World of Patterns

World of Patterns

A Global History of Knowledge

Rens Bod

Translated by Leston Buell

Johns Hopkins University Press
Baltimore

This publication has been made possible with financial support from the Dutch Foundation for Literature.

Nederlands letterenfonds
dutch foundation
for literature

The Open Access edition of this publication has been funded by the Dutch Research Council (NWO).

This book was previously published as *Een Wereld Vol Patronen: De Geschiedenis van Kennis*. © Rens Bod, Uitgeverij Prometheus, Amsterdam, 2019.

Printed in the United States of America on acid-free paper
9 8 7 6 5 4 3 2 1

Johns Hopkins University Press
2715 North Charles Street
Baltimore, Maryland 21218-4363
www.press.jhu.edu

Library of Congress Cataloging-in-Publication Data

Names: Bod, Rens, 1965– author.
Title: World of patterns : a global history of knowledge / Rens Bod ; translated by Leston Buell.
Description: Baltimore : Johns Hopkins University Press, 2022. | Includes bibliographical references and index.
Identifiers: LCCN 2021020604 | ISBN 9781421443447 (hardcover) | ISBN 9781421443454 (ebook) | ISBN 9781421443461 (ebook open access)
Subjects: LCSH: Knowledge, Theory of—History. | Intellectual life—History.
Classification: LCC BD161 .B538 2022 | DDC 121—dc23/eng/20211104
LC record available at https://lccn.loc.gov/2021020604

A catalog record for this book is available from the British Library.

If scholars had not recorded their thoughts over the centuries, the foundations of knowledge would have collapsed and their conclusions would have been lost. For any branch of knowledge to exist, it must be derived from history.

AL-MASUDI, *THE MEADOWS OF GOLD AND MINES OF GEMS*, 10TH CENTURY

Contents

Preface

The Wonder of Knowledge

The idea that the world can be understood through patterns and underlying principles is one of humankind's most important insights and perhaps its most successful survival strategy. The search for patterns and principles started at least 40,000 years ago with striped patterns scratched on mammoth bones, leading to modern-day knowledge disciplines. What paths has human knowledge taken to grow from these humble beginnings, via many detours and dead ends, to today's understanding of nature and culture? This book answers this question and shows what role patterns and principles have played in different regions and cultures. I discuss not only the study of nature (the natural sciences) but also the study of culture (the humanities), medicine, mathematics, jurisprudence, and a number of fields of knowledge that we no longer consider "science."

My previous book, *A New History of the Humanities* (2013), focused on the fields of knowledge that deal with the study of culture: the humanities. There I started in classical antiquity, when the notions of patterns and principles had already partially crystallized. However, I overlooked the fact that these notions have a history of their own. In the current book I broaden my perspective and take a step back in time: to understand how the notions of patterns and principles have developed in different places around the world since the Stone Age. In this way I hope to find an answer to the question that has intrigued me for years: How did what we know now originate and grow? I had wanted to write such a book for some time, but 10 years ago the task seemed too ambitious. So, as a sort of exercise, I began work on a slightly less ambitious project: a global history of the humanities, something that was also sorely lacking. The resulting book, *A New History of the Humanities*, published with Oxford University Press (originally published in Dutch as *De vergeten wetenschappen*), had a greater impact than I could have hoped. The book was translated into Chinese, Polish, Ukrainian,

Korean, Armenian, and Italian, and the history of the humanities has grown from a nonexistent field into a discipline with its own journal (*History of Humanities*), an annual conference (The Making of the Humanities), a book series, an international society, courses at universities in many parts of the world, and university chairs. In the Netherlands, for example, both the Dutch Research Council (NWO) and the Royal Netherlands Academy of Arts and Sciences (KNAW) referred to *De vergeten wetenschappen* when they wanted to highlight the importance of cross-fertilization between the sciences and the humanities.[1] And the popular science magazine *Scientific American* dedicated an opinion article to my book in its June 2015 issue, which concluded that "regardless of which university building scholars inhabit, we are all working toward the same goal of improving our understanding of the true nature of things, and that is the way of both the sciences and the humanities, a *scientia humanitatis*."[2]

I resumed my initial project of creating a global history of knowledge in January 2014. I was stimulated by the establishment of the Vossius Center for the History of Humanities and Sciences at the University of Amsterdam, where we have brought in researchers as fellows since 2016. The fruitful interactions with these fellows and with my two codirectors—Julia Kursell and Jeroen van Dongen—have inspired me on many occasions. Now that this book is finished, I realize how strange it is that no work has previously integrated the histories of science and the humanities.[3] While some historians, such as George Sarton (1884–1956), have made an impressive attempt to arrive at a global history of knowledge disciplines,[4] they have failed in that endeavor, sometimes because they died before they could finish their work, sometimes because of their limited access to resources outside Europe, and especially because of their partiality to the natural sciences.[5] The history of science has long consisted mainly in the history of Western natural sciences.[6] Their fruitful interaction with the other disciplines, both in the West and in other parts of the world, has been overlooked.[7] With this book I show what the history of disciplines, and with it the history of knowledge, looks like when we remove the natural sciences and the West from their central position. Such a history takes into account as many disciplines from as many regions and cultures as possible on an equal footing.

Despite the wide variety of disciplines I discuss, it became clear to me in the course of writing that there was also a certain unity in that variety. But it took me until the end of my research to comprehend that unity. So I then decided to rewrite the book from scratch, successively becoming enamored with a region, culture, school, or historical person. I still cannot get over the fact that the

16th-century Indian Kerala school produced so many new mathematical and astronomical insights that are unknown to the general public. That the many female scholars and scientists, regardless of their region, have been kept under wraps in historiography for so long. That the practice of inoculation was invented not in Europe but in China. And that jurisprudence—from the Roman Empire to the Ottoman Empire—was the model for many other disciplines. These examples may be known to specialists in the relevant fields, but they have never been brought together in a general history of knowledge.

Not everyone will find everything to their liking in this book: I was forced to make choices and focus on a dozen or so disciplines that occur in most regions from ancient times—astronomy, mathematics, mechanics, medicine, linguistics, history, musicology, philology, jurisprudence, and art theory. These disciplines have by no means remained stable since antiquity, but they do show a high degree of continuity with regard to their subject matter (see the introduction). I also make excursions into several other disciplines, such as botany, zoology, geography, logic, poetics, philosophy, astrology, magic, and alchemy. I should emphasize that I am not a specialist in the history of most of these disciplines, so suggestions and criticism are welcome. You can send them by email to rens.bod@gmail.com, and I will gratefully acknowledge your comments in any revised edition. You can also follow developments about this book at http://devergetenwetenschappen.blogspot.com.

World of Patterns

Understanding the World through Patterns and Principles

This book is about the human quest for patterns and principles in the world that surrounds us. Where can we find the first evidence for this quest, and how has human knowledge evolved in different regions and cultures? Many philosophical works have addressed the question of what human knowledge is and how we know what we think we know, but none has attempted a general *historical* overview. This book aims to provide such an overview.

Patterns

At all times and in all cultures, people have sought and found patterns in the world around them.[1] My story is about the often successful, sometimes fruitless, but always impressive human search for patterns and the systematic knowledge derived from them. Knowledge of solar, lunar, and planetary motion; knowledge of language and music; knowledge of animal behavior; knowledge of plant cultivation—all this knowledge consists not in a simple summation of facts but in regularities that allow one to bring individual facts together. This is what we understand by systematic, or pattern-based, knowledge.

Not all knowledge is pattern based. Knowledge of the names of members of a family, for example, does not constitute a pattern. However, if there is a certain order or regularity underlying them—such as the way certain names are passed down—then knowledge of the regularity in question is a form of pattern-based, systematic knowledge.[2]

Pattern recognition and interpretation is one of the most essential skills of *Homo sapiens*. Humans seek, recognize, and interpret patterns—both in nature and in culture. In this book I argue that humans have always been pattern-seeking, interpreting creatures. While pattern recognition also occurs in animals and even in plants,[3] what humans do with the patterns they discern differs radically from how other living beings do this (see below under "Principles"). This brings us to the question of what exactly a pattern is. Intuitively speaking, a *pattern* is an observed regularity that contains an element of repetition. It is not necessarily immutable but may be subject to exceptions and variation. While there are more precise definitions of a pattern,[4] most concern quantitative patterns,[5] whereas this book also deals with qualitative patterns, such as social, historical, cultural, and narrative patterns, which are usually not quantitative. Moreover, existing definitions do not do justice to variable patterns. For this reason, in this book we will first allow our notion of pattern to crystallize before attempting to refine our definition of the concept. For the time being, the word "pattern" will be an umbrella term encompassing a range from the most unstable regularity to the most absolute.

Not only do patterns allow observed phenomena and events to be categorized together; they can also say something about phenomena and events that are as yet unobserved and thus unknown. Patterns have predictive power. Take for instance the pattern of the rise, peak, and decline of states, which was discovered in antiquity. This pattern was discerned in previous events by the Greek historians Herodotus and Thucydides. It was also described by the Chinese historian Sima Qian, the Arab historian Ibn Khaldun, and the Italian historian Giambattista Vico. The pattern makes generalizations about the past, but it also makes a claim about the future for states that do not yet exist. The same applies to patterns in nature, such as the rising and setting of the sun, moon, and planets, events that were recorded in Neolithic stone circles thousands of years ago and were also recorded in detail on clay tablets by Babylonian astronomers. I do not claim that all quests for patterns have been successful—I discuss quests that either ended in failure or led to patterns that later proved to be invalid, such as the Babylonian search for a link between planetary motions and the price of grain.

Besides the notion of pattern, there is a notion of what is unique or exceptional, what does not follow a pattern. We will see that the exceptional cannot exist without a pattern; an exception is something that deviates from a pattern, connecting the two notions intimately (see the conclusion).

Principles

Patterns in themselves do not explain anything; that is what principles aim to do. Whereas patterns are observable, principles concern the underlying relations between things that are not directly observable. Principles are theoretical, making them more difficult to verify or refute than patterns are. In a sense, patterns involve knowledge that is more certain than the knowledge described by principles. In her book *How the Laws of Physics Lie* (1983), the philosopher of science Nancy Cartwright shows that phenomenological (pattern-based) laws in physics often make better predictions than fundamental (principle-based) laws.[6] Her point applies not only to physics (or economics, which Cartwright has also dealt with) but to all sciences and all forms of systematic knowledge. Nevertheless, principles do go a decisive step beyond patterns: principles reveal an underlying regularity with which they attempt to explain the "superficial" patterns.

I will argue that it is a thoroughly human trait to explain patterns, and even understand them, using deeper principles. For example, the regularity of how names are passed down can be understood using principles of kinship. Legal rules can sometimes be traced to deeper principles such as the retaliation, or *talio*, principle that we find in many legal systems. The patterns in the planetary motions have been reduced to underlying principles multiple times and in different ways (see chapters 3.2, 4.2, 5.2). And as we will see, people have also sought principles underlying patterns in the human body, language, music, art, literature, and more.

The search for patterns and their underlying principles can be found all over the world. The concepts of principles and patterns thus appear to be universal, which is further supported by the very similar linguistic contexts in which the words for these concepts appear in different languages. For Chinese and English, for example, this can be verified using the bilingual *Thesaurus Linguae Sericae*.[7]

Principles are at their best when they can predict new patterns, just as patterns can predict new phenomena or events. But when exactly is it a principle that we are dealing with? In this book I use the word "principle" when it can cover multiple patterns at the same time. This notion of principle differs somewhat from that in my previous book, *A New History of the Humanities* (2013). In that book, principles were primarily methodological, such as the principle of a system of rules in linguistics or the principle of numerical relations in music theory. In the current book I refine my definition to a generalization that underlies patterns—a more expansive interpretation that incorporates the notion of methodological principle.

Principles themselves can often also be generalized to increasingly deep principles (such as the notion of "universal laws" in natural science). But I will continue to refer to deeper principles simply as "principles," regardless of how comprehensive they are. So in this book, a principle is nothing more than a statement that applies to more than one pattern. And while we will see that people are often convinced that principles explain patterns, I agree with the philosopher of science Alan Musgrave that the opposition between a description and an explanation is rather illusory: "We explain one thing *by* describing another."[8] But Musgrave's observation does not detract from the human practice of using generalizing principles to explain observed patterns. Although pattern searching can also be found in other living beings, at this point in time, the search for underlying principles appears to be uniquely human.

Relations between Patterns and Principles

Can patterns be formally derived from principles, or do principles only loosely generalize over patterns? People have been asking these questions since the 4th century BCE. There appear to be all kinds of relations between principles and patterns: from logical inferences and procedures with more or less formal rules, to informal relationships and preconditions. I will not go into more detail here about the nature of the possible relations between patterns and principles in different disciplines, periods, and regions, let alone into the question as to whether these relationships themselves also exhibit patterns. That would make our story unnecessarily abstract at this point. We will come back to this in the chapters that follow when we have considered sufficient historical material.

Polycentric and Comparative: The Problem of Global Historiography

This book discusses the history of patterns and principles not only in the natural sciences but also in other disciplines, such as jurisprudence, medicine, mathematics, philology, musicology, and art history. In this way I examine what the history of the sciences and humanities looks like when we treat the natural sciences on an equal footing with other domains of knowledge.[9] When we do so, a new perspective on knowledge emerges, providing insight into the intertwined nature of widely divergent disciplines such as astronomy and philology. With this book I also endeavor to show what the history of knowledge looks like if we do not presume a single center of activity, such as Europe or the West, but instead assume that there are multiple centers, including from Asia (e.g., the Mughal Empire, China), Africa (e.g., the Songhai Empire, Ethiopia), the Arab world, Oceania (e.g., Tonga), and the pre-Columbian Americas (e.g., the Incan and Mayan Empires). I do not limit myself to discussing these centers separately but also explore the extent to which knowledge was exchanged between them.[10] While the many knowledge activities differ from each other, they can be compared with respect to patterns and principles.

But how do we determine whether there are patterns and principles in the works of the past in a particular region? Should we use our own definitions given above, or should we only attribute the terms "patterns" and "principles" to historical actors when they have used these terms themselves? The problem is that while most past scholars and scientists searched extensively for regularities and generalizations, they did not always refer to them as patterns and principles. We encounter a whole array of terms, such as "law," "rule," "motif," and "regularity"—in a variety of languages—but it is more often the case that the results are presented without any such terminology at all. For an overarching history like this one, I do not require that the historical actors themselves have used certain terms in order to claim that they employed the concepts those terms denote. For example, no 16th-century philologists or 17th-century physician used the term "empirical cycle" (see chapter 5.1), but that does not mean that they did not apply such a cycle in their research. To the contrary, the empirical cycle was used widely starting in early modern times, from Europe to China. For a history of knowledge spanning many centuries, it is counterproductive to limit ourselves to using so-called *actors' categories*. After all, not all anachronisms are misleading, as historian Nicholas Jardine has convincingly argued.[11] Limiting ourselves to actors' categories is appropriate for a biography or a history of knowledge for a specific period, which delves into the world of a single individual with his or her specific concepts and idiosyncratic terms. Although I discuss biographical details in this book, this is not microhistory but rather an attempt to unite the micro level of concrete historical events with the macro level of long-term developments.

As mentioned, this book presumes the existence of multiple knowledge centers on all inhabited continents. Furthermore, with my polycentric approach, I focus not only on regions with a written culture but also on regions without writing, not only in prehistory but also in later periods, such as in the Incan Empire, where information was recorded in the form of cords with knots (*quipu*), and in Melanesian and Polynesian civilization, where knowledge was recorded in architectural structures. Even in classical Greece, knowledge was reflected not only in writing but also in visual and material sources, such as in art theory, where the principles of correct proportions were recorded in sculpture (for instance, the Canon of Polykleitos). I also discuss the historical pictorial narratives of the Incas, Aztecs, and Mixtecs. And where possible, I use oral sources, such as for Xeer jurisprudence in Somalia and for historiography in the Gonja kingdom.[12] But by far the greatest emphasis is on written material, simply because written sources are the most informative for our purposes. However, troves of written material remain largely inaccessible. Take Timbuktu's ancient manuscripts for example, of which a mere 6,000 of the approximately 700,000 books are accessible (see chapter 5.1). Most of the manuscripts are in private households, and an unknown number were destroyed during recent assaults. However, the manuscripts that made it into the Ahmed Baba Institute testify to an extraordinary wealth of ideas, insights, and discoveries in many areas.

Writing a polycentric, global history of knowledge remains thus an enormous challenge. While it is relatively easy to treat the different disciplines on an equal footing, this is much more difficult, if not impossible, for the different centers. And that is not only because in some knowledge centers many of the sources are inaccessible, but also because in many cases the sources have not been deciphered (such as the genealogical and astronomical texts from Easter Island), and even more often because the available sources in a certain knowledge center do not always concern all fields of knowledge. Thus, while my history is polycentric, it is not always "equicentric."

Historical Generalizations and Trends

In any historiography—and especially in a history spanning many centuries—we must ask whether the historical cases assembled provide sufficient evidence for the conclusions reached. First, there is the problem of unknown information. However, this is not the biggest stumbling block, as long as we remain open to the possibility that its future emergence could change or refute the previously drawn conclusions. The second, bigger problem is that we have to select from the *known* facts and cases. Although I try to do justice to the various fields of knowledge from as many knowledge centers as possible, there can never be an impartial selection of the possible cases. Even though I cover some 20 disciplines (see below under "Knowledge Activities and Disciplines"), it is still quite likely that I have missed certain patterns, principles, and their mutual relationships. To keep myself on my toes, I looked not only at quests for patterns, principles, and relationships but also at cases where there were no such quests or where they were even rejected (see, e.g., chapter 3.3).

Making generalizations about different historical events is even more challenging. Doing so was considered suspect[13] and was associated with positivist historiography from a period prior to the discipline's professionalization.[14] Later theoreticians of history considered generalizations to be out of the question, and that was the historiographical maxim of the 20th century. Yet generalizations were formulated by all the major 20th-century historians: from Johan Huizinga, who saw the returning practice of games and play as a general pattern in the history of all human cultures,[15] to Fernand Braudel, who did not hesitate to propose universal patterns in the emergence of capitalism.[16] So although generalizations never really left us, they fell out of favor for a long time. However, for the past couple of decades historical generalizations have been back on the agenda, although now more than earlier the historian is required to be as skeptical as possible, and every generalization must be examined with an extremely critical eye.[17] It is important to make the range of historical cases as broad as possible and to propose each generalization only as a provisional hypothesis—or rather as a "tendency" that can be substantiated or disputed by further cases.

Discovering a counterexample does not mean that a generalization we have found should automatically be rejected. Indeed, in history there are hardly any absolute regularities. The best we can find are historical trends, or "historical lines," to put it in the terms of historian Jan Romein.[18] None of these trends or lines is absolute. There will always be exceptions.[19] Yet we will see that the tendencies I propose are "risky": they state that there is a certain line in the search for patterns and principles and that this line can be broken if new cases suggest a different trend. The simple (and admittedly caricatured) Popperian refutation by a single counterexample does not hold for history. As historians, we must face the colossal problem of data that are incomplete and even corrupt. But we can work with the data we have, where a generalization or tendency we find can say something not only about the past but also about the future.

In this book I therefore discuss both regularities and exceptions in the history of knowledge, always questioning whether there really is a tendency and how it relates to the exceptions found. A tendency within a single discipline and within one specific culture is easier to find than a tendency that spans multiple disciplines, a longer period, or even multiple cultures. These latter trends are extremely rare: I have found only a few. The notion of "tendency" is not formally different from that of "pattern." However, I reserve the word "tendency" for a regularity that I have found in the past, while I reserve the word "pattern" for a regularity found by a historical actor, regardless of what term he or she uses for it.

So it is not my goal to achieve a complete overview of the history of knowledge, something that is in any case unattainable. Instead, I strive for a history that highlights the diversity of knowledge disciplines as much as possible so as to give the broadest support possible for my hypotheses about the development of human knowledge, even if that support is not complete.

The Past as Empirical World: Digital History

I see my historiography pertaining to *digital history*, in which digital data and resources are used for historical research. I have already mentioned the *Thesaurus Linguae Sericae*, but I have also used many other digital corpora and thesauri, such as the *Old Babylonian Grammatical Texts*, the *Thesaurus Linguae Grecae*, the *Thesaurus Linguae Latinae*, the *Corpus Iuris Civilis*, the *Hadith* collections, facsimiles of the Mesoamerican Codices, and the CKCC corpus (containing letters of 17th-century Dutch scholars). In addition, I have also made extensive use of specific corpora that focus on one particular scholar or scientist, such as the digitized *Opera Omnia* by Desiderius Erasmus and the digitized works of Kepler (*Herausgabe der Werke von Johannes Kepler*).

Use of such digital files is not strictly necessary, and in this book I will mainly refer to the sources themselves. However, searching through large amounts of resources has become much easier and faster since they were digitized. And what is more, the digital analysis of sources also leads to a different working procedure. For example, I am an advocate of what is called *distant reading*.[20] Lexical and syntactic tools—from so-called *topic modeling* to *parsing*[21]—are used to quickly search texts for content and see whether patterns and principles are used. The historian then decides whether to follow up this *distant reading* with a *close reading*, where the texts are studied in detail. In this way, a much larger number of texts can be gone through in a given amount of time. *Distant reading* is similar to the time-honored technique of "diagonal reading," in which a researcher scans a text without reading it in detail. But there is an important difference: thanks to the *topic modeling* tools available, it has now become possible to search text files for frequently appearing topics without human intervention,[22] whether that be the topic of cyclicity in historical narratives or that of geometric models in astronomical texts. The locations of these *topics* in the texts can be collected automatically, after which the historical handiwork can begin, without a preceding round of diagonal reading. I dare say that I could not have completed this book in a single lifetime without digital tools. Having said this, however, my ultimate concern was not the digital aspect of my approach but the historical narrative that resulted from it, which is human work from start to finish.

I realize that I have some advantage over most other historians in that I have been working in computational linguistics for a quarter of a century. For example, thanks to the analysis and parsing techniques developed by my research group, I was able to analyze most sources syntactically, making it possible to search not only for lexical patterns in texts but also for syntactic patterns.[23] This sort of syntactic search made it possible to include "long-distance" relationships between words, rather than just looking at words that were adjacent to each other (for further explanation, see chapter 2.1).

Knowledge Activities and Disciplines

I will generally refer to the knowledge activities I cover in this book as "disciplines" or "sciences," but we must realize that these categories are recent.[24] Until the 18th century the word "science" was rarely used in the sense of a discipline or field. "Science" simply meant knowledge or knowing.[25] It is especially in the course of the 19th century that we see its gradual transformation into an institutionalized discipline. Ideally, we should use the local designations of the day for the various knowledge activities in this book, but that is easier said than done. For example, the study of art (or art history) was classified by the Roman author Pliny mainly under "mineralogy and application of materials."[26] This book would become unreadable were we to limit ourselves to the local, historical terms for the various knowledge activities. Where possible I will mention the historical or regional terms for a particular knowledge activity and then replace them with what I believe to be the most coherent terms. In some instances that will be the historical term, while in others it will be a modern-day term.[27]

When I use the word "discipline" in this book, I am not referring to the notion of an academic discipline but to a collection of activities that share an *object* of study, such as language, numbers, nature, diseases, the cosmos, law, music, art, the past, the plant kingdom, or the animal kingdom. The disciplines that are the focus of this book can all be traced back to ancient times and can also be found in most regions: astronomy, mathematics, mechanics, linguistics, history, musicology, philology, medicine, jurisprudence, and art theory. These 10 disciplines are fairly representative for a knowledge history spanning many centuries; they include the study of both nature and culture, as well as medicine and mathematics, two disciplines that cannot be unambiguously categorized under either the study of nature or of culture. I also discuss logic, poetics, botany, zoology, geography, theology, philosophy, astrology, magic, and alchemy—sometimes even quite extensively—but these other 10 disciplines are not the focus of this book.

Of course, my selection of disciplines does not do justice to domains of knowledge that have emerged only more recently. As I will explain in chapter 5, the detailed part of my historiography covers the period from the Stone Age to the 18th century. For the period from 1800 to 2000, I provide only a rough sketch for most disciplines. However, in the conclusion, I briefly touch on some new disciplines that have emerged in the 20th and 21st centuries. We will see that the disciplines I discuss can hardly be said to have remained stable over the centuries: all disciplines have undergone far-reaching transformations in which their boundaries have shifted several times.[28]

CHAPTER ONE

The Awareness of Patterns

Prehistory

2.5 Million Years Ago–3000 BCE: All Regions

The search for patterns is as old as humanity itself and probably even older. More than 2.5 million years ago, *Homo habilis* began making increasingly complex stone tools. And some 500,000 years ago, *Homo erectus* scratched a geometric zigzag pattern on a shell, the meaning of which is unclear.[1] The control of fire also dates from this time. These events occurred in the Paleolithic, or Old Stone Age, which runs from about 2.5 million years ago to 12,000 years ago. This period is followed by the New Stone Age, or the Neolithic, when people began practicing agriculture and keeping livestock. While *Homo erectus* spread from Africa across Europe and Asia more than a million years ago, *Homo sapiens* was still hiding out in a remote corner of Africa. Around 70,000 years ago this "wise man" also began to populate the other continents. Around this time something special happened: *Homo sapiens* manifested an outburst of creativity. We see this in the form of the many cave paintings, records of the phases of the moon, increasingly refined tools, and the development of a precursor to writing. Some historians even speak of a cognitive revolution in the late Paleolithic and attribute it to genetic mutation, for lack of a better explanation.[2]

What exactly happened during this cognitive revolution or "leap" is unclear, as is the answer to the question of whether it might be better to think of it as a development rather than as a leap. But what is undeniable is that a certain acceleration occurred at that time, one that was crucial for the history of knowledge. At the end of the Paleolithic, *Homo sapiens* was the only one of the many human species remaining, together with the last few Neanderthals, who disappeared some 34,000 years ago. For this reason, when I discuss humans in this book, I am referring primarily to *Homo sapiens*, unless otherwise indicated. In our study of the search for patterns in the Stone Age, we are, of course, dependent on unwritten sources.

1.1 The Paleolithic: From Primal Human to Jack-of-All-Trades

The Oldest Shared Pattern in the World

It was in the summer of 1983 when I was traveling around southern Europe as an 18-year-old that I first heard about the cave paintings at Altamira. These prehistoric caves had been closed to the public for years but could finally be seen again, though not for long, unfortunately: the air exhaled by visitors proved to be so harmful to them that nowadays people have to make do with a replica. So I arrived at the right time and was overwhelmed by the staggering number of images of bison, deer, horses, and wild boar (figure 1).

The paintings are so realistic that when they were discovered in 1878, archaeologists didn't want to believe that they dated from the Paleolithic.[3] In their opinion, a "primitive" human would have been unable to produce this sort of artwork, and the person who had discovered the drawings was accused of forgery. But similar caves were soon found in Spain and France. Radiometric dating methods have now established that these paintings range from 20,000 to 40,000 years old.[4] We also know how they were made. Each image was painted in three phases: the figures were first scratched into the rock with a sharp object, then they were outlined with black charcoal, and finally they were colored with ocher.

What is particularly striking is the systematic way in which the bison, horses, wild boar, and other animals are depicted. The prehistoric painters followed a particular pattern in how they represented the animals' positions and actions: they are painted in profile, that is, from the side. The animals are depicted with all of their legs, their tail, and—when applicable—with both horns. The painters apparently wanted to show as much of the animal's body as possible, which required

Figure 1. Profile images of bison in different positions in the caves of Altamira, between 35,000 and 20,000 years ago. CreativeCommons, photo by Matthias Kabel, 2005; https://commons.wikimedia.org/wiki/File:Reproduction_cave_of_Altamira_01.jpg.

drawing it from the side rather than from the front. We find the same pattern in other Paleolithic paintings, such as the well-known Apollo 11 caves in Namibia,[5] the Sulawesi caves in Indonesia (figure 2),[6] and the Cueva de las Manos in Argentina, although the latter is not as old (between 13,000 and 9,000 years), considering that *Homo sapiens* didn't arrive in the Americas until around 13,000 years ago.

In addition to the sideways portrayal, we also find a pattern in the cave paintings known as "twisted perspective": the heads of the animals are shown in profile, but the horns are shifted in relation to each other, or twisted, making them clearly distinguishable. Apparently the horns were too important to be overlooked, unlike the eyes, for example. We also encounter images of people, especially their hands. Such stencils of hands can be found all over the world. They were probably made by blowing or spraying liquid ocher over a hand, as in the Sulawesi cave paintings in figure 2. In addition, we sometimes see representations of depth illusions. An example of this is in the Altamira cave, where a relief of the rock face was used to evoke depth. The contours of animal bodies follow the bulges in the wall, rendering the images three-dimensional. The result is of unsurpassed beauty. Picasso allegedly had this to say on a visit to the cave: "After Altamira, all is decadence."[7]

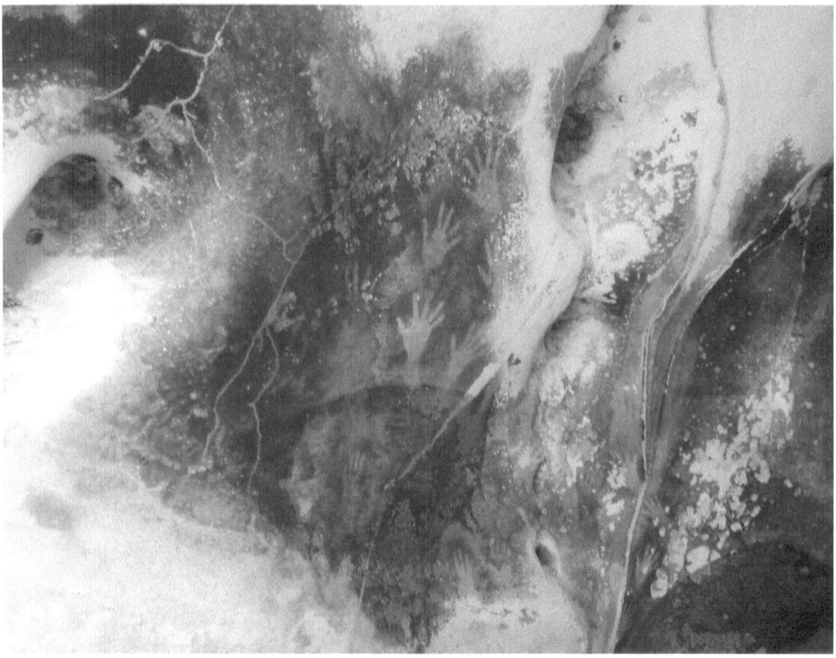

Figure 2. Cave painting from Sulawesi (Indonesia). CreativeCommons, photo by Cahyo Ramadhani, 2014; https://commons.wikimedia.org/wiki/File:Hands_in_Pettakere_Cave.jpg.

While all the patterns we are discussing here are unambiguously present, they remain implicit and were only identified by people studying them. There are no inscriptions or texts that mention the pattern of the side view or twisted perspective. All we have is the images themselves.

Astronomical Knowledge: Explicit Patterns

In other forms of prehistoric knowledge, patterns can also be found that are more explicit, such as the oldest known observations of the lunar cycle. These have been transmitted through inscriptions on thousands of bone fragments of reindeer and mammoths on which people kept track of the phases of the moon. For example, the long lines on the mammoth tusk from Gontzi (figure 3) seem to refer to the days with the new moon and full moon, while the short lines refer to the days in between. Bones and tusks of this kind have been found at various sites in Africa and Europe and are between 15,000 and 40,000 years old.

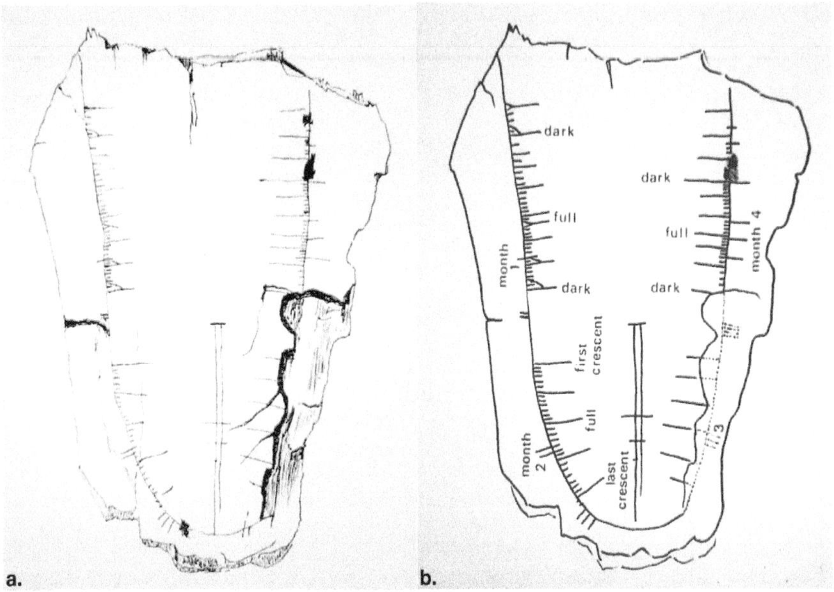

Figure 3. Moon observations engraved on a mammoth tusk from Gontzi, Ukraine. This specimen is 15,000 years old, where (a) is the original and (b) is a diagrammatic representation. From James McClellan III and Harold Dorn, *Science and Technology in World History: An Introduction*, 2nd edition (Johns Hopkins University Press, 2006, p. 15); used with permission.

Although the lunar interpretation of these dash patterns is generally accepted by archaeologists and archaeoastronomers, we have no conclusive evidence that they actually concern the lunar cycle.[8] Yet the indications are strong. To start with, moonlight was of great importance to Paleolithic humans, who depended on it for the nocturnal hunt. Second, the cycles in the dash patterns are subdivided into smaller cycles, which—though not always consistently—represent the first and last quarters of the lunar phases (the half-moons). But even if the pattern were to refer to something completely different, for example, to a woman's menstrual cycle, it is still an explicit representation of a pattern. In addition, the count appears to have been tallied using dashes, making the pattern one of the oldest quantitative representations known.

A different lunar cycle pattern appears to be engraved on the 30,000-year-old Blanchard bone from Sergeac in France (figure 4). Here it is the lunar phases themselves that appear to be depicted. With a little bit of effort, we can make out the waxing and waning moons, but there is no clear cycle in the lunar phases on the

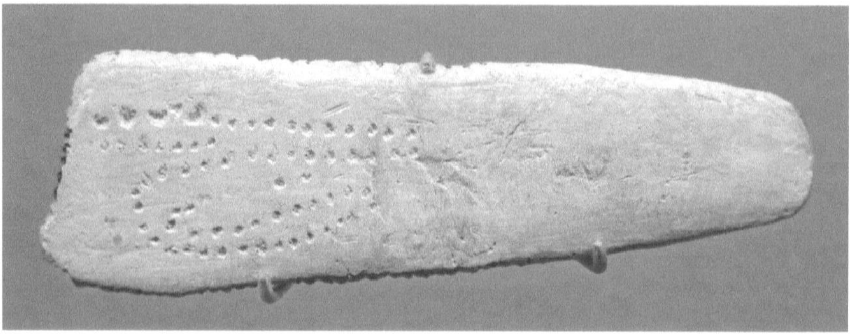

Figure 4. Moon observations on the Blanchard bone, Sergeac, France, 30,000 years old. CreativeCommons, photo by Don Hitchcock, 2014; https://en.m.wikipedia.org/wiki/File:Blanchard_plaque.jpg.

bone. Moreover, the images include all sorts of details whose meaning is unclear. This has led some archaeologists to suggest that the Blanchard bone had a primarily decorative character and does not represent observations of the moon.[9]

However, most Paleolithic lunar observations resemble the Gontzi dash patterns. Viewed from a certain distance, this system bears a degree of similarity to that of cave paintings. Of course, the dash patterns are abstract while the animals are figural. But both cases involve the noting and recording of shapes, be they of animals or of the moon. It has also been suggested that certain constellations of black dots in the Lascaux cave paintings show similarities to star constellations in the sky, especially the Pleiades. This would mean that the Lascaux paintings are the first in which art and the cosmos come together.[10]

In any case, we can state that the conscious or unconscious search for patterns starts with Paleolithic humans and is at least 40,000 years old. We find this search among groups of people who were sometimes more than 12,000 kilometers apart and who were not in contact with each other. This suggests that Stone Age people already had this pattern-seeking creativity and took it with them when they left Africa to spread to other parts of the world.

Knowledge of Domestication: From Unconscious to Conscious Patterns

Paleolithic humans also gathered systematic knowledge in a completely different field: domestication, or knowledge concerning animal breeding and the cultivation of plants. Domestication involves selecting animals and plants for certain characteristics so that their offspring are better suited for human needs. Domestication can be used to cultivate tastier and more robust crops and to breed tamer and stronger animals.

It is almost certain that the oldest form of domestication—from wolf to dog—took place *unwittingly*.[11] We know this through genetic research into the proto-dog, which split off from the wolf at least 33,000 years ago (but possibly as early as 100,000 years ago).[12] This split was the result of *self-domestication*: over many generations, some wolves who were less afraid of humans than others gradually evolved into dogs through a process of "self-selection" by following people and eating their food scraps at campfires. This gave proto-dogs an advantage over their more fearful counterparts. People discovered that these animals could warn them of danger, help them hunt, and even serve as food in times of scarcity. A symbiotic relationship between humans and proto-dogs developed following an unconscious domestication pattern. But as soon as people became aware of this pattern and its benefits, they adopted it for breeding other animals as well, such as the further breeding of the dog itself (about 13,000 years ago). The dog was followed first by the goat, the sheep, and the pig (around 8000 BCE), and then later by the cow and the horse (6000 and 4000 BCE, respectively). But by that time we are in the Neolithic Age, with a pastoral peasant culture, and the unconscious selection pattern has become conscious.

Similar processes have occurred in the domestication of wheat and other crops.[13] When the grain is ripe, wild wheat falls to the ground and goes to seed, but some of the grains remain on the stalk. This wheat remaining on the stalk could be harvested more easily and hence came to be domesticated—through an unwitting process. For its survival, this wheat depends on the farmer harvesting it and sowing it again. Here too, the unconscious pattern must have become a conscious one, as evidenced by the domestication of many crops in the later Stone Age. The first crop to be domesticated was probably rye (around 12,500 BCE), followed by other cereals such as wheat (9500 BCE) and peas (around 9000 BCE); fruit trees would follow millennia later. Awareness of the domestication pattern will become the driving force behind the "Neolithic revolution" that I discuss below, but the first seeds of this pattern were sown in the Paleolithic.

With domestication we find a transition from an unconscious pattern to a conscious one: human interaction with plants and animals triggers selection processes that lead to adaptations in them, after which this pattern is used to domesticate other species.

Awareness of the domestication pattern initially yielded no knowledge of the underlying principle.[14] The underlying process was only discovered some 150 years ago by the likes of Gregor Mendel and Charles Darwin, while an understanding of the process in terms of genetics is even more recent. Darwin dubbed deliberate selection carried out by people "artificial selection," contrasting it with his famous notion of "natural selection." But the underlying principle is the same: since the second half of the 20th century, we have known that this deeper selection principle on which domestication depends is based on gene mutations and on new combinations of existing genes. So it can take a long time for a pattern to be reduced to its underlying principle; in the case of the domestication pattern, this took some 30,000 years. Of course, it is still possible that Stone Age people also formulated their own principles for the domestication pattern, but they are not known.

Knowledge of Technology and Culture

It goes without saying that the development from the chipped stone to the more advanced hand ax was also accompanied by a search for a pattern, namely for the best possible tool for a given purpose. But the hand ax may alternatively have been the result of a happy coincidence or some individual's brilliant insight.[15] This also applies to the making of spears, harpoons, arrows, and bows, and for the centuries-long improvement of the oil lamp, the oldest of which (ca. 15,000–10,000 BCE) was found in the caves of Lascaux and consists of no more than a stone dish filled with animal fat and a wick made of plant fibers.

Human control of fire is much older. Archaeological finds show that *Homo erectus* was already occasionally making fire a million years ago, and that around 450,000 years ago they were doing so systematically, just like the Neanderthals.[16] From that time on, a new social pattern probably arose with evening campfires accompanied by communal meals and other social activities.[17] This led to tighter group cohesion and a better understanding of others. Research into contemporary hunter-gatherer societies tells us that cultural transfer and group bonding does take place around campfires.[18]

With one particular development, the pattern seems obvious: the survival strategies used in the ever-colder locations to which *Homo sapiens* migrated over the course of the late Paleolithic. For this emigration from Africa to Eurasia, increasingly sophisticated techniques were developed to make clothing that better held heat in. The most important tool was a needle made of bone or ivory, which was gradually refined to seal animal skins off as well as possible from the cold. After 40,000 BCE this technique had been developed to the point that people around 15,000 BCE were able to survive at temperatures of −50°C (−58°F). This is an impressive feat of survival, one indebted to a unique human adaptation pattern: the more extreme the environment (in terms of temperature or some other factor), the more refined the technique (in this case the needle). It is a form of adaptation without genetic mutation.

The knowledge that Paleolithic humans possessed of their natural surroundings must have been tremendous. Stone Age humans could distinguish edible from inedible fruits, they knew the growth behavior of every plant, the course of rivers and streams, and the burrows of predators and prey, and they were experts at following animal tracks.[19] It is obvious that Stone Age people used patterns for this. For example, knowledge about the growth behavior of plants and the habits of animals is almost by definition pattern based because it makes generalizations about individual plants and animals.

So Stone Age people had knowledge of many things and of many patterns, but what we do not find is knowledge of underlying principles that generalize over patterns. This doesn't necessarily mean that they lacked this knowledge. But there is no indication that principle-based knowledge existed in the Paleolithic Age. And perhaps humans did not need such knowledge to survive.

Yet Paleolithic humanity must have had rules of law, rules for living together, for kinship, for rituals, for play, and for burying the dead, rules we could term "man-made patterns" or "cultural patterns," though it is difficult to make a sharp distinction between patterns developed by people themselves and those found in the natural environment surrounding them (see the discussion in the conclusion). The numerous Paleolithic burial sites are also subject to a pattern: the tombs that have been found are aligned with the course of the midwinter or midsummer sun.[20] In addition, burial gifts in the Paleolithic are evenly distributed practically everywhere we look, indicating a fairly flat social structure.

The current view is that Stone Age people were *animists*, just like the hunter-gatherers that remain today. According to animism, everything in the world has a soul: not just people but animals, plants, stones, mountains, and rivers as well, and even natural phenomena such as thunder and lightning. This abundance seems to indicate a world that lacks an underlying unity, since every object or being has its own soul or spirit. Although the evidence for animism from the Stone Age is paper thin, such an animistic worldview would fit well with the knowledge that Stone Age people had of their environment. They knew a great many patterns—botanical, zoological, geographical, artistic, astronomical, technological, and social—but all these patterns did not form a coherent whole.

Behold Paleolithic humanity, creatures that over the course of 2.5 million years evolved from primordial *Homo habilis* to the *Homo sapiens* jack-of-all-trades. These humans became aware of the many patterns in the world around them and stored them in their ever-expanding brains (there are indications that the human brain was at its largest in the late Paleolithic, only to decrease again in size starting in the Neolithic).[21] But an awareness of patterns is not the same as an awareness of deeper underlying principles. A search for the "one in the many" was neither natural nor necessary for Stone Age peoples.

1.2 The Neolithic: From Jack-of-All-Trades to Specialist

The Neolithic Revolution and the Inequality Pattern

The greatest change ever to occur in the history of humanity was undoubtedly the transition from a food-gathering culture to a culture based on food production.[22] After leading a nomadic life for 2.5 million years, humanity transitioned to sedentary life almost everywhere in the world. On a macrohistorical scale, this transformation took place at lightning speed: at around 10,000 BCE, people began producing food, including wheat, barley, and peas, in the Fertile Crescent (a contiguous area in the Middle East that includes parts of present-day Egypt, Israel, Palestine, Jordan, Kuwait, Lebanon, Syria, Iraq, Iran, and Turkey). Wherever a farming culture is established, the hunters and gatherers come under pressure. Around 7000 BCE, agriculture and animal husbandry spread from Anatolia to Palestine and Iran. At around the same time, people in Central America also transitioned to an agricultural lifestyle, with maize being the oldest crop (ca. 7500 BCE), even though there was no contact between the Old and New Worlds. Around 3500 BCE, we find agriculture and animal husbandry almost everywhere in the world. A radically different society emerged, where people settled in villages and lived in houses. Humanity itself became "domesticated" and formed a hierarchical society with much greater social inequality.

A great deal has been written about the causes of this Neolithic revolution and the emergence of social inequality. It is generally assumed that it was a shortage of land or food in combination with population growth that drove people to agriculture and livestock farming. But the unconscious domestication of plants and animals described above is also cited as the cause. Social stratification and inequality are a product of the Neolithic revolution, but it's not entirely clear which of the two is the cause and which is the effect. Moreover, the archaeological discoveries are ambiguous. All that can be deduced from research into the distribution of burial gifts, for example, is that social inequality thrives in a society with a permanent place of residence and a surplus of food.[23] But this should not be taken to mean that social inequality arises only with the introduction of agriculture and animal husbandry. Food surpluses can also occur in a sedentary hunter-gatherer culture where food is stored, leading to unequal distribution of wealth.[24] But such sedentary hunter-gatherer cultures always turn out to be in transition to pastoral or peasant life, and this transition from hunter to farmer brings further inequality. The possession of livestock leads to inherited wealth. Additionally, shepherds and farmers who specialized were more successful in expanding their livestock. This led to even more inequality, resulting in further specialization. While everyone seems to do almost everything in the Paleolithic, in the Neolithic we find specialized craftspeople, such as potters, masons, and weavers.

Like the domestication pattern, these patterns of specialization and (increasing) inequality were initially unconscious: they were no more than a by-product of the transition from a hunter-gatherer culture to a settled (peasant) existence. But as soon as Neolithic peoples became aware of the economic benefit of specialization, this initially unconscious specialization pattern came to be pursued consciously, along with the inequality pattern. So here again there is a process from an unconscious pattern to a conscious one.

This process is a recurring *meta*-pattern and therefore constitutes a recognizable trend in the history of knowledge: initially certain processes arise "organically," such as the coevolution of humans and dogs, the random selection of plant characteristics (such as with wheat), and the increase in social inequality. But from the moment that people become aware of this pattern, they can either embrace the pattern or reject it. So far, we have seen only the embrace of patterns once people become aware of them, but we also find instances of rejection in this book. However, the pattern of inequality seems to have been embraced everywhere—at least by those at the top of the inequality curve. Rejection may have been impossible for those at the bottom of the inequality curve. For those at the top of the inequality curve who would have been able to reject it, there was no advantage in doing so.

This is what is called a *positive feedback loop*, where the pursuit of a certain pattern leads to another pattern, which in turn reinforces the earlier pattern. Or to be more precise, positive feedback occurs when A leads to more B, which in turn leads to more A.[25] In the case of the inequality pattern, A is the specialization that leads to more inequality (B), which in turn leads to more specialization (A) (bearing in mind that specialization is not the only cause of inequality). If no adjustments are made from the outside, this process can go on indefinitely. The rich get richer and the poor get poorer—a meta-pattern known in modern sociology as the Matthew effect, based on the parable of the talents in the Gospel of Matthew: "For to all those who have, more will be given, and they will have an abundance; but from those who have nothing, even what they have will be taken away."[26]

It isn't until the Bronze Age (ca. 3000–800 BCE; see chapter 2) that the population increases so much—and the concomitant inequality—that we can speak of the emergence of states with profound social stratification: with a monarch at the head, ruling over a priestly class, an army, and subjects, with enslaved people at the bottom of the heap. Moreover, each social layer is itself layered: subjects can be rich or poor traders, weavers, or farmers, just to name a few stations in life. Whereas someone in Mesopotamia in 6000 BCE could still be born into an egalitarian society, a few thousand years later he would come into the world as a crown prince, subject, or enslaved.

The inequality pattern is one of the most persistent patterns in human history. Despite the many attempts to contain it, this pattern has not died out since it first emerged in the Neolithic.

Knowledge of Technological Production Patterns

The sowing, growing, harvesting, and processing of crops required new tools. The typical tools from the Neolithic are the mortar and pestle, the digging stick, the (hook) plow, and the sickle. Milk production (from cows, sheep, goats, and horses) also led to a search for ways to make raw milk keep longer by processing it, since raw milk goes sour quickly.[27] Thus we see the successive development of cheese (starting in 7000 BCE), butter (6500 BCE), and yogurt (2000 BCE).

Together with the development of new food products, a demand for storage also arose. Pots, barrels, and jars heralded the beginnings of pottery technology, in which water is extracted from molded clay by firing it, turning it into earthenware. Clay was also baked in the Paleolithic but mainly for art objects. The Neolithic furnaces could be fired up to around 900°C (1650°F), a temperature sufficient for earthenware but not for metalworking. At the end of the Stone Age, we do see the use of *unprocessed* copper for making axes, spearheads, arrowheads, and other weapons. It was this sort of ax that was carried by Ötzi, the iceman mummy of fame, who dates from 3300 BCE and was found in the Italian Alps in 1991.[28]

Wool and cotton were also first produced in the Neolithic. Textile production can be described as a serial pattern: (1) sheep are sheered or cotton is harvested, (2) thread is spun, (3) looms are constructed, (4) cloth is woven, (5) cloth is dyed, and (6) garments are made.[29] The first five of these steps were totally unknown in the Paleolithic. So, the transition from Paleolithic to Neolithic also entailed a transition from short to long production patterns, sometimes even serial patterns, which often consisted in subpatterns. It is partly these series-based production patterns that contributed to increasing specialization and social inequality.

Despite all the technological innovations, Neolithic people were no healthier than their Old Stone Age ancestors. Quite the opposite: the production and storage of grain and milk products led to a diet that was less varied and to people who were both unhealthier and smaller, as can deduced from the skeletons found. This unbalanced diet resulted in more illnesses, while leading to more births because more children could be fed at the same time. The net effect was population growth and the emergence of the first cities, such as the walled city Jericho, which dates as far back as 7000 BCE.

Horizon Astronomy and Stone Circles: Construction Patterns

The New Stone Age shows increased knowledge of the movements of the sun and moon. This is evident from the thousands of rings of standing stones dating from around 7000 BCE that were constructed at various sites throughout the world. A stone henge consists of a group of large upright stones, called megaliths, arranged in the form of a circle or an ellipse. The number of megaliths per henge can vary considerably: from 4 to over 60. There can also be stones that lie sideways on the upright stones. The largest concentration of stone circles can be found in the British Isles and in Brittany. More than 1,000 have been found there, of which Stonehenge (ca. 2500 BCE) is the most famous.

It is usually thought that these stone circles served a religious purpose, but many also seem to have had an astronomical or calendar function.[30] For example, when we look at Stonehenge from the center of the circle, the midsummer sun rises exactly behind the so-called Heel Stone (the top right stone in figure 5). The monument is also aligned with the midwinter sunrise, which like the midsummer sun corresponds to a solstice. The directions of sunrise at the beginning of spring and autumn are also indicated, on the equinoxes, when day and night

Figure 5. Midsummer at Stonehenge. CreativeCommons, *Nordisk familjebok*, 1918, p. 27:115; https://commons.wikimedia.org/wiki/File:Stonehenge_vid_midsommar_1700_f_Kr,_Nordisk_familjebok.png.

are the same length. In addition, Stonehenge marks the more complex movements of the moon along the horizon. According to some archaeological astronomers, the central observation position is not at the center of the Stonehenge circle but at the Heel Stone itself.[31] This would mean that Stonehenge focuses on the observation of the midwinter sunset rather than the midsummer sunrise. This is not entirely unlikely, since in many cultures the winter solstice is a metaphor for death and rebirth and is more important than its summer counterpart.

Whatever the case may be, everything indicates that Stonehenge shows a one-dimensional projection on the horizon of the movements of the sun and the moon. This is also referred to as horizon astronomy, in which patterns in the movements of celestial bodies are recorded over time along the horizon. For us, these patterns remain implicit in the circular stone construction. So, the evidence for these patterns is indirect: we do not have inscriptions about what a ring of megaliths means or how it was used. And we know even less about what the builders of these stone circles thought about the sun and the moon. Did they consider these heavenly bodies to be gods? And was Stonehenge a holy place to them? All we know is that the monument marks the movements of the two most prominent celestial bodies.

If we compare the search for astronomical patterns in the Old and New Stone Ages, what is striking is that in the Paleolithic, as far as we know, the only search was for the lunar cycle, while in the Neolithic there was a search for a much larger number of patterns in movements of the sun and moon. This difference in the number of patterns sought and found is not surprising, considering that systematic knowledge of the seasons is vital for a food-producing culture, whereas it is of no importance, or at least of much less importance, for the hunter-gatherer life.

The Neolithic Revolution Depicted in Painting

The transition from a nomadic existence to a sedentary one can also be seen in painting. One of the murals from the Neolithic Anatolian settlement of Çatal Höyük (ca. 6,150 BCE) arguably represents the world's oldest depiction of a landscape.[32] In the foreground there are the rectangular houses of the town, while in the background a volcanic mountain rises that could be identified as Hasan Dağ. If this interpretation is correct, this wall painting would be the oldest known map of a settlement.

In other murals at Çatal Höyük, we encounter hunting scenes, which were also popular in the Old Stone Age. Although agriculture and animal husbandry were widely practiced there, hunting remained an important food source, as shown in the scene in figure 6 (from around 6000 BCE). However, this hunting scene is very different from the Paleolithic paintings in figures 1 and 2. We see a group of hunters working with something that looks like a red bull. The painter depicts the weapons in detail, including bows and arrows. But the most impressive aspect is the multitude of positions and actions of the hunters. Some of them are running, others are shooting, and yet others are watching the hunting scene. As in Paleolithic painting, a method is used in which only the most important parts of the bodies are shown: whereas the animals are depicted from the side, with humans only the head is shown from the side, while the torso with arms and legs are shown from the front.

Figure 6. Hunting scene from Catal Höyük, around 6000 BCE. CreativeCommons, photo by Omar Hoftun, 2013; https://commons.wikimedia.org/wiki/File:Mural_from_%C3%87atalh%C3%B6y%C3%BCk_excavated_by_James_Mellaart_showing_neolithic_hunters_attacking_an_aurochs_(Bos_primigenius)..jpg.

So there seems to be a tendency running throughout the entire Stone Age: people and animals are depicted in such a way that the most important parts of their body are visible. For the depiction of people, this led to a front-on representation, while animals were drawn in profile. Furthermore, in Çatal Höyük we encounter a shift to a more narrative structure: whereas Paleolithic paintings mainly depict single animals, Neolithic paintings show an entire scene.

On the Cusp of the Bronze Age and Early Antiquity: Knowledge of Writing

If there is anything in which systematic knowledge in the later Stone Age is essentially different from that of the early Stone Age, it is in the early development of writing. Writing was not a sudden invention; it started with the first ideograms and pictograms. Ideograms express ideas or concepts, and if these signs resemble a physical object, they are referred to as pictograms. Combinations also occur.

The oldest of these characters are found in China (around 7000 BCE) and are known as the Jiahu characters.[33] However, their meaning is unknown. There are also the Vinča characters from Romania from around 6000 BCE, which we are also unable to interpret, although it is assumed that they relate to rituals.[34] The same applies to the Kish tablet from Sumer from around 3500 BCE, which contains the oldest form of proto-cuneiform script.[35]

Like ideograms, pictograms are not suitable for expressing sentences consisting of multiple words. This is because in addition to content words—such as nouns and verbs—human languages also contain function words, such as articles, conjunctions, and demonstrative pronouns, which do not refer to physical objects as expressed by pictograms. One of the revolutionary developments in writing was the insight that function words could also be represented with signs. In this way, any sentence in a language could be expressed by a series of characters, representing variously content words or function words. Statements, reports, stories, hymns, laws, treaties, contracts, and so forth were recorded verbatim—the first time being in Sumerian script, called cuneiform because of its characteristic little "wedges," *cunei* in Latin (see chapter 2). Although starting around 3000 BCE we are actually talking about the Bronze Age or early antiquity rather than the Stone Age, the transition from ideograms to alphabetic writing starts as early as the late Neolithic. At around 3400 BCE, we find a shift in cuneiform script from ideograms and pictograms to logograms (signs or characters representing a word or phrase). In addition, there was also a shift in the cuneiform script to phonograms: signs expressing sounds, much like the Latin alphabet, which developed later. The oldest phonograms were inspired by logograms, the sound of the phonogram corresponding to the first or last sound of the word to which the original logogram referred. One of the places where we see this is in the development of the precursors to the Latin alphabet, such as in Phoenician, where the first letter, *aleph* (today's letter *a*), originally meant "ox," and the second letter, *beth* (our letter *b*), initially meant "house."[36]

Although most writing systems (like cuneiform) have followed the process from ideograms, pictograms, and logograms to phonograms, this is not the case with all writing systems. Furthermore, a writing system does not indicate which combinations of signs produce well-formed words or grammatical sentences. The question of whether there is an underlying system (a grammar) that can predict the correct combinations of characters would not appear until centuries later (see chapter 3.1).

In any case, Sumerian cuneiform script was an overwhelming success: practically all peoples who came into contact with Mesopotamian civilization adopted the idea. However, it should also be mentioned that other peoples have developed writing systems independently of Sumerian, an example of which is the Zapotec script that we encounter around 600 BCE in Central America.

1.3 Conclusion: Stone Age Patterns from All Regions

Science and scholarship are usually thought to have had their beginnings in ancient Babylonia or even later, in classical Greece. However, the search for systematic knowledge appears to be thousands of years older than that, as we see in dash patterns representing the lunar cycles and the early development of writing. Moreover, the oldest remains of this search are in places tremendously far from each other. We conclude from this that people did not develop their pattern-seeking practices in Europe, Asia, Oceania, or America, but that they must have taken them with them when they left Africa. And many of the patterns found are still in use. These Stone Age patterns "of lasting value" include the pattern of domestication from the Old Stone Age and the patterns in the early development of writing in the New Stone Age. Domestication served as the engine for many later developments: breeding and growing led to a food-producing culture that became the driving force behind new technology and increasing specialization accompanied by social inequality.

From Unconscious Patterns to Conscious Ones

The transition from unconscious to conscious patterns is a recurring process and, as such, constitutes a tendency in prehistory. We encountered this with the domestication of plants and animals, as well as with the emergence of inequality. Domestication of the proto-dog initially took place unconsciously; afterward the technique was used consciously to domesticate other animals and plants. The pattern of inequality was also created unconsciously as a side effect of domestication and of the resulting transition to a sedentary culture, after which the pattern was maintained by those who benefited from the inequality. A process from unconscious patterns to conscious ones may also have taken place in the development of writing.

We can also describe the process of shifting from passive recognition of patterns to the conscious search for them as a transition from mere *perception* to *apperception* (conscious perception) of patterns. The perception of patterns often occurs unconsciously and is not unique to humans: almost all animals perceive and use patterns.[37] But it is highly questionable whether animals are consciously looking for patterns. And we have never detected deeper principles with animals, although we have yet to hear the last word on this question.

Implicit versus Explicit Knowledge

All Stone Age patterns are more or less implicit. For example, the pattern of side-view representation of animals can be deduced from the data only indirectly. The patterns of the lunar cycle engraved in bones are much less implicit since dashes are used to tally. However, no pattern is completely explicit unless it is described or explained as such. For this reason there is no evidence for the apparent transition from implicit to explicit knowledge in the Stone Age, and this naturally also applies to the aforementioned process from unconscious patterns to conscious ones. It isn't until the Bronze Age, or early antiquity, that patterns are explicitly described for the first time.

No Awareness of Principles

In the Stone Age we perceive an awareness of patterns but no awareness of principles that generalize over patterns. The many Stone Age patterns do not show any further coherence, and this seems to correspond to the survival strategy of the Paleolithic human as a kind of a jack-of-all-trades. The Neolithic transition from jack-of-all-trades to specialist—or the transition from some knowledge about many things to an abundance of knowledge of some things—could have a parallel in the transition from the search for patterns to the search for principles: in both transitions we see a shift from many to one. But it isn't until the Bronze Age that we find the first concrete principles (chapter 2).

CHAPTER TWO

The Explosion of Patterns and the Awareness of Principles

Early Antiquity

3000 BCE–600 BCE: Fertile Crescent, China, India, Europe

The transition from what we call prehistory to the period that we usually term antiquity coincides with the transition from the Stone Age to the Bronze Age. The invention of writing is usually considered to be a turning point. However, this invention takes place at different times depending on the region: at around 3200 BCE in Mesopotamia, around 3000 BCE in Egypt, around 1200 BCE in China, and between 1000 and 600 BCE in Mesoamerica.

We divide antiquity into two periods: early antiquity, from 3000 to 600 BCE, with its focus on the search for patterns (this chapter), and classical antiquity, from 600 BCE up to 500 CE, in which the main game is the search for principles (chapter 3). In early antiquity, Mesopotamia, especially the kingdom of Babylonia, is the region that devoted the most attention to systematic knowledge. From 3200 BCE, the first Mesopotamian civilizations used cuneiform script for administrative purposes. This writing system developed and spread quickly, and by 2700 BCE it was being used throughout the region to record agreements, contracts, and treaties on clay tablets, written by specially trained clerks. Starting around 2600 BCE, clay tablets were being used to record laws, dictionaries, celestial observations, astrological signs, chronicles, meteorological observations, mathematical calculations, and medical diagnoses. And thus began what we can call the first knowledge disciplines. It is estimated that between one and two million clay tablets have been excavated, of which 100,000 have been deciphered and published to date.

These and other sources offer us a glimpse into early antiquity, a world many times more detailed than prehistory. Massive data collections were assembled in Babylonia in particular. But while in Babylonia there was an explicit search for patterns, the same can hardly be said of a search for underlying principles. In this chapter, in addition to Mesopotamia, we will also look at Egypt, China, and India, while Europe will come up a few times as well. In other regions, such as pre-Columbian America and Oceania, science and scholarship do not blossom until after 500 CE (see chapter 4).

2.1 Linguistics: Babylonia's Unique Case

Nothing is so obvious as language: it is part of our daily existence, but we are usually not conscious of it. Although the Mesopotamians were able to write starting around 3200 BCE, the study of language in Babylonia—that is, the collection, analysis, and interpretation of language data—didn't begin until around 1600 BCE. But that is still a thousand years earlier than anywhere else in the world.

Today, linguistics has a somewhat ambiguous reputation. On the one hand, it is one of the most thriving disciplines in the humanities; on the other hand, there are so many different schools that detractors claim that there are more linguistic theories than there are linguists.[1] But there are occasions where linguists from the most diverse schools look for consensus. It isn't at large conferences that this happens but at intimate locations, such as at Villa Serbelloni in Bellagio, Italy, which now belongs to the Rockefeller Foundation. It was at the beginning of this century that I, as a young computational linguist, was participating in a discussion on the question, Are there linguistic phenomena that are recognized by all linguists and that are common to all languages? After a day of discussion, it was agreed that the phenomenon of *discontinuous relationships* (also called *structural relationships*) was a serious candidate.

Let me use an English sentence to illustrate this phenomenon: *The dog on the hill barked*. In this sentence, there is a connection between *dog* and *barked* but not between *hill* and *barked*, even though these last two words are directly adjacent to each other. That is, it is that dog who barks, rather than the hill, even though in the sentence the word *hill* comes between *dog* and *barked*. But no native speaker of English would parse the sentence incorrectly, interpreting the hill as the thing barking. One might hypothesize that this is because, from a semantic perspective, a hill cannot bark. But that can't be the reason, because even if we take a sentence like *The young dog next to the old dog barked*, then *barked* is interpreted as referring to *the young dog*, rather than to *the old dog*. So apparently, words in a sentence (such as the subject and the predicate) don't necessarily have to be contiguous to be in a relationship; they can just as easily be discontinuous. In fact, relationships within a sentence can span over arbitrarily long distances, as in the sentence *The dog under the tree next to the house on the hill barked*. For this reason, this phenomenon is rightly regarded as one of the most important characteristics of human language: not only in English but in all known languages, relationships between words and between phrases can be discontinuous. It now appears that this particular characteristic of language was first described in Babylonia around 1600 BCE but in relation to different parts of a word rather than individual words in a sentence.[2]

Babylonia: Discontinuous Patterns within Words

The circumstances under which the Babylonian study of language came about are similar to those that would occur later elsewhere in the world:[3] people wanted to preserve old literature written in a dying language. In Babylonia this ancient literature, exemplified by the famous *Epic of Gilgamesh* from the 21st century BCE, was written not in their own language, which was Akkadian, but in Sumerian. In the 3rd millennium BCE, a cultural symbiosis between the Sumerians and the Akkadians had taken place, where Sumerian had a major influence on Akkadian, especially with regard to pronunciation and loanwords. This is all the more notable considering that the two languages were not related: Sumerian is a so-called linguistic isolate, having no known relatives, whereas Akkadian is the oldest known Semitic language. In Babylon's heyday, around 2000 BCE, Akkadian gradually came to replace Sumerian, but the Babylonians wanted to retain their knowledge of the language because it was used in ceremonial, literary, scientific, and scholarly works.

Where did the Babylonians need to begin if they wanted to save a language that was not their own? A dictionary was considered the first requirement. But the way words were used in their linguistic context had to be recorded as well, both in Sumerian and Babylonian, so that they could serve as an aid for translation. Conjugations, inflections, and compound words—that is, the morphology—also had to be recorded for both languages. The precise rules for word order—the syntax—were considered of less importance. This latter point is not surprising considering that the most striking language patterns are the regularities in the conjugations and inflections of words; the Sumerian verb *gar* (to place) has at least 227 different forms.[4]

What is remarkable is that the Babylonians reported a phenomenon that is still a current issue in linguistics even now after 36 centuries have passed: *discontinuous relationships*.[5] Although the examples given above of discontinuous relationships concerned relationships between independent parts of a sentence, such as the subject and the verb, discontinuous relationships can also exist within compound words—not just in Sumerian but also in modern Germanic languages such as English. Word compounds in English can be easily constructed by sticking together existing words. So, by combining the words *freedom* and *proponent*, we can create the compound phrase *freedom proponent*, which we would interpret as a person who advocates freedom. We call the relationship between the two parts of this compound phrase adjacent or continuous: nothing intervenes between the two words in this compound. But we can also create a compound with the same relation between those two words even if we don't make them adjacent. Let's imagine that rather than *freedom* in general, what is at issue is more specifically *freedom from violence*. We could then refer to someone who advocates for this type of freedom as a *freedom from violence proponent*, which is thus still a sort of freedom proponent and not a proponent of violence, even though the words *violence* and *proponent* are adjacent to each other while *freedom* and *proponent* are not. So apparently, a nonadjacent relationship is sometimes required to construct an acceptable compound in English.

Something similar applies in Sumerian, but within verbs. For example, the OBGT VI clay tablet lists the following conjugations and constructions for the verb *gar*, "to place" (table 1).[6]

Table 1 Transcribed forms of the Sumerian verb *gar*, with translations into Akkadian, on the OBGT VI clay tablet

	Sumerian	Akkadian	
VI § 2:	gar-**bí**-íb	šuškin	(make **someone** place it)
VI § 4:	gar-ra-**ni**-íb	šuškiš**šu**	(make **him** place it)
VI § 19:	gar-**mu**-ub	šuškin**anni**	(make **me** place it)
	ga-**ri**-íb-gar	lušaškik**ka**	(make me place it for **you**)

Note: For each example, the parts in boldface (added by me) show how particular parts of the word in Sumerian are translated.

What is striking is the rich word structure of Sumerian and Akkadian. An entire verbal cluster, including personal pronoun and direct object, can be expressed with a single word. There is no real equivalent to this in English, but similar cases do exist in other European languages, such as Italian and Spanish. Take the single Italian word *diciamoglielo*, for example, which means "let's tell it to him." This single word expresses a combination of a verb in the imperative (*dire*), with an implicit subject "we" (*diciamo*), an indirect object (*gli*), and a direct object (*lo*).

What we find in Sumerian but not in a language like Italian is the use of *infixes*, internal affixes that lead to discontinuous relationships or dependencies. These are the infixes *bi*, *ni*, *mu*, and *ri*, which must always be placed inside the verb form *garib*, for example, *gar-**bí**-íb* "make someone place it" and *gar-ra-**ni**-íb* "make him place it." The verb is split in two, as it were, and another word (in this case a personal pronoun) is placed in the middle of the verb, resulting in a noncontinuous dependence on the outer parts of the verb and leading to a new meaning of the word as a whole. It is quite possible that the Babylonian lexicographers became aware of this insertion pattern only when they compared these constructions in Sumerian and Akkadian as translations (as in table 1). It must have been a eureka moment for them when they realized that a meaningful unit changes in the middle of a word, while the parts to both sides of it remain the same.

However, no further explanation of the pattern is to be found on any of the clay tablets. In contemporary terminology, we could present what is going on as follows: if x, y, and z stand for parts of a word or sentence (let's call them "linguistic units") and if we use a subscript i to indicate a relationship between these units, then we can see the discontinuous pattern as $x_i\, y\, z_i$. This representation summarizes discontinuous relations not only in Sumerian but in other languages as well, as long as x, y, and z can stand for linguistic units of arbitrary size: ranging from sounds, syllables, and words to entire phrases.

There are also linguistic clay tablets that show no sign of a search for regularities. This is the case, for example, with fixed verbal expressions and sayings in Akkadian and Sumerian (OBGT VII–X). English examples would include phrases like *to kick the bucket* and *to throw in the towel*. These expressions can't be translated literally into other languages without losing their meaning. In most cases these constructions are not based on rules or patterns. The Babylonians listed such exceptional cases in detail, in addition to the pattern-based constructions, and thus documented both *patterns of similarity* and *patterns of dissimilarity*. That makes the Babylonians the first linguists to discover non-rule-based translatability of idiomatic expressions. Not only were Babylonian lexicographers interested in regularities; they were equally passionate about recording all possible irregularities and differences between Akkadian and Sumerian. The distinction between patterns of similarity and patterns of dissimilarity will go on for millennia in the search for knowledge, in domains ranging from linguistics to philology and from Chinese astronomy to Roman law (see chapter 3).

No Linguistics Elsewhere in Early Antiquity

The linguistics we encounter in Babylonia is unique. As far as we know, for a thousand years nowhere else in the world do we encounter any investigation into the regularities or irregularities in language. This contrasts with disciplines such as mathematics, astronomy, and legal studies, where this sort of investigation does take place in other regions during early antiquity. It isn't until classical antiquity, around 600 BCE, that we first find linguistic activities outside of Babylonia—in India, China, and Greece.

2.2 Mathematics: First Awareness of a Principle?

The basis of arithmetic is counting. But what do we actually know about counting? And what are the underlying rules? The oldest surviving counting patterns in the world are the dash patterns of lunar cycles from the Paleolithic (see chapter 1.1).[7] But this is a case of tallying, not of counting. When did people make the transition from tallying to counting and subsequently to searching for patterns in numbers and shapes?

Babylonia: The Result Is More Important Than the Road There

All our knowledge about Babylonian arithmetic and geometry is extracted from 400 clay tablets, the oldest of which dates from the 3rd millennium BCE.[8] The first thing that strikes us is that the Babylonians do not have a counting system based on 10s (the decimal system), which seems an obvious choice since they counted on their fingers. Nor did they have a system based on 12s. This "duodecimal system" is something we encounter in many other places and that could also be considered obvious, based as it is on the number of phalanges (finger bones) of the four fingers of a single hand, yielding a total of 12, with the thumb being used as a pointer. But the Babylonians opted instead for a sexagesimal system—that is, a system based on 60—which they adopted from the Sumerians, who may have developed it by folding the 10-based and 12-based systems together.[9] The advantage of a sexagesimal system is that 60 is divisible by a large number of other numbers (2, 3, 4, 5, 6, 10, 12, 15, 20, and 30), meaning that a fraction often results in a whole number. This property facilitated all kinds of arithmetic operations that were necessary for the growing trade in Mesopotamia. But it had the additional benefit of facilitating the calculation of time, because the Babylonians' basic unit of time—the hour—could be divided into equal parts of 30, 20, 15, 12, 10, 6, 5, 4, 3, and 2 minutes. This 60-based system was adopted in large parts of the Hellenistic world. Carried along with the Romans and the later European expansion, use of the sexagesimal system stretched across the globe, not for counting but for measuring time and angles. The hours, minutes, and degrees we use today are still expressed in the sexagesimal system, a notable anomaly in a culture where the decimal system is dominant.

The Babylonian number system is also the oldest *positional notation*, where the value of a digit is determined not only by the digit itself but also by its position in the number. This insight is still relevant today: almost all contemporary number systems use positional notation. For example, in our decimal system, the numeral 7 in the number 73 has the value not of 7 but of 70. Without positional notation, calculating with large numbers would become extremely impractical. The Babylonian number system was based on a systematic pattern: it was a digit's position in a number that determined its magnitude, with each step to the left signifying one power greater. Equipped with their positional notation, the Babylonians could more easily discover regularities in numbers, as compared with the Egyptians and especially with the Romans later on. Obviously, this positional notation pattern is not a "natural" pattern; it is an artificial pattern devised by people, but it is just as relevant to our search for patterns in history.

The oldest clay tablets with series of numbers date from around 2600 BCE and consist of arithmetic tables such as multiplication tables.[10] There are also tablets from the same period with exercises that seem to come straight from Mesopotamian schools and arithmetic lessons. Around 2000 BCE we also find clay tablets with squared (from 2^2 to 59^2) and cubed (from 2^3 to 31^3) values. In addition to these arithmetic tables, the Babylonians devised methods for solving quadratic equations. These solutions are again presented in the form of tables, reversed from the previous tables to find the square roots needed to solve the equations. The Babylonians were also able to solve a number of cubic equations without using any algebraic notation. Solving equations must have been cumbersome, but it was quite important for all sorts of practical problems, such as determining the dimensions of a rectangular strip of land along the Euphrates if only the area was given.

The most fascinating of all mathematical clay tablets is Plimpton 322 (figure 7), which dates from the time of Hammurabi (ca. 1800 BCE), the first king of the Babylonian Empire. For a long time it was thought that this tablet contained nothing more than accounting data, until mathematicians in the 1940s discovered that the numbers corresponded to Pythagorean triples, that is to say, integers a, b, and c, such that $a^2 + b^2 = c^2$.[11] These numbers represent the two legs (a and b) and the hypotenuse (c) of a right triangle, respectively. To illustrate, the set (3, 4, 5) constitutes a Pythagorean triple because $3^2 + 4^2 = 5^2$ or $9 + 16 = 25$; other such examples include (5, 12, 13) and (8, 15, 17). These three triples consist of small numbers that could be discovered by systematically trying out different combinations. But the list on Plimpton 322 includes much larger Pythagorean triples, such as (3456, 3367, 4825) and even (13500, 12709, 18541); see table 2, which shows the triples in decimal notation.

An underlying principle is required to generate these complex Pythagorean triples; one cannot simply discover them through trial and error, because the

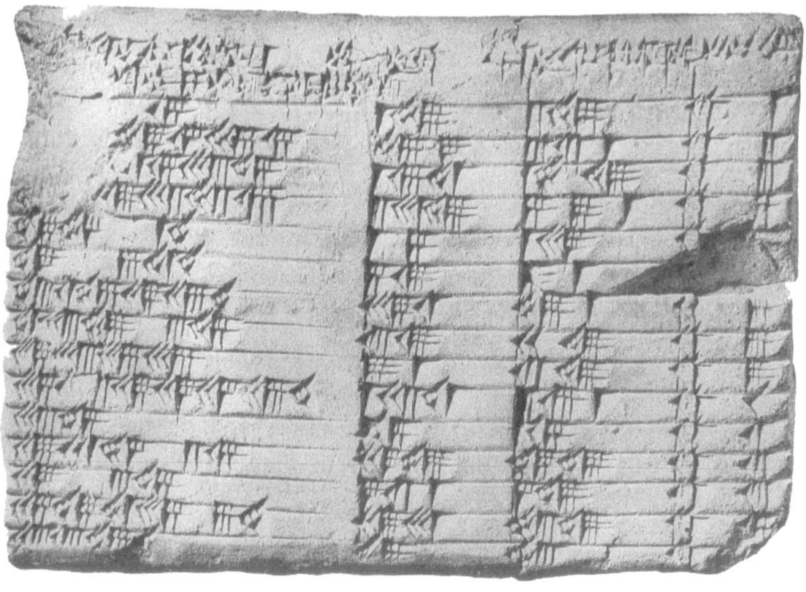

Figure 7. Tablet Plimpton 322 with Pythagorean triples. Photo author unknown, https://commons.wikimedia.org/wiki/File:Plimpton_322.jpg.

Table 2 15 Pythagorean triples on the Plimpton 322 clay tablet

a	b	c
120	119	169
3456	3367	4825
4800	4601	6649
13500	12709	18541
72	65	97
360	319	481
2700	2291	3541
960	799	1249
600	481	769
6480	4961	8161
60	45	75
2400	1679	2929
240	161	289
2700	1771	3229
90	56	106

number of combinations of numbers for all possible triples is too large to yield a successful triple in a reasonable amount of time. It is not known what principle or method was used, but mathematicians who studied Plimpton reason that the procedure may not have been very different from the method described many centuries later by the Greek mathematician Euclid (ca. 300 BCE). Euclid's method works as follows: choose random integers p and q with $p > q$ and let $a = 2pq$, $b = p^2 - q^2$ and $c = p^2 + q^2$, then $a^2 + b^2 = c^2$ applies.[12] This can be easily verified by entering the values above for a, b, and c in $a^2 + b^2 = c^2$ and working them out. While the procedure itself is simple, discovering it requires considerable mathematical insight. For this reason, it was long thought that such an underlying principle for generating Pythagorean triples was not discovered until the time of Euclid. But we now know that such a method must date from at least 1,400 years earlier.[13]

If Babylonian mathematicians really did use this method or some similar one, why didn't they explain it or write it down anywhere? One of the greatest enigmas of the Babylonians is that they describe the result but not how they obtained it, although it could be that all the methodical clay tablets have simply disappeared (that, however, is unlikely). But what is certain is that the Babylonians must have been aware of an underlying principle, because without it, discovery of the larger Pythagorean triples is infeasible. Awareness of mathematical principles must therefore have begun with the Babylonians no later than 1800 BCE. This is the second important turning point in the history of systematic knowledge: the first turning point was becoming aware of patterns in the Paleolithic (see chapter 1.1). To the best of our knowledge, it is around 1800 BCE that the second turning point takes place: the awareness of principles.

Do we know anything about the author of the Plimpton tablet? Was it really a product of a mathematical genius? The Plimpton tablet is anonymous and thus fits in with the Babylonian tradition in which anonymous literary masterpieces were usually dedicated to deities or kings. Intriguingly, however, the structure of the Plimpton tablet resembles that of a school tablet: it repeats the same mathematical pattern 15 times, each time with a different triplet. This also explains the seemingly random choice of the triples. For this reason, the tablet itself was probably not the result of some deep mathematical thought, but an answer key that could be used to check the answer to a math problem without having to carry out the same calculations over and over again.[14] There must have been some older mathematical insight underlying this tablet, of course, but no record has been found. The most likely candidate is tablet YBC 7289 from around 1800 BCE.[15] This tablet bears a diagram showing that the Babylonians also knew the geometric implication of $a^2 + b^2 = c^2$ (1,200 years before Pythagoras; see chapter 3.4). The diagram consists of a square with diagonals, and according to the Pythagorean theorem, the length of the diagonal is the length of the side multiplied by the square root of 2. An accurate approximation of $\sqrt{2}$ is given along one of the diagonals (in sexagesimal notation). So, in addition to arithmetic, the Babylonians were also active in the field of geometry.

Egypt: Elaborations of Solutions

The oldest Egyptian papyri on arithmetic and mathematical topics date from the Twelfth Dynasty (1990–1800 BCE). Many of these papyri contain mathematical problems with solutions, often in the form of questions and answers, which therefore seem to be written for students. What is striking is that the Egyptians, more than the Babylonians, were interested in elaborating a solution, although, just as with the Babylonians, we never see any explicit mention of mathematical principles. Egyptian arithmetic was based on a decimal system, which spread farther through later Hellenistic mathematicians and today is the most widely used numerical system in the world.

One of the most important mathematical documents from ancient Egypt is the Rhind Papyrus, dating from around 1650 BCE.[16] It contains 87 problems concerning multiplication, division, and linear equations, plus the oldest mention of prime numbers. We also see an approximation of π as 3.16045, a margin of error of 1 percent. Later papyri from around 1300 BCE contain solutions for quadratic equations, like those we found in Babylonia half a millennium earlier. And we also see the use of Pythagorean triangle geometry in agriculture.

Math in Other Regions

Mathematical sources from early Chinese antiquity are few and far between. Mathematics first begins to flourish in China in the classical period (see chapter 3.4). In contrast, we have mathematical sources from India that may come from early antiquity, but these sources are very difficult to date and may actually be much more recent. There are several sutras that mention prime numbers, cubic roots, and irrational numbers. In the *Sulba Sutras* (thought to have originated between 800 and 500 BCE), the square root of 2 was approximated to no fewer than 5 decimal points. A methodical principle is also explained for Pythagorean triples, a few centuries before Euclid.[17] So everything seems to indicate that Babylonian and Indian mathematicians, as well as their later Chinese counterparts, all discovered the Pythagorean theorem independently.

2.3 Astronomy and Astrology: Patterns in Planetary Movements and Eclipses

The step from mathematics to astronomy was a small one in ancient times. This had everything to do with the importance of calendars, which are both arithmetical and astronomical. The star-studded sky was studied all over the world, but the Babylonians were the first to apply mathematics to it. Knowledge of the stars and planets was considered important for agriculture; for predicting the weather, calamities, and wars; and for identifying omens. Like other ancient peoples, the Babylonians believed that the gods spoke through celestial phenomena to communicate information about the future. Accordingly, celestial observations were primarily intended as a means to better understand the world around them, but the Babylonians also discovered astronomical patterns in these observations.

Babylonia: From Wandering Stars to Patterns in "Big Data"

One of the oldest astronomical clay tablets dates from the time of King Ammizaduga, a great-grandson of the great King Hammurabi (17th century BCE). The Venus tablet of Ammizaduga records the rise and set of the planet Venus with accompanying astrological interpretations concerning climate, disease, war, and love. The tablet contains the oldest known formulation of a regularity in the movement of a planet.[18] We have already seen that the recurring patterns of the moon and the sun in the sky have been observed since prehistoric times (see chapter 1), but planets were of a different nature: they were wandering stars that moved irregularly between the fixed stars. As we now know, their erratic movement arises from the fact that planets have their own orbit around the sun that differs from that of the earth. This makes their movement across the firmament look like complex loops with backward motions. In reality, of course, the planets do not make retrograde loop motions; they revolve around the sun. But the Babylonians were not aware of all this: they observed the wandering movements on the firmament without an underlying geometric model. And at first glance, especially over short periods of time, these movements seem irregular and erratic. But the Babylonians continued making observations over long periods: the Venus tablet of Ammizaduga contains observations that span no fewer than 21 years. As a result, Babylonian astronomers discovered that the rise and set times of Venus repeat themselves almost exactly every eight years. They had discovered regularity in the apparent irregularity!

And that is what opened the floodgates: Babylonian astronomers began their quest to discover as many patterns in the firmament as they could. For example, the series of clay tablets known as MUL.APIN, whose name refers to the constellation the Plow, contains tables with the rise and set of planets, the duration of days as measured by water clocks, and the rise and set of the moon.[19] What makes MUL.APIN particularly special is that patterns derived from these tables can be used to predict future rises and settings of celestial bodies. For example, the MUL.APIN II (ii 43–iii 15) tablet includes the following description for calculating the visibility of the moon (showing the numbers in the sexagesimal system): "4 is the visibility coefficient for the moon; one multiplies 3 minas by 4 and gets 12, the visibility of the moon. One multiplies 40 ninda, the difference between daytime and nighttime, by 4, and gets 2;40, the difference in visibility." We won't go into the exact meaning of *mina* and *ninda* here (which is still a point of discussion among historians).[20] What is important is that the above description explicitly reports on a pattern that can be used to calculate what is called the visibility of the moon and the difference in visibility. Lunar visibility is described as "visibility coefficient of the moon $\times$ 3 minas," making this text fragment one of the most explicit patterns of early antiquity. The MUL.APIN II tablet thus differs essentially from the mathematical Plimpton tablet (see above). On Plimpton the Pythagorean triples are listed without the mathematical rule (the Pythagorean theorem), whereas MUL.APIN contains both the tables and the mathematical rule: the data and the pattern appear together.

We can best compare this form of astronomy with contemporary *data science*: large amounts of data are searched for patterns using specific procedures (algorithms) but without appealing to a deeper theory or underlying principles. Although the precise procedures used are unknown, what is clear is that the Babylonian astronomers searched for recurring sequences in the observations they recorded. For example, they found recurring (numerical) sequences in the number of days that a planet is visible, in the variation in how long a day was throughout the year, and in the dates of lunar and solar eclipses. We must bear in mind that these regularities were seldom or never seen as a result in themselves but in terms of their consequences for life on earth, especially for the king and the state. It is perhaps because of their effect on earthly life that many of these astronomical patterns were represented in causal form. The 70-part tablet series Enuma Anu Enlil (In the days of Anu and Anlil) contains an amazing 7,000 celestial phenomena with their consequences for daily life, as seen in the example below:[21]

- If the moon is visible on the first day of the month, then there is reliable language and the country will be happy.
- If the appearance of the moon bears a crown, the king will reach his highest status.
- If a lunar eclipse occurs in the evening during the month of Ajaru, the king will die.

What is striking is that these predictions follow an *if–then* pattern. They have the form "*if* <sign and/or date> *then* <event>." They seem to express a causal relationship, with a given phenomenon leading to a certain consequence. We have not encountered causal relationships with patterns that express mathematical relationships, even though they also have consequences for daily life. Consider the use of the Pythagorean theorem in agriculture to determine the size of a piece of land. But these patterns were not given a causal interpretation, in contrast to astronomical patterns, which were thought to have a direct causal relation to an event on earth that we might therefore call astrological patterns (bearing in mind that the Babylonians made no distinction between astrology and astronomy). For example, the third line in the list above states that a lunar eclipse in the month of Ajaru is the cause of the king's death, or at least that is the omen for it.

Contrary to what we would expect with these kinds of astrological predictions, the patterns in question are in all likelihood empirically substantiated.[22] The third line, for example, was probably deduced from some event in the past where a king actually died during an evening lunar eclipse in the month of Ajaru. This type of deduction is based on the following reasoning: if omen A is associated with event B at some point in the past, then the repetition of A will lead to the repetition of B. This reasoning does not mean that the king will die during every evening lunar eclipse in the month of Ajaru. The *if–then* construction should not be read as a logical implication.[23] Only under identical circumstances (ceteris paribus) will the king die. But identical circumstances occur only rarely if at all, and some special circumstance can always be found to prevent the rule from applying. But the Babylonians never left the matter up to chance: when a lunar eclipse was predicted in the month of Ajaru, they sought a replacement for the king. After the lunar eclipse, they made the replacement disappear, possibly by poisoning. And in this way fate was deceived.

Astronomical observations reach a peak starting at the end of the 8th century BCE. These observations come from what is known as the Astronomical Diaries, which constitute the world's longest tradition of continuous scientific observation: the clay tablets consist of tables with the positions of the sun, moon, and planets (*ephemeris*), day after day, month after month, year after year—for eight consecutive centuries. There are also tablets with procedural texts that represent the mathematical rules for calculating new ephemeris. These rules describe a pattern of solar, lunar, and planetary movements that recur within a certain period.[24] This period can range from a month (such as with the moon) to many years (as with Venus and other planets), and the events can consist either of the rise of a celestial body or of solar or lunar eclipses following an 18-year *Saros* cycle. The arithmetic rules are interesting not only because of their content but also because of the form in which they are cast. They again follow an *if–then* pattern and can be represented schematically as:

if <date and period> *then* <event>

This pattern bears an uncanny resemblance to the astrological patterns in the Enuma Anu Enlil, which, as discussed earlier, have the following form:

if <sign and/or date> *then* <event>

The main difference between the two patterns is that an event in the Astronomical Diaries takes place on the firmament (rise and set of celestial bodies, eclipses of the moon and sun), while an event in the Enuma Anu Enlil can (also) take place on earth. So the *if–then* pattern applies to both astronomical and astrological phenomena, whether they describe ephemeris or omens. This pattern should not be confused with the notion of a principle. A principle can predict patterns or even explain them. The *if–then* pattern above is no more than a format in which the concrete patterns are cast or expressed. Nothing is known about the principles used by the Babylonians for recognizing and predicting astronomical patterns, but they must have searched for matching numerical sequences in their observations. Only then could the long-term patterns be found in the recurring times of the rise and set of Venus and other planets.

The importance of Babylonian astronomy can hardly be overestimated. It is the first time in history that long-term patterns were derived from such whimsical planetary movements—quite an impressive achievement. It is thanks to these discoveries that later Greek astronomers were able to develop their models of the cosmos (see chapter 3). In addition to these patterns, the Babylonians also had a form of cosmology, in which heaven and earth were presented as a spatially round whole. But somewhat surprisingly, this cosmology was not used as an explanation for celestial phenomena.[25]

Egypt: Astronomical Data with Few Patterns

Egyptian astronomy seems to be of a completely different nature: the search for patterns is much less prevalent here—which is remarkable, considering that patterns were the order of the day in Egyptian mathematics. There are no systematic descriptions of planetary motions, eclipses, or other astronomical phenomena. The positions of the pyramids do seem to be aligned with the polar star and the midwinter sun, but this view is controversial.[26] Starting in 2000 BCE we also find tables with star positions on the inside of the wooden coffins, which incidentally do not mention any other regularities in the data. The great discoveries of the Alexandrian astronomers will wait until the Hellenistic period in Egypt.

India, China, and Other Regions

To the best of our knowledge, Indian astronomy begins with the *Vedanga jyotisha*, a text of uncertain date, with estimates ranging from around 1400 BCE until the last century BCE.[27] The *Vedanga jyotisha* contains information about the sun, the moon, and calendars, such as lunar months, solar months, and leap months. There are several indications that early Indian astronomy is of neo-Babylonian origin, suggesting that composition of the *Vedanga jyotisha* did not start until after 600 BCE.[28] Although the older Indus civilization (ca. 3300–1300 BCE) produced many technological innovations, we know very little about any in the field of astronomy and other sciences.

In China, astronomy has its beginnings in the Shang dynasty (ca. 1600–ca. 1046 BCE). Star catalogs were compiled, and hundreds of observations of solar and lunar eclipses were recorded, which Chinese astronomers probably used to predict new eclipses. No precise patterns were reported of the sort we find in Babylonian astronomy, but we will encounter these in abundance in classical antiquity (see chapter 3.2).

In Europe we find examples only of nonwritten astronomy, such as the bronze sky disk from Nebra (Germany). This disk, approximately 30 centimeters in diameter, is from around 1600 BCE and contains a representation of the starry sky with the moon, sun, and some stars, including the Pleiades. The sunrise and sunset can be simulated by means of circular arcs. The sky disk is therefore the first portable astronomical instrument in Europe.[29]

In pre-Columbian America, astronomy came to fruition after 500 CE, especially in the Mayan Empire (see chapter 4).

2.4 Jurisprudence: From Legal Rules to Legal Principles

In the modern-day classification of academic knowledge, it is hard to conceive of a step larger than the one from mathematics and astronomy to jurisprudence, or legal studies. Whereas mathematics is precise and unambiguous, jurisprudence has the reputation of being inexact and polysemous. But in early antiquity that couldn't be further from the truth: there is no other early antique discipline where the rules are formulated as unequivocally as in Mesopotamian law. The most important difference is that the mathematics of early antiquity is numerical and quantitative, while jurisprudence is linguistic and qualitative.

Sumer: If–Then *Patterns and the Retaliation Principle*

The oldest known legal text with rules of law dates from the 21st century BCE and is attributed to King Ur-Nammu of Ur. It is with this monarch that the last major Sumerian heyday began, which would last for almost a century. Ur-Nammu was responsible for the construction of temples, city walls, ziggurats, and an extensive system of irrigation canals. Ur-Nammu's laws, drafted in Sumerian, contain 57 legal rules, of which about 40 have survived, as below (where the numbering corresponds to that on the clay tablets):[30]

1. If a man commits a murder, the man must be killed.
2. If a man steals something, the man must be killed.
3. If a man kidnaps someone, the man must be jailed and pay 15 shekels of silver.
5. If a male or female slave marries a native [free] person, the slave person must give the firstborn to [his or her] enslaver.
6. If a man violates the rights of another and deflowers the virgin wife of a young man, then they must kill that man.
25. If a man's female slave compares herself to her mistress and talks to her impudently, then her mouth shall be cleaned out with 1 quart of salt.

The legal rules thus spell out what needed to be done for which offense. They take the form of a causal *if–then* pattern that we can represent as follows:

if <offense> *then* <punishment>

We have already encountered causal patterns in Babylonian astronomy (above), but the Sumerian legal pattern predates it by a few centuries. This may be an indication that the notion of astronomical causal law is modeled on the notion of legal causal law.

What makes Ur-Nammu's rules of law particularly fascinating is that they are based on a deeper principle, the *principle of retaliation*, or *talio principle* (from the later Latin *ius talionis*, that is, "the right to retaliation"). This is also called the "eye for an eye" principle.[31] This principle would serve as the basis of justice for many millennia, and even today the talio principle is used in countries where murder is punishable by death.

However, it is unlikely that the principle of retaliation was devised by either Ur-Nammu or his counselors. Many of the legal rules, as well as the underlying legal principle, must have existed before they were written down. Even before the codification of these laws, it was customary in Ur to kill a murderer. The codification established an existing practice and was therefore primarily *descriptive*; it was not yet *prescriptive*. But after this initially descriptive codification, laws became prescriptive: they no longer harked back to an older practice but started to serve as the basis of normative jurisprudence. We see here a *transition from descriptive to prescriptive*: initially, pattern-seeking activities represent what is being observed, after which these observations can be used prescriptively. This process from descriptive to prescriptive will prove to be one of the most common meta-patterns in the history of systematic knowledge, but it is here that it appears for the first time.

Babylonia: Laws of Hammurabi, Replacement, Satisfaction, and Retaliation

Three centuries after Ur-Nammu, the principle of retaliation was elaborated further in the famous laws of Hammurabi (1760 BCE). This most extensive legal system from antiquity before the Romans consisted of 281 numbered laws, largely preserved on a black chiseled stela measuring more than two meters high that can now be admired in the Louvre.[32] At the time of Hammurabi, these kinds of stelae were displayed in public so that no one could resort to ignorance of the law as an argument for acquittal. The numbered laws run from 1 to 282, but law 13 has been omitted (13 was already considered an unlucky number among the Babylonians), and laws 66–99 on the stela are no longer legible. Moreover, law 182 literally refers to law 181. This cross-reference would put in motion a legal tradition running from the 18th century BCE through the present day: in all contemporary legal systems, laws refer to each other, and these references define the structure of the law.[33] Hammurabi's laws constitute an extremely detailed system of rules, which he wanted to use to consolidate the existing order.

At first glance, many of Hammurabi's laws are very similar in form to those of Ur-Nammu. Both systems use the pattern "*if* <offense> *then* <punishment>" and the retaliation principle. Yet there is an important difference: with Hammurabi the eye-for-an-eye principle is implemented much more consistently—sometimes to the point of absurdity. While with Ur-Nammu this principle in its literal sense really applies only to rule 1, Hammurabi follows it much more closely. The following laws are examples of this, where the numbering corresponds to that on the stela:

196. If a man blinds another man's eye, his eye shall be blinded [an eye for an eye].
197. If he breaks a man's bone, his bone shall be broken.
200. If a man knocks out a tooth of a man of his own rank, his tooth shall be knocked out [a tooth for a tooth].
229. If a builder builds a house for someone and does not build it well, and the house that he built collapses and kills the owner, the builder shall be put to death.
230. If it kills the owner's son, the builder's son shall be put to death.
231. If it kills a slave of the owner, then he shall pay the owner of the house slave for slave.

These laws refer not only to "an eye for an eye" and "a tooth for a tooth" (lines 196 and 200) but also to "a son for a son" (230). Yet even with Hammurabi it is not the case that all laws follow the principle of retaliation. A few other principles seem to also play a role. The first among these is the *replacement principle*: contrary to the retaliation principle, under rule 231, the master builder's slave is not killed if the occupant's slave dies when the built house collapses, but the master builder's slave is replaced or reimbursed instead. Enslaved people were seen as replaceable in Babylonia, so the principle of "an eye for an eye" was unnecessary. In contrast, a son of the occupant was not replaceable if he were killed, so according to the retaliation principle the master builder's son had to be killed. But replacement was stipulated wherever it was deemed possible. Finally, there was a *satisfaction principle*: even when replacement was not possible, in many cases one could buy one's way out of the literal application of the eye-for-an-eye principle by means of financial compensation, such as in rule 209:

209. If a man hits a free-born woman resulting in her losing her unborn child, he shall pay 10 shekels for her loss.

In Hammurabi's legal system there are different laws for different social groups: the nobility, free men, free women, unborn free children, and enslaved people. But the aforementioned three principles of replacement, redress, and retaliation remain fully in force, and they also seem to be prioritized in a certain way: in the event of damage caused by an offense, it is first determined whether replacement is possible; if not, then redress is in order; and finally, eye-for-an-eye retribution should be applied. We could also merge the two principles of replacement and satisfaction into a more general principle of *compensation*. And if we also replace the word "offense" with "legal act" (since many laws are not about punishment but about a "legal act"), then the following principle may apply in Hammurabi's legal rules:

if <legal act> *then* <compensation> *otherwise* <retaliation>

In this way it becomes clear that Hammurabi's laws can prescribe either retaliation or compensation, which in modern terms would both fall under civil law. In this way, Hammurabi's legal code rolled up criminal and civil law all into one.

Could Hammurabi's principles also be used to deduce the rules of law themselves, in the same way we saw in mathematics, where Pythagorean triples could be generated from a single methodical principle? That isn't as easy as it might seem. Deriving or "predicting" rules of law from the aforementioned three principles would require knowledge of the customs and social order in Babylonian society. For example, we cannot use general principles to determine whether someone is replaceable. Enslaved people were replaceable, and the loss of an unborn children could be bought off (line 209), but the killing of a woman could not be bought off. This act was not punished with the death penalty, nor was the offender's wife killed, rather it was offender's *daughter* who was to be killed.

In addition to the king, there were four types of men in Babylonia: the nobility, freeborn, freedmen, and slaves.[34] Each of these groups had its own (mostly unwritten) rules of retaliation, satisfaction, or replacement. In addition, there were women, daughters, and born and unborn babies, who had their own rules. The pattern of inequality that we first encountered in the Neolithic (see chapter 1.2) was institutionalized with Hammurabi's laws and etched in stone. A certain sort of equality existed only within a given social class and gender. Although the relationship between the underlying principles and the rules of law may have been obvious to the Babylonians, to us they are not. For this reason, the three principles alone are insufficient to predict rules for new cases.

Legal clay tablets from the Old Babylonian period reveal that new cases did occur. One of these tablets deals with the adoption of a foundling, with the judge stipulating that should someone in the future try to claim the boy, a barrel of 20 liters of human milk must be paid as compensation.[35] Apparently, the judge estimated that raising a child was equivalent to 20 liters of breast milk. This example makes it clear that in the absence of a rule of law the judge had to rely on a legal principle to justify his decision. In this example, the relevant principle was the principle of satisfaction. But just how large that compensation should be could not be deduced from this principle, so the court ruling was an estimate of what was deemed "reasonable" compensation in Babylonia. Thus, the predictive power of Hammurabi's law system is limited and less significant than the rules and patterns in linguistics, mathematics, and astronomy. On more than one occasion, one had to resort to an analogy with previous similar cases to arrive at a decision.

However, the principles of replacement, satisfaction, and retaliation do reflect the framework within which the rules and judgments could be formulated. Not all rules or judgments were possible: the three ordered principles had to be met. We'll call such a system of principles a *declarative system*: declarative principles indicate the preconditions within which the derived laws or rules are drawn up. This differs from a *procedural system* of principles in which there is a deductive procedure used to derive statements from principles (as we saw in mathematics when generating Pythagorean triples). Legal systems don't seem to include deductive systems of this sort, in which rules and statements are predicted on the basis of principles, although legal scholars from the 17th and 18th century CE have made a serious attempt at this (see chapter 5).

Were Hammurabi's laws descriptive, just like Ur-Nammu's? It would appear that they were. In his prologue to the stela, Hammurabi explains that he wants to bring the various Sumerian city-states under his authority. This is why he attempts to standardize the law with this legal code. The laws describe a legal practice that already (partially) existed and was observed. Hammurabi—or his legal scholars—adopted both the ancient Sumerian legal habits of Ur-Nammu and those of the Semitic people (including the Akkadians) and forged them into a coherent whole.

Egypt and Other Regions

The Egyptian legal system is even older than its Babylonian counterpart, but there are virtually no surviving sources that describe its legal practice. We know that the oldest Egyptian civil law dates to around 3000 BCE and that it was based on the concept of *Maat*, which stood for truth and justice.[36] Maat was also a deity who had brought order out of the chaos in the universe. Although little is known about Egyptian jurisprudence itself, it appears that it was founded on an underlying principle—that of truth and justice, personified in Maat.

In early antiquity, it is difficult to find a detailed jurisprudence in other regions as well. Although the Chinese legal system goes back to at least the 11th century BCE (Zhou dynasty), we must largely reconstruct this from the much later *Book of Documents*, the *Shujing*, which was not assembled until the 6th century BCE. This legal system focused on notions of first birthright and respect for the elderly. In India, the most important text is the *Dharmasastra*, which consists of legal traditions that go back to at least 600 BCE but may actually be much older. I will discuss this in the next chapter.

In the Hebrew Bible we find rules of law that closely resemble Babylonian law, including the eye-for-an-eye principle. According to Mosaic law, this principle is even universal. The biblical book of Leviticus (from around the 6th century BCE) also contains many moral rules, as well as rules concerning worship and purity. The rules reflect the view of humankind and the world found in the Creation story in Genesis.

2.5 Medicine and the Role of Magic: Diagnosis, Prognosis, and Treatment

The earliest evidence we have of attempts to treat ill health originate from prehistory. Instances of skull trepanning, insofar as they were medical in nature, date back to the Old Stone Age, and in the New Stone Age it was customary to carry medicinal herbs, as Ötzi the iceman shows (see chapter 1.2). The first medical handbooks date from early antiquity, and we find these mainly in Babylonia and Egypt.

Babylonia: If–Then *Diagnoses and Forecasts*

In Babylonia, the oldest known text is the *Treatise on Medical Diagnoses and Prognoses* from ca. 1600 BCE. It covers 40 clay tablets and contains some 3,000 descriptions of diseases and their course. The *Diagnostic Handbook*, *Sakikkū*, came five centuries later. We also know the author of this handbook, Esagil-kin-apli, who was employed as a scholar by King Adad-apla-iddina (1067–1046 BCE).[37] This handbook contains not only descriptions of diseases but procedures for making diagnoses as well. The Babylonians were able to distinguish numerous ailments, ranging from strokes and epilepsy to disorders of the eyes, ears, skin, and heart. In addition, the manual provides instructions for their treatment, including the following typical diagnoses and forecasts:[38]

- If someone's face is covered with a yellow ointment, his lips covered with a film, his eyes secrete a yellow substance, and his right eye squints, he will die.
- If his face is distorted, and his tongue is yellow, and his body is yellow, then he is sick in his stomach, and he will die on the third day (at the latest).
- If [a man] has a diseased anus, crush 5 *silas* of linseed, strain it, and soak it in milk, tie it on the chest and shoulder for 14 days, and he will get better.

Just like in jurisprudence and astronomy, diagnoses, prognoses, and treatments follow an *if–then* pattern. Within this pattern we can distinguish two basic patterns in the Diagnostic Handbook. The first pattern makes a diagnosis based on symptoms, sometimes indicating a prognosis:

if <symptoms> *then* <diagnosis and/or prognosis>

The first two lines above are examples of this pattern. The other basic pattern determines a treatment based on a diagnosis, sometimes also indicating the prognosis:

if <diagnosis> *then* <treatment and/or prognosis>

The third line above is an example of this basic pattern. Combinations of the two basic patterns also occur, although no treatment could be determined without diagnosis or symptoms. The patterns in the Diagnostic Handbook are arranged systematically, with the body being discussed from head to foot and from left to right, and even in a certain color sequence. This system is reinforced by specific procedures, such as for making a diagnosis, which can take place only after the symptoms have been identified. It is tempting to see a formal system of reasoning in the set of causal patterns. We will come back to this issue after we have considered Egyptian medicine.

Egypt: Founder of Babylonian Medicine?

Egyptian medicine is older than its Babylonian counterpart: the Kahun Papyrus from around 1800 BCE is the oldest surviving medical document. It contains descriptions of women's illnesses, elaborations on fertility and infertility, and the first description of a contraceptive: a pessary made from crocodile manure mixed with herbs and honey.[39] The later Ebers Papyrus from ca. 1550 BCE is at least 20 meters long and lists 700 medicines and 800 pharmaceutical recipes. Some of these drugs indicate proven procedures. For example, the prescription for night blindness was fried ox liver, which is rich in vitamin A, a deficiency of which can indeed cause the disease.

While the above papyri are still far from systematic, in the Edwin Smith Papyrus from around 1500 BCE we find a detailed system of procedures with causal *if–then* patterns.[40] Like the Babylonian Diagnostic Handbook, this papyrus consists of observations of symptoms followed by diagnosis, prognosis, and possible treatment. Every medical examination starts with a clinical observation: "If you examine a man who . . . ," followed by a diagnosis. Unlike the Babylonian handbook, the Egyptian papyrus distinguishes among three types of clinical treatments: (1) a certain cure ("a condition that I will treat"), (2) a possible cure ("a condition that I will fight"), and (3) a hopeless case ("a condition that cannot be treated"). In this way, consideration is always given to the question as to whether treatment is possible in the first place. A typical example from this Edwin Smith Papyrus is the following case:[41]

> If you examine a man who has a crack in his cheek and you find a swelling, raised and red, on the outside of the crack,
> you will say [about him]: Someone with a crack in his cheek. A condition that I will treat.
> You shall bandage it with fresh meat on the first day. His treatment consists of waiting until the swelling has subsided. Then you will treat him with fat, honey, and a pillow every day until he is healthy.

A different example from this papyrus shows that in some cases it is better not to do anything at all (case 5):

> If you examine a man with a gaping wound in his head . . . , bleeding from both nostrils and both ears and suffering from stiffness in his neck so that he is unable to look at either of his shoulders and his chest,
> then you will say [about him]: Someone with a gaping wound in his head. A condition that cannot be treated.
> You will not bandage him up, but rather you will bind him to the stakes, until the period of his injury passes.

In contrast to the Babylonian Handbook, the Edwin Smith Papyrus uses a medical assessment to determine whether a treatment is useful at all. The underlying medical pattern here can be represented as follows:

if <symptoms> *then* <diagnosis, prognosis, evaluation, and possibly treatment>

This is a broadening of the patterns in the Babylonian handbook, which are actually simplified versions of the Egyptian pattern. Given that the Edwin Smith Papyrus is a few centuries older than the Diagnostic Handbook, the question arises as to whether any exchanges occurred between Egyptian and Babylonian medicine, the former influencing the latter. We know that Egyptian doctors were very highly regarded in neighboring countries. Pharaoh Ramses II, also known as Ramses the Great (13th century BCE), sent one of his physicians to the court of the Hittites in Babylonia.[42] In addition, the Babylonian text appears to be derivative of the Egyptian papyrus, in terms of both content and approach. It is therefore likely that Babylonian medicine had its origins in Egyptian medicine or was influenced by it, although the Babylonians processed this influence in their own way.

Some attribute the Egyptians' extensive medical knowledge to medicine's ritual use in the mummification of the dead, claiming that the removal of organs from the body for this purpose promoted knowledge of human anatomy. However, research has shown that ritual mummification bore no relation to medicine: the two activities were performed by distinct groups of people. Egyptian—and also Babylonian—medicine did influence the later Greek doctors: many of the causal patterns are found in the Hippocratic corpus (see chapter 3).

The Role of Magic and the Possible Origin of the Pattern Search

In the world of early antiquity, magical[43] and medical treatments existed side by side. Diseases were attributed not only to natural causes such as cold, drought, poisoning, malnutrition, and infection, but also to supernatural ones such as demons, gods, spirits, spells, and the violation of taboos.[44] Magical spells, talismans, incantations, and amulets were used against the latter. Where no natural cause or treatment could be found, recourse was taken to the supernatural. Illness and death were matters too critical to be left to chance. Particularly great importance was attached to omens: they conveyed the counsel of the gods. For example, studying the intestines (and especially the liver) of sacrificial animals was a widespread practice for identifying omens. These signs showed a tendency toward a system, as a way to expose the fabric of the world. This also applied to the use of invocations, prayers, and incantations, which often followed the same underlying basic pattern (see above in this section and the section on astronomy):

if <X happens> *then* <Y may be the case>

Following this pattern, an offering, prayer, or incantation will not necessarily result in an event such as a healing, but it may. Magical knowledge did not differ from other forms of systematic knowledge.[45] The manipulation of the supernatural was an integral part of systematic knowledge: everything that happened needed to be understood and preferably controlled.

Underlying Principles and the Role of Logic and Reasoning

At first sight, diagnosis on the basis of the aforementioned *if–then* rules would appear to be a relatively simple activity: on the basis of the symptoms observed, one seeks the corresponding rules that provide the diagnosis, prognosis, and associated treatment. However, putting this into practice is tricky, since a given set of symptoms can sometimes indicate different conditions, and the same condition can manifest different symptoms that do not always occur together. Then it is up to the art and skill of the physician to find the most appropriate *if–then* rules for such inconsistent symptoms and make the most likely diagnosis. Such a probabilistic-logical reasoning practice is used in all disciplines that deal with uncertainty and incomplete information. But the details of such a system of reasoning are not always clear. So, at first sight, the system of medical *if–then* rules appears to merely be a *procedural system* (where there is a clear, deductive procedure for each case; see above on law). But in practice, what we have here is an imprecise system in which reasoning is based on similar cases.

How the *if–then* rules came about is not known. If we limit ourselves to making diagnoses and forecasts, then the relevant rules are probably the result of determining recurring patterns in the course of previous ailments and disorders. But the origin of the derivation of treatments is harder to identify. Many papyri and clay tablets prescribe treatments that are not only ineffective but flat-out harmful. The treatment rules may have arisen through trial and error in conjunction with traditional or ritual practices. These customs were never subjected to further testing. With today's knowledge, we understand that failure to perform certain treatments would have neither improved nor worsened the patient's condition, even though in the case of harmless treatments we cannot exclude the placebo effect.

Babylonian and Egyptian medicine is virtually devoid of medical theories. The most developed is that of the Egyptians: they believed that people were born healthy but the body was susceptible to disturbances, which could be caused not only by demons but also by intestinal infections. Just like the Nile, the intestinal tract had to be regularly unblocked. The Greek historian Herodotus tells us that the Egyptians kept three days a month free for intestinal flushes with enemas and laxatives.[46]

2.6 Historiography: First Evidence of Recording the Past, No Patterns?

Where do the notions of "historiography" and "past" come from? Before about 2500 BCE, no notion of the past can be found on clay tablets or papyri. This does not mean that people were not giving thought to the past, but the first indisputable proof of recording the past is found on Sumerian king lists from around 2500 BCE, such as the *Chronicle of the Single Kingdom*.[47] It consists of short summaries of the names of kings and their victories and defeats. In addition to summaries, the later Weidner Chronicle, originating from the dynasty of Sargon from around 2000 BCE, also gives historical explanations, such as for the fall of Akkad, which was thought to be the result of a divine punishment—a form of explanation that was used on multiple occasions.

The oldest Egyptian texts about the past also date from around 2500 BCE, the most impressive example being the Palermo Stone: the stone contains lists that date back to around 5000 BCE. The Palermo Stone would be used centuries later by the Egyptian-Hellenistic historian Manetho for an overview of all Egyptian kings until that time. The rediscovery of Manetho's list in the 16th century became a sensation in Europe, perhaps even leading to a revolution: it was learned that there were pharaohs who had lived more than a thousand years before the then-accepted Judeo-Christian date of Creation (see chapter 5.1). In addition to the Palermo Stone, there are Egyptian annals, chronicles, king lists, and biographies, but surprisingly, we do not find a search for patterns.[48]

Historiography is quite different from other disciplines in early antiquity. Unlike linguistics, mathematics, astronomy, law, and medicine, a search for explicit patterns is absent in historiography, other than possibly the implicit pattern of the rise, peak, and decline of successive rulers and states. However, this pattern can be recognized only with a certain amount of effort—it is mentioned explicitly no earlier than the 5th century BCE, by Herodotus (see chapter 3.3).

2.7 Other Disciplines: Successful versus Failed Patterns

The search for patterns in early antiquity was not limited to the disciplines mentioned above but could be found in almost all knowledge activities. In what follows, I discuss the search for patterns in some of them.

Technological Patterns

The Babylonians and Egyptians were masters in the field of searching for technological patterns. Many patterns can be found in the Mesopotamian irrigation systems. Since agriculture in Mesopotamia depended on water levels, the construction of irrigation canals began early on. One of the patterns that was discovered was that the soil became more saline with increasing irrigation.[49] This led the Mesopotamians to search for new land, but because of a chronic lack of arable soil, they started growing crops that were more resistant to salt, such as barley instead of wheat. The empirical insight that irrigation leads to salinization is still one of the most widespread and problematic patterns in intensive agriculture.

Other Mesopotamian technological developments concern metalworking, glass-blowing techniques, and the manufacture of textiles. The discovery of the wheel is usually also attributed to the Mesopotamians, but this discovery was made simultaneously in the Caucasus and central Europe (around the 3rd millennium BCE).[50]

Religious Patterns

Patterns can also be found in religion: while initially each Mesopotamian village had its own deity, the deities of individual villages were merged into a kind of pantheon when city-states emerged. And just as a hierarchy of important and less important villages and towns emerged, a hierarchy of greater and lesser gods emerged, with pedigrees and all. When the first kingdoms came into existence, a kind of kingdom-wide god was put forward as the most important of all gods, such as Marduk for the kingdom of Babylonia.

The Mesopotamian religion did not have a book or founder, something we do find in the later Jewish, Christian, and Islamic religions, nor were Egyptian religious practices based on a holy book. There were, however, deeper concepts in Egyptian religion, such as *ba*, which can be considered the personality of an individual and his or her physical manifestation after death. Alongside *ba*, there was the concept *ka*, the soul. To attain eternal life, the *ba* and *ka* had to be reunited after death into *akh*, an immortal spirit.

Successful and Failed Patterns in Early Economic Thinking

The Babylonians can rightly be called the inventors of accounting. The notion of an accounting balance, a systematic distinction between debit and credit, is first found on clay tablets dating from the Third Dynasty of Ur, roughly the same time as the clay tablets recording the law of King Ur-Nammu (21st century BCE; see above).[51] The oldest mention of the link between grain yield and labor input—expressed in women's working days—dates from that time. The Babylonians also searched for patterns in commodity prices for centuries. The Astronomical Diaries, which were kept starting in the 8th century BCE (see above), contain not only tables, ephemeris, and movements of celestial bodies, but also data on the weather; water levels; and the cost of food, herbs, and wool, whose market prices showed an erratic and mysterious course.[52] The Babylonians hoped to discover a relation between the heavenly bodies, the weather, the water levels, and these prices. They probably never found such a relation: among the thousands of clay tablets with astronomical and economic data, not a single tablet reports a pattern. As a byproduct, the Babylonians did find the famous patterns in planetary movements and the rules for predicting ephemeris, as well as solar and lunar eclipses. They searched for patterns in the cost of merchandise but found them in the movements of celestial bodies—the possibly oldest known example of serendipity.

2.8 Conclusion: Patterns in Nature and Culture Compared

If we try to get a general picture of the search for patterns in early antiquity, it is striking that this activity is most prevalent in Babylonia, except for medicine and possibly historiography, where Egypt was the trailblazer. Finding patterns depends on the availability of data: written observations of phenomena in both nature and culture, from planet positions to verb conjugations.

Babylonian "science" is often dismissed as no more than "the enumeration of all natural and cultural entities."[53] Although the Babylonians were indeed intensively engaged in enumerating and classifying data, this was primarily a means to a higher goal: uncovering regularities in that data to bring the world under control. The same also applies to the Egyptians, Chinese, and Indians.

Table 3 Patterns in Babylonian and Egyptian disciplines

Discipline	Patterns
Linguistics	$x_i y z_i$
Mathematics	$a^2 + b^2 = c^2$
Astronomy	*moon_visibility_coefficient* × 3 *minas* *if* <sign and/or date> *then* <event> *if* <date and period> *then* <event>
Law	*if* <offense> *then* <punishment> *if* <legal act> *then* <compensation> *otherwise* <retaliation>
Medicine	*if* <symptoms> *then* <diagnosis and/or prognosis> *if* <diagnosis> *then* <treatment and/or prognosis> *if* <symptoms> *then* <diagnosis, prognosis, evaluation, and possibly treatment>

Moreover, nature and culture were not viewed as separate categories; they were instead considered to belong to the same, undivided reality that was subject to patterns and regularity. The question that now arises is whether we can also derive more general tendencies from the many patterns discussed in this chapter (table 3) in that undivided reality.

Quantitative versus Qualitative, Causal versus Noncausal

We can observe in table 3 that certain patterns, such as those in mathematics and (partly) in astronomy, are quantitative and numerical, whereas other patterns, such as in jurisprudence and medicine, are qualitative and expressed in words. The linguistic pattern lies somewhere in between: it is relational in nature and expresses a discontinuous connection between linguistic units. In the chapters that follow, we will see how this distinction between quantitative and qualitative comes back in later periods and consider whether we can speak of a long-term tendency in the history of knowledge.

What also stands out is that the qualitative patterns are causal (the *if–then* patterns), while the quantitative patterns are noncausal. Causal patterns can be understood as follows: If event A at any time(s) in the past is associated with event B, then A can be regarded as a cause of B—which is represented in both Babylonia and Egypt by *if A then B* statements. Causality does not seem to play a role in quantitative patterns (e.g., relations between numbers or between visibility and the visibility coefficient of the moon). The origin of the *if–then* pattern is unknown, although the oldest surviving occurrence dates back to the laws of Ur-Nammu (21st century BCE). In the Mesopotamian world, the *if–then* pattern runs like a common thread through almost all disciplines, whereas in Egypt it occurs only in medicine. The pattern does not appear to return in this form in later periods.

Parallel Discoveries of Patterns

One of the most fascinating aspects of the human search for patterns is that the same patterns have been discovered in different regions without any contact between them. For example, the pattern of Pythagorean triples was discovered both in India and China as well as in Babylonia, and prime numbers were known in both Egypt and India. In addition, knowledge of the sun, moon, and calendars, as well as of lunar eclipses and the seven planets was present in all regions. We have also encountered parallel patterns in prehistoric times, where the side-view representation of animals and the use of hand stencils in cave paintings can be found in various places in the world. In addition, patterns of the solar and lunar movements, as shown in stone circles, also occur in practically all regions. A big exception is linguistics in early antiquity, which is found only in Babylonia.

Predictive Power of Patterns

All patterns have predictive power (see the introduction), but this is strongest in mathematics and astronomy, where the patterns numerically indicate the solution to a quadratic equation or the visibility of the moon. Patterns in language and medicine also have considerable predictive power, although in their qualitative form it is not as great as in mathematics and astronomy. With respect to jurisprudence, legal rules have significantly less predictive power, since one of the principles must be invoked for each new case, while there is no formal way to trace a particular court ruling back to a principle. Historiography is a dubious case: while we do not find any explicit patterns, the implicit pattern of the rise, peak, and decline of states has considerable predictive power, but it takes some doing to distill this pattern from the chronicles.

Process from Descriptive to Prescriptive

We have seen that some patterns are descriptive (as in mathematics) while others are prescriptive (as in jurisprudence), but we have also seen that this distinction is often difficult to make because even in jurisprudence the patterns (the rules of law) have a descriptive origin. Codification of the Babylonian legal rules started with the description of an existing legal practice, and once written down, these rules were subsequently employed prescriptively. This process is not limited to law; it can also take place in language, history, and even in astronomy: once certain patterns have been observed and written down, it is particularly difficult to think outside these patterns, let alone "break through" them, even when there is preponderance of empirical evidence that one should do so. I have also discussed the process from descriptive to prescriptive in my previous book *A New History of the Humanities*, where I attribute it to Pliny's art theory and Aristotle's poetics. Now it appears that this process can be found 1,500 years earlier in Babylonian jurisprudence.

From Unconscious to Conscious Principles

Although patterns proliferated in early antiquity, principles are used only sparsely and mainly in jurisprudence. There are, however, indications of a certain awareness of principles. For example, the underlying mathematical principle for generating new Pythagorean triples must have been known to the Babylonians, as well as the legal principles of replacement, satisfaction, and retaliation that determine the penalty or compensation in new situations and finally, the principle of probabilistic reasoning in medicine, without which no diagnosis can be made with incomplete information. But it remains to be seen whether people were actually aware of these principles or whether it was just a traditional and ingrained way of working ("implicit" principles).

CHAPTER THREE

The Explosion of Principles and the Awareness of Deduction

Classical Antiquity

600 BCE–500 CE: Greece, Roman Empire, China, India

Classical antiquity differs radically from early antiquity: the search for principles is a central issue almost everywhere. Starting with the first established principles, the relationship with patterns also comes to the fore. How do the two relate to each other? Can patterns be derived from principles, or conversely, can patterns be reduced to principles? Or are principles just descriptions that make loose generalizations about the patterns? These kinds of questions are addressed in various regions—from China and India to Greece. It would appear that Greece was the first to embark on this search, but that is far from certain, due to the fact that many works have been lost, especially in the great Chinese book burning of 213 BCE.

The Greeks were in an exceptional situation. The Greek natural philosophers, historians, philologists, physicians, and logicians were either self-employed, often supporting themselves by giving private lessons, or they were wealthy enough not to have to work for a living. This allowed them to pursue science and scholarship without a practical purpose. And so it happened that the Greeks sometimes paid more attention to theoretical principles than to empirical patterns. In addition to Greece, India and China also played a leading role. In India a sort of theoretical linguistics was developed that is unparalleled in the history of knowledge. And in China, underlying principles were sought in astronomy and mathematics but especially in the field of logic, where the Chinese established the oldest known laws of reasoning.

3.1 A Tale of Two Principles: Thales and Panini

The world's oldest knowledge principles have their origins in Thales and Panini, around 600 BCE. Whereas the Greek philosopher Thales posits a basic principle for nature—that it is composed entirely of water—the Indian linguist Panini puts forth a general principle for language—that it is recursive. At first glance, there would appear to be a world of difference between the two: the first principle applies to all of nature, whereas the other applies "only" to language. But that depends on how one views language and nature. In Panini's world, language was of the utmost importance imaginable, considering that all ritual, philosophical, and cosmological knowledge was expressed in language; so knowledge of the principles of language provided access to knowledge of the world as a whole. But Panini had nothing to say about nature. For the Greek natural philosophers, it is not language that was the central issue but nature, so it was knowledge of the principles of nature that provided access to knowledge of the whole world. But Thales and his colleagues had nothing to say about language.

Unfortunately, none of the original texts by either Panini or Thales has survived. But while we are acquainted with Panini's work through copies, for Thales we have only statements ascribed to him by later philosophers. Some historians contend that this fact calls the reliability of Thales's work into question. For instance, Aristotle (384–322 BCE) argued that earlier philosophy was a laborious process slowly inching toward the truth as revealed by none other than him,[1] so he probably presents the early natural philosophers as much more naive than they actually were. But in spite of issues like these, we will have to make do with the sources we have at our disposal.

Who were Thales and Panini? Thales lived in Asia Minor (Miletus), whereas Panini was a Brahman from Vedic India (Gandhara). Nothing more is known about Panini's life, but there is no dearth of anecdotes about Thales. The Greek philosopher was aware of Egyptian and Babylonian mathematics and astronomy, he has multiple mathematical discoveries to his name (see below), and he successfully even predicted a solar eclipse that was decisive for the outcome of a war. Thales also demonstrated that philosophers were not only excellent thinkers but that they were also smart enough to earn great sums of money. Thales managed to amass a fortune by buying up olive presses and subsequently renting them out for a hefty price.[2]

The most famous statement ascribed to Thales is that all of nature is composed of a single material substance: water.[3] Assuming such a primary substance is not as odd as it might seem at first glance. In fact, the modern idea that the entire world is composed of elementary particles, or even of strings, is actually an extrapolation of Thales's principle. Neither is Thales's assumption that the primary substance must be water as strange as it may appear to us from our 21st-century perspective, considering that water occurs in all three phases in everyday life: solid (ice), liquid (water), and gas (steam). Thales assumed that the earth itself was floating on liquid water. This allowed him to explain the existence of earthquakes without the need to ascribe them to the whims of the gods: they were caused by waves crashing against solid ground. According to Aristotle, Thales's methodology lies in the idea that the study of nature can be based only on nature itself rather than on the supernatural.

But Thales could not use his principle to make any concrete predictions. Although his contemporary, the Greek historian Herodotus, reports Thales predicting the solar eclipse that occurred on May 28, 585 BCE,[4] this prediction had nothing to do with his principle of a primary substance. Perhaps Thales had knowledge of the Babylonian Saros period, which describes the regularity with which solar eclipses occur (see chapter 2.3). However, his principle of a primary substance could not be used to make concrete predictions, and neither was that his intention. What mattered to him was the idea that a multitude of phenomena could be explained using a single principle,[5] and this quest for the one in the many turned out to be extraordinarily productive.

In contrast to Thales, Panini was interested not in the world of nature but in the world of language. His most important hypothesis was that all language is based on the principle of recursion. The central question for Panini was how all possible linguistic utterances—in Sanskrit in his case—could be described using a finite number of rules. Since there is no limit on a sentence's length, the number of utterances is unbounded. For example, a given sentence can contain one or more embedded clauses. An example of this in English is *The man who hit the dog is a criminal.* This sentence can be lengthened and complicated by adding more and more embedded clauses, such as *The man who hit the dog that was bitten by the cat is a criminal*, or even longer: *The man who hit the dog that was bitten by the cat that ran down the street is a criminal.* Panini's insight was that this productive feature of language, that one embedded clause can be contained in another, can be ascribed to the principle of recursion. Sentences can also be simplex, of course, without recursion, as in *The man is a criminal*, which can be seen as a sort of zero-recursion. The pattern of discontinuous relations, which had already been discovered by the Babylonians in their translations of words (see chapter 2), can also be described in terms of recursion. We saw this at the beginning of the chapter with the compounds *freedom proponent* and *freedom from violence proponent*, where the insertion of *from violence* results in a discontinuous relation between *freedom* and *proponent*. This insertion process is also a form of recursion: just like sentences, words can in principle also be made longer and longer.

Panini's grammar of Sanskrit contains a whopping 3,959 rules, which he claims can describe the entire classical Sanskrit language.[6] While this may seem like a lot of rules, it is unlikely that any language can get by with fewer. Furthermore, Panini's rules concern language as a whole—from pronunciation, morphological word formation, and syntactic sentence structure to the semantic assignment of meaning and pragmatic functions of language. Additionally, Panini formulated his rules in such a way that they could be used as an algorithm: the rules can be applied to a string of words to determine whether it constitutes a grammatically correct sentence. Panini also introduced a number of meta-rules into this system to deal with exceptions or with (potentially) contradictory rules. Exceptional cases in language have priority over general cases. A good example of this concerns the conjugation of verbs, where a special case, such as the past tense of an irregular verb like *bring*—in this case *brought*—takes priority over application of the general past tense rule, which would lead to the incorrect form *bringed*. Panini handled such exceptional cases with his meta-rule: "If two rules contradict each other, the latter rule prevails."[7] Panini devised his grammar in such a way that an exceptional rule was listed after the general rule, so that the relevant meta-rule was sufficient to account for the exception. Meta-rules are actually "principles," considering that they apply to the grammar as a whole. However, these meta-rules are first and foremost procedural and do not generalize over linguistic phenomena, whereas the principle of recursion does.

Why were Panini and Thales interested in a single basic principle for all patterns? Examining Panini's grammatical text, all we learn is that he devised his system of rules in the most concise way possible, regardless of how great their number. This concise structure certainly reinforced acceptance of Panini's system in India. Unfortunately, we know even less about Thales's motives, but what is clear is that he considered it of the utmost importance to group various phenomena together. This idea even became a constant in all Greek scholarship (see below). Perhaps it is because Thales and Panini had to compete with the worldviews of their own time—the Homeric and the Vedic—that they wanted to present their ideas as forcefully, and thus as succinctly, as possible.

After Thales and Panini

The pupils of Thales and Panini remained true to the idea of the basic principle. But something we find with the Greek natural philosophers but not with the Indian linguists is a constant critiquing of earlier principles. For example, Anaximander argued that the world was composed not of water but of an infinite, eternal, and indefinable mass, which he called *apeiron*.[8] The earth floated and remained in place thanks to its equal distance from other objects in the universe. He also presented a form of evolutionary thinking in which the earth first existed in a liquid state and subsequently was dried by the sun; it was after this that fish emerged from the water, followed by all other animals, and people evolved from those fish. In contrast, Anaximander's pupil Anaximenes posited that the primary substance was air. He did not believe that the earth floated but rather that it was suspended in the air.[9]

We also encounter the idea that the world is composed not of a single primary substance but of multiple primary substances, or *elements*. Empedocles argued that the entire cosmos consists of earth, water, air, and fire. However, the most radical vision came from Leucippus and his pupil Democritus, who in the 5th century BCE argued that all matter was composed of tiny, indivisible particles that move around in a vacuum: *atoms*. Later atomists such as Epicurus (342–270 BCE) and Lucretius, a Roman (ca. 99–55 BCE), made frantic efforts to prove atomic theory. For example, Lucretius reasoned in his didactic poem *De rerum natura* that the current state of the universe could be explained from an entirely atomist standpoint. All objects and phenomena consist of nothing but combinations of atoms, which by their constant and eternal motion can create new objects by clumping together. Even human free will could be explained by the fact that atoms move randomly rather than in a deterministic way. Just like the primary substance hypothesis, atomism remained extremely speculative, and predicting new phenomena was beyond its reach.

Panini's thinking fared quite differently in India. His pupils not only followed his idea of a basic principle; they also stayed true to the principle of recursion. For example, the commentaries by Katyayana and Patanjali are useful for understanding his method, but we don't find any criticism of Panini in them, let alone a rejection of him.[10] Some improvements are proposed, discovering and fixing inconsistencies in the Paninian system of rules. Panini's grammar also provided a model for languages like Tamil and Tibetan, demonstrating that his principle was not limited to Sanskrit but could be used to describe other non-Indo-European languages as well. Panini's formalism, and the predictions that proceeded from it, worked too well to be rejected.

The Greeks do not seem to have been acquainted with Panini's work, either via Alexander the Great or through other routes. Given their penchant for the one in the many, they would certainly have valued Panini. In contrast to its Indian counterpart, Greek linguistics concerned itself mainly with studying word formation and word categories, in addition to the philosophy of language. The oldest surviving grammar—the *Technè grammatiké*—is a school grammar by Dionysius Thrax (1st century BCE) comprising a scant 30 pages. It discusses pronunciation, nouns, verbs, articles, prepositions, adverbs, conjunctions, and the various meters of Greek poetry.[11] The system of rules is limited: Dionysius does not go beyond a description of the conjugations and declensions. While linguists after Dionysius do show an interest in syntax, a principle-based rule system that can predict whether a given sequence of words is a grammatical utterance in Greek or Latin is lacking. For instance, Apollonius Dyscolus (2nd century CE) describes the case system of Greek, noting—with surprise—that the subject is sometimes in the nominative but at other times in the accusative.[12] But Apollonius doesn't provide the underlying rule for this phenomenon; he merely discusses it using examples. For this reason, not only is his grammar rule based; it is also partially *example based*: where he cannot identify any rules, he discusses linguistic phenomena using examples, failing to generalize over the examples by identifying some pattern (let alone a principle). Later Roman linguists, such as Varro, Donatus, and Priscian, were less interested in syntax and focused mainly on word forms and their semantic functions.[13] Although Priscian introduces the notion of rule (*regula*), his grammar mainly concerns word structure.

3.2 The Convergence of Principles and Patterns: Astronomy

The question of what underlies visible patterns has rarely been worded as aptly as by Plato (ca. 380 BCE): "We may regard those patterns in the sky as the most beautiful and precise patterns in the material world, but they are simply visible things, and therefore, in terms of purity, they still lag far behind the mutual relationships between the movements of those celestial bodies that can be expressed in pure numbers and pure figures. The latter, of course, can only be comprehended using reason and intellect, not by eyesight."[14] In my book, the "mutual relationships" mentioned by Plato are considered principles, and according to him they are not visible but can be expressed in "pure numbers and pure figures." Here Plato builds on Pythagoras (ca. 570–495 BCE), who is said to have suggested that all celestial bodies move in perfect circles (around a central fire) and that their mutual distances can be expressed as ratios of the first four integers—1, 2, 3, and 4—as a cosmic harmony.[15] In all its simplicity, this idea proved so powerful that it determined not only Greek history of knowledge but large parts of its Western counterpart as well.

The idea that the divergent planetary motions can be explained with a single mathematical model instead of with a multitude of patterns (like the Babylonians did) may seem revolutionary, but the idea of unity in the multitude was not new. It was actually a continuation of Thales's idea that there is one single principle for all of nature. But as beautiful as Pythagoras's starting point of pure numbers and circles may be, it was far removed from the planetary motions actually observed. Plato was aware of this and remained primarily philosophical in his reflections. His so-called theory of ideas was partly based on Pythagoras's philosophy.[16] According to this doctrine, the visible world is only a shadow of the real world—the world of ideas. This world was mathematical and could be ascertained only by reason.

It took a mathematician, Eudoxus of Cnidus (ca. 410–347 BCE), to develop a geometric system that could describe the observed motions of the planets, sun, and moon. The biggest problem for such a system was the apparent retrograde loop motion that had already been described by the Babylonians (see chapter 2.3). This apparent motion is the result of the celestial bodies revolving not around the earth but around the sun, and in ellipses rather than in circles—but this was unknown to the Greeks.

Eudoxus energetically addressed Plato's question, "By the assumption of what uniform and orderly motions can the apparent motions of the planets be accounted for?"[17] Eudoxus insisted on the idea of explaining the complex planetary motions using circular orbits. He introduced a model that consisted of concentric spheres, rotating celestial orbs with the earth at the center. In Eudoxus's system, the spheres have sloping axes and can move both clockwise and counterclockwise so that backward motion of the celestial bodies can be accounted for. A given celestial body can have multiple spheres. For example, at least two orbs were needed for the sun: one for the fast daily rotation and another for the annual motion of the sun in the opposite direction. At least three orbs were needed for each of the planets. In total, Eudoxus designed 26 spheres to simulate the patterns of all the celestial bodies. For the first time there was a single geometric model that could account for the motions of the various planets.

Eudoxus's model marked the beginning of a long tradition. For example, his student Callippus (ca. 370–300 BCE)[18] added seven spheres to better describe the motions of the inner planets and the variable length of the seasons.[19] Aristotle (384–322 BCE), who also worked in Athens, even increased the number of spheres to around 55. He also explained his comprehensive notion of the cosmos: a two-world model with a heavenly mechanics and an earthly counterpart.[20] According to Aristotle, all natural motions in the sky were circular revolutions caused by an unmoved mover, while on earth all natural motions tend toward rest, toward the center of the earth, unless an object is kept in motion by an external force. The heavens above the moon were perfect and unchangeable, while the sublunary sphere of the earth was imperfect and subject to change. We will delve further into Aristotle's theory of motion below.

From Athens to Alexandria

As influential as it was, Eudoxus's model had serious shortcomings. Spheres could not explain why planets have variable speed and brightness. Since the spheres were concentric, each planet was always at the same distance from the earth, making the variation in brightness inexplicable. It was thanks to the brilliant mathematician Apollonius of Perga (ca. 262–190 BCE), who was also the discoverer of the conic sections (see below), that two new concepts were introduced into astronomy. The first was the notion of an eccentric circular orbit, the *eccentric*,[21] where the center of a planet's orbit has been shifted slightly relative to the earth. With his eccentric orbit, Apollonius could explain an important variation in the brightness of a planet but not its backward motion (which is the result of the earth's annual orbit around the sun). To this end, Apollonius introduced the notion of the epicycle. An epicycle is an auxiliary circle whose center lies on the circle of the orbit. This orbit, which is also referred to as a *deferent*, bears the epicycle, so to speak, which in turn carries the planet. The combined model of eccentric and epicycle explains both a planet's backward motion and its periodic increase and decrease in brightness.

Apollonius probably presented his eccentric-epicycle system as a conceptual model. In fact, the precision of his model still lagged behind the accuracy of the planetary patterns discovered by the Babylonians (see chapter 2.3). These patterns were known to the Greeks thanks to the Chaldean astronomers who lived under the empire that Alexander the Great had conquered. Although this empire fell apart after Alexander's death, Hellenistic culture in the fields of art and science persisted, with Alexandria as its intellectual center. Slowly but surely, the Hellenistic astronomers became aware of the gap between the precision of their theoretical, principle-based models and the patterns observed by the Babylonians.

Hipparchus (ca. 190–120 BCE) knew both the geometric models of his Greek predecessors and the arithmetic techniques that the peoples of Mesopotamia had developed over the centuries.[22] He advocated that Greek astronomers pursue the same precision as the Babylonians. To make that possible he designed a method for depicting the celestial sphere on a flat surface to test the Greek models against the observed positions and motions of the planets. This method, known as stereography, led to the invention of a new instrument that is also attributed to Hipparchus—the astrolabe—which can be used to calculate the location and height of a celestial body. Hipparchus's work heralds the beginning of the use of star coordinates and of his famous star catalog. On the basis of extensive observations of the orbit of the sun and the moon, Hipparchus constructed improved eccentrics and epicycles, resulting in a geometric model that for the first time matched the solar and lunar patterns found by the Babylonians. Although Hipparchus failed to achieve the same for the planetary patterns, it is thanks to him that the notion of "exact prediction" was brought to the attention of Hellenistic astronomers and that the same empirical demands were made on both principle-based and pattern-based knowledge.

Where Hipparchus failed, the Alexandrine Claudius Ptolemy (ca. 100–170 BCE) would succeed. These two astronomers are separated by a full three centuries, but hardly any astronomical developments are recorded during this time—with the notable exception of an astronomical calculator known as the Antikythera mechanism (ca. 100 BCE).[23] In his impressive *Almagest*, Ptolemy added a third geometric principle, the *equant*, which describes the variable speed of the planets in the sky.[24] Although planets do not have a constant velocity, Ptolemy discovered that they do have a constant *angular* velocity relative to an imaginary point (the equant), which is slightly shifted relative to the center of the orbit.

Together with the geometric principles of eccentric and epicycle, Ptolemy used his equant to describe and predict the motion of the planets with a precision never seen before, surpassing the Babylonians. His *Almagest* gives an overview of ancient astronomy, complete with propositions, derivations, and observations. Not only does it describe the three basic principles of the planetary motions (eccentric, epicycle, and equant); it also shows how to use these principles to infer the apparent planetary motions (the patterns) and predict the planet's positions. Astronomy had become a deductive science modeled on mathematics (see below), where every planetary pattern could be fitted to the Ptolemaic model based on these three geometric principles.

Since deriving the celestial patterns was a very laborious task, Ptolemy also published the so-called *Handy Tables*, which included only the tables and rules of thumb needed to predict the celestial phenomena. Although Hellenistic astronomers generally sought to establish a principle-based theory, many were pragmatic enough to summarize their theories in the form of tables so that, as with the Babylonians, predictions could be made quickly without further insight into the deeper underlying relations. These tables would become the model for the later astronomical tables in the Islamic world and the European Middle Ages (see chapter 4.2).

Ptolemy was skeptical about his own model: in the 13th book of the *Almagest* and in his later *Planetary Hypotheses*, he explains that although his system was simpler than the model of the spheres—based as it was on only three principles—it should primarily be used to make calculations and not to describe reality.[25] According to Ptolemy, one of the most important applications of his model was to astrology. In his *Tetrabiblos* (Four books), he discusses the principles of horoscopes and the influence of heavenly bodies on earthly events.[26] The *Tetrabiblos* became the most influential work in Western astrology for more than 1,500 years, lending a legitimacy to astronomy that would last for centuries.

Process from Misalignment to Alignment between Principles and Patterns

The story above has been told many times, and in more detail than here, but what come to the fore in our story are initial conceptual models with little predictive power developing into more refined, mathematical models that approach the accuracy of the Babylonian pattern-based approach and ultimately surpass it. This shows the role that the Greeks assigned to theory: their theoretical principles initially stemmed from predetermined ideas about the world, such as the belief that the motion of the firmament could be described using perfect circles and pure numbers. When it became clear that these theoretical principles did not (fully) correspond to empirical reality, the Greeks also became aware of a misalignment between principles and patterns. For Plato and Eudoxus, this sort of a misalignment was not the primary source of concern, as long as the model provided a "good" explanation and insight into underlying relations. But later astronomers such as Hipparchus and Ptolemy did not settle for that and insisted that one should be able to use the principles to make "precise" predictions about these patterns. But exactitude is not feasible in the empirical sciences, as we know today. Observations always involve a measurement error, no matter how accurately we conduct them. Nevertheless, the scientific ideal was established: however convincing theoretical principles may appear, they need to be assessed for empirical adequacy.

However, empirically adequate principles do not necessarily reflect reality. Principles can have great predictive power but still be removed from reality. For example, with its notions of eccentric, epicycle, and equant, the Ptolemaic model accurately predicts the planetary motions in the sky as they appear, but not as they actually are. Planets do not revolve around the earth in looping motions; they revolve around the sun in ellipses.[27] So just how "principled" were the Greek models? Wasn't Greek astronomy just an exercise run amok to save the phenomena using perfect circles alone at all costs, where a new epicycle—or even a new concept such as the equant—was introduced for every deviation in planetary motion or brightness? The equant in particular was a thorn in the side of many later astronomers. Today, the introduction of the epicycle has become proverbial for bad science. But that would be selling the Greeks short, because they had made an admirable achievement, successfully reducing the vast multitude of heavenly patterns to a much smaller number of principles.

What we have seen is *a process from misalignment to alignment between principle-based and pattern-based knowledge* in Greek astronomy. Theoretical principles are initially less accurate than patterns, but principles can be modified and expanded to potentially match or even exceed the accuracy of the patterns. However, this does not mean that Greek astronomy can be said to have had an "empirical cycle" (see chapter 5), where the modified or expanded principles are repeatedly tested against the patterns, followed by feedback to the principles, and so on. While such an empirical cycle may have taken place, there is no evidence to support it. All we know is that Hipparchus and Ptolemy wanted their principles to correspond as closely as possible to the patterns observed. So in this chapter I speak only of a convergence, or a process from misalignment to alignment, of principles and patterns, rather than of a cycle.

Moreover, the development from conceptual to predictive models applies only to the geocentric systems discussed above. There were also alternative models in the Hellenistic period. Based on the motions of the two inner planets, Mercury and Venus, Heraclides Ponticus (ca. 387–312 BCE) suggested that their orbits had the sun as their center rather than the earth.[28] This seems to be a step in the direction of a heliocentric system, but in Heraclid's model both the sun and the remaining planets continue to revolve around the earth, with the two inner planets, Venus and Mercury, additionally revolving around the sun. We often see these two planets near the sun just after sunset or just before sunrise, so the idea that they revolved around the sun wasn't really so preposterous. This model met with some approval by later Greek astronomers, such as Theon of Smyrna and Chalcidius. As far as we know, the more extreme, heliocentric model in which all the planets, including the earth, orbit the sun, was first proposed by Aristarchus of Samos (ca. 310–230 BCE).[29] Although Aristarchus seemed ahead of his time in this regard, his heliocentric model failed to yield better predictions. Eccentric circles and epicycles were still needed to make the model correspond to the planetary motions, though Aristarchus himself did not employ them. His model remained conceptual and theoretical, just like that of Heraclides. Since Aristarchus's worldview deviated from what one could see on the firmament, he was generally criticized, his only adherent being Seleucus of Seleucia.[30]

Chinese Astronomy: Cosmological versus Arithmetical Models

While the Greeks focused on geometric principles for celestial patterns, such as epicycles and equants, Chinese astronomers searched for arithmetical principles for these patterns. Historical research into mathematical astronomy in China has been long neglected,[31] but that has changed considerably in recent decades.

The first major Chinese thinker, Confucius (551–478 BCE), was not a system builder in the way that Pythagoras or Aristotle were.[32] What was central to Confucian thought was the worship of the cosmos by worshiping its parts. The notion of God or the "prime mover" cannot be found in China. In a sense, the Chinese God was heaven itself, and the emperor was the son of heaven and the head of the state religion. There were various cosmological schools, however. For example, the Gai Tian school represented heaven as a hemisphere over a dome-shaped earth. In contrast, the Hun Tian school held to a spherical earth floating in a celestial sphere (somewhat similar to the Greek tradition). And the Xuan Ye school placed the celestial bodies in an empty and infinite space.[33] However, these conceptual models were far removed from the patterns observed in planetary motions, and as far as we know, no attempts were made to describe and predict the celestial patterns using the ideas advocated by these schools.

The oldest surviving Chinese document with a systematic description of planetary motion is the *Wu xing zhan*, or *Prognostics of the Five Stars* (meaning the planets).[34] This anonymous text was discovered when a Han tomb dated to before 168 BCE was opened in 1973. Although some earlier documents describe planetary patterns, such as the *Huainanzi* (The master of Huainan) and the *Shiji* (Historical reports) by Sima Qian (see below), these works cannot be used to calculate the positions of the planets, something that can be done using the *Prognostics of the Five Stars*. The text contains tables with observations of the rise and set times and positions of Saturn, Jupiter, Mars, Venus, and (to a lesser extent) Mercury, always mentioning their cyclic patterns explicitly. In addition, the text also contains a predictive model for each planet: their initial state in 246 BCE is chosen as where the cyclical planetary motions begin and from which the positions of the planets can be calculated at later dates. The year 246 BCE does not appear to have simply been pulled out of a hat: it corresponds to the first year of the reign of the Qin king who would become the first emperor of a united China a quarter of a century later.

Computational Rules of Liu Hong: Algorithm as Principle

The approach in the *Prognostics of the Five Stars* is elaborated into a principle-based system by Liu Hong (ca. 129–210 CE). He refines the assumptions of initial condition and cyclicity and poses these de facto as the underlying principles of the planetary patterns. Liu's theory was so advanced that it was immediately adopted by the imperial government of the Eastern Han dynasty. In the text *Qian Xiang li* (Qian Xiang calendar), Liu sets forth a theory of the motion of the moon relative to the ecliptic (the orbit of the sun).[35] He provides a number of calculation rules that can predict the solar and lunar motions in the sky with unprecedented precision. These rules take the form of a procedure or algorithm used to derive these patterns in a step-by-step fashion using an initial state and a number of constants. Liu Hong's theory consists of three parts:

1. An initial state: a time when all elements in the system have simple initial values. This is what Liu calls the moment of "ultimate origin," which according to him corresponds to January 21, 7172 BCE.
2. A collection of constants for the various elements in the system.
3. A calculation procedure or algorithm to predict the state of all elements in the system at any given moment after the initial state using (1) and (2).

So, Liu gives three basic principles—an initial state, a collection of constants, and an algorithm—which together can be used to calculate not only the position of the moon but those of the other planets as well. If we know the initial states and constants, we can calculate the state at any later point in time. At first glance, it may seem strange to consider the initial state and the set of constants as principles, but they do meet our definition of a principle as set out in the introduction: the initial state and the constants are valid, together with the algorithm, for several celestial patterns. In addition, Liu's principles have a surprising parallel in modern physics and astronomy: as much later was put into words by Pierre-Simon Laplace (see chapter 5.3), in physics it is assumed that if the initial state and constants of a system are known, all later states can be calculated.

Liu's system is principle based just like that of Ptolemy, but the two could hardly be more different. While for Ptolemy the principles consist in the geometric notions of eccentric, epicycle, and equant, Liu's principles are entirely arithmetical: an initial state, constants, and an algorithm. And where the Greeks constructed a cosmological model based on circular motions, Liu's principles are based on recurring variations in the data but make no statement about the shape of these motions. The Greek geometric approach was based on a specific, *geocentric* structure of the cosmos, while the Chinese algorithmic method neither adopted a particular cosmological model nor excluded its possibility. That is why it is usually said that the Greeks provided explanatory models of planetary motion while the Chinese did not.[36] However, to what extent is this position true? When the Greeks tried to construct explanatory models based on spheres, as did Eudoxus, Callippus, and Aristotle, they proved largely inadequate from an empirical standpoint. And where their model was (more) empirically adequate, as with Hipparchus and Ptolemy, it was neither explanatory nor intended as such, a point that Ptolemy himself emphasized. Liu's cyclic analysis explains just as much or as little as Ptolemy's equant analysis. However, in the long run, the Greek geometric approach has proven more fruitful than the Chinese arithmetical approach: the discussion of geocentric versus heliocentric models was immensely important for bringing together celestial mechanics and terrestrial mechanics using a single gravitational theory (see chapter 5.3).

However, there is another problem with Liu's calculation procedures: although he had more observation data at his disposal than Ptolemy did, it is unclear how he came upon his rules of calculation. They have an ad hoc character, with all sorts of unspecified factors having to be multiplied and with many undefined terms having to be added up to finally arrive at the time and position of a phenomenon, such as a full moon. Liu's method even resembles data-oriented "modeling," where simple rules of thumb are derived from large quantities of positions observed—a Herculean task but one that does not provide any deeper insight. All this does not detract from the importance of Liu's rules of calculation. We can still use his algorithm quite easily, with beautiful sine-curve-like graphs emerging from his rules of calculation, which predict not only the lunar positions to within less than 1 degree of accuracy (an improvement on the *Prognostics of the Five Stars*)[37] but also the times of the phases of the moon, plus the lunar and solar eclipses. With its great precision, Liu's model is on par with those of Hipparchus and Ptolemy, so it is quite unfortunate that his other work, *Qi yao shu* (Art of the seven planets), used to calculate the positions of the other planets, has been lost. However, Liu's algorithmic method does not stand alone: his Chinese predecessors, as well as some of his successors, used roughly the same three basic principles of initial conditions, constants, and algorithms.[38]

Convergence of Principles and Patterns versus Exceptional Phenomena in China

Liu's death roughly coincides with the fall of the Han dynasty (220 CE), which marks the beginning of a period of great unrest in China. One of the most important astronomers after Liu Hong is Yu Xi (4th century CE), but his emphasis is more on collecting data and observations than on improving arithmetic rules. However, everything suggests that, just as in Greece and Alexandria, we can also speak of a convergence between principles and patterns in Chinese astronomy. Such a convergence can be seen in the succession of the first arithmetic models, such as the *Prognostics of the Five Stars*, leading to the better predictions in Liu Hong's *Qian Xiang Calendar*, among others. Here we see a first indication of a possibly more general trend.

After Yu Xi, however, we see no further improvement in mathematical astronomy in China, a situation that is no different from that in Europe after Ptolemy. There was increasing skepticism among Chinese astronomers regarding principle-based algorithms. The relevance of patterns was undisputed, but cosmic reality was considered by many to be too complex to be defined in terms of mathematical algorithms. This skeptical attitude had the side effect that abnormal phenomena that did not appear to show any patterns were also recorded in China.[39] These irregular phenomena, such as sunspots, supernovas, and comets, were hardly studied by the Greeks, if at all (also because comets were regarded as atmospheric phenomena). In China, starting in 28 BCE, smoked crystal was used to observe sunspots, which were reported each year with great precision in the Imperial Annals.

India: Astronomical Melting Pot

Indian astronomy consists of a combination of Babylonian, Greek, and Vedic knowledge. The Babylonian calendar system reached India in the late 5th century BCE as a result of the conquest of northwestern India by the Persians. The Greek theory of the spheres also reached India, possibly even before the conquests of Alexander the Great. These divergent forms of knowledge, together with the mythical Vedas, were forged into a cosmological model of the world consisting of *Puranas* covering vast periods of millions of years. These Puranas were subdivided into *Yugas* and provided an overview of the cosmos from creation of the world to its destruction.

As mentioned earlier, the oldest astronomical text from India, the *Vedanga jyotisha*, is difficult to date, with estimates ranging from 1400 BCE to the 1st century BCE (see chapter 2.3). Dating Indian texts is complicated considering that we have no reliable Indian chronologies until the 12th century CE. Some attribute this absence to the vast extent of the Puranas, which span millions of years and lay all human history to waste. Whatever the case, in an astronomical text dated roughly between the 1st and the 5th centuries CE, the *Paitamahasiddhanta* constitutes a model of planetary motions that is amazingly successful in accounting for the effects of the Ptolemaic equant without using the equant itself.[40] This is based on a double epicycle model, which is explained in more detail in later Indian works. We will come back to this in the next chapter. The insight that a double-epicyclic model makes the equant superfluous demonstrates the great ingenuity of Indian astronomers. It also shows that these astronomers were not satisfied with either the empirically inadequate pre-Ptolemaic models (such as that of Hipparchus) or with ad hoc solutions such as the equant.[41] The Indian astronomers went their own way, looking for new principle-based models that were empirically adequate. So it seems that in India too a search was underway for a convergence of principles and patterns.

3.3 Prior Convergence between Principles and Patterns: Musicology, History, Poetics, Art Theory, and Philology

At this point it may seem the convergence between theoretical principles and empirical patterns—and the process from misalignment to alignment—should preeminently concern the exact sciences, where concrete predictions can be made and tested. But nothing could be further from the truth. We also find this convergence in a number of disciplines or knowledge activities that we today associate with the humanities, especially in the study of language, music, art, and text. Moreover, it is not the case that these disciplines tried to imitate astronomy. To the contrary, the pursuit of convergence between principles and patterns took place earlier in the study of music, language, and text and would only later be applied in the study of celestial bodies and nature. In this chapter I also discuss disciplines where a convergence between principles and patterns is harder to clearly identify or simply failed to occur, such as in historiography and poetics.

Musicology: Principles for Harmonic Patterns

In the study of music we find a search for the underlying principles of consonant intervals starting in the 6th century BCE. Consonant intervals are harmonies in which the separate tones dissolve into one another, or blend, so to speak. Next to consonant intervals, there are dissonant intervals where the individual notes are in sharp contrast. Consonant versus dissonant intervals can be established by plucking a string, followed by plucking half its length, or two-thirds, or three-quarters. You then hear an octave, a fifth, and a fourth, respectively, while dissonant intervals, such as the second, correspond to ratios such as 8:9.[42]

Perhaps the question as to what principles underlie the pattern of consonant intervals was already proposed by the Babylonians, but no explicit musical principle emerges from the surviving clay tablets.[43] According to tradition, Pythagoras was the first to introduce a "law" for the consonant intervals, which coincided with the ratios between the first four integers: 1, 2, 3, and 4.[44] The sum of these four numbers is 10, the sacred Pythagorean number. The dissonant intervals, unlike the consonant intervals, could not be constructed as ratios of these integers and were characterized by more complex ratios. According to Pythagoras, it was this mathematical principle, which formed the basis for cosmic harmony (see above), that explained the phenomenon of consonant intervals. This led him to categorize the study of music as a mathematical discipline, along with astronomy, arithmetic, and geometry. In later antiquity, these disciplines, together with the linguistic disciplines of grammar, logic, and rhetoric, would constitute the liberal arts, or *artes liberales*—the standard curriculum of every free man.

But although Pythagoras's principle did an excellent job of predicting the consonant intervals of octave, fifth, and fourth, it encountered problems with intervals such as the third and the sixth. In so-called just intonation, a third is generally perceived as consonant, whereas in the Pythagorean system it corresponds to a rather complex ratio (64:81). This ratio is even more complex than the dissonant second: 8:9. Pythagoras rejected the third as a consonant interval, possibly for numerological reasons.

The first to refute Pythagoras's principle was Aristoxenus of Tarentum (late 4th century BCE), a student of Aristotle. Because of his in-depth empirical study of both harmony and melody, Aristoxenus is sometimes called the first musicologist.[45] He believed that intervals, including the consonant intervals, should not be assessed on the basis of simple integer ratios but on the basis of human hearing. Pythagoras put principle above pattern, but Aristoxenus advocated the inverse. First, the consonant intervals had to be determined empirically, and only then could mathematical principles be established to generalize about them. In this way, Aristoxenus was for musicology what Hipparchus was for astronomy: the theory had to correspond to the observations. Aristoxenus lived more than a century before Hipparchus; thus, the explicit pursuit of alignment between principles and patterns appears earlier in the study of music.

Unfortunately, Aristoxenus's works, the *Harmonic Elements* and *Rhythmic Elements*, survive only in fragments. From these fragments comes the insight that music consists in regularities that can be discovered by studying pieces of music themselves instead of starting from mathematical ideas. Ultimately, this allows one to discover not only the deeper principles of harmony but also the principles that underlie melodies: which sequences of notes constitute acceptable melodies in the Greek musical idiom (a kind of grammar for music). Aristoxenus discovered a number of theorems in the area of harmonic intervals, although he did not find a simple mathematical principle that could predict the consonant intervals.[46] Nevertheless, Aristoxenus never gave up his pursuit of an underlying mathematical principle. He was even strongly opposed to the work of his non-mathematical contemporaries, the so-called harmonists, who abandoned all musical regularity. Although Aristoxenus acquired followers of the likes of Kleionides, Aristides Quintilianus, and Psellos, they often ignored the empirical foundation of his work.[47] This has led to many misunderstandings affecting how Aristoxenus was received: mathematicians Euclid (3rd century BCE, see below) and Claudius Ptolemy (whom we have already encountered as astronomer) wrongly labeled his work anti-mathematical. In the 6th century CE, the two approaches—the Pythagorean and the Aristoxenic—are discussed in Boethius's (480–525 CE) *De institutione musica*. But until the early modern period, we encounter no further search for harmonic principles, although we do find a search for melodic principles (see chapter 4.4).

From a distance, the history of musicology shows a surprising affinity to that of astronomy. The consonant intervals initially seemed to be described fairly adequately by the principles of Pythagoras, but after Aristoxenus's empirical criticism—analogous to that of Hipparchus in astronomy—the Greeks became aware of the misalignment between principles and patterns, after which a search took place for them in music theory.

In India and China, a convergence of principles and patterns in music theory is less apparent. In India we find in the *Natya shastra* of Bharata Muni (ca. 1st century BCE–1st century CE) a description of the principles of consonant intervals, but these are limited to the octave and the fifth, which can be expressed as simple ratios (1:2 and 2:3).[48] In China, consonant intervals are described as a unity between micro- and macrocosm. In the Confucian *Book of Rites*, the *Liji*, the relationship between music and reality is explained using the pentatonic Zhou scale (named after the Zhou dynasty): "The basic note *kung* represents the sovereign; *shang* (one note above *kung*) represents the minister; *chiao* (one third above *kung*) represents the people; *chih* (one fifth above *kung*) represents national affairs and *yu* (one sixth above *kung*) represents things. If these five notes are not in disarray, harmony will prevail in the country."[49] The most important musicological discovery in China, however, is found in the *Huainanzi* (c. 139 BCE), which was compiled at the court of the Prince Liu An (179–122 BCE). It provides the oldest known analysis of a full 12-tone tuning with approximations of up to six digits and two decimal points.[50] But we have not been able to find a convergence between principles and patterns, let alone a process from misalignment to alignment.

History: Historical Patterns and Their Underlying Principles

The course of history shows no stable patterns. Yet the oldest historical pattern—that of the rise, prosperity, and decline of states and civilizations—goes back to the Babylonian historiography of early antiquity (see chapter 2.6). The first to mention this pattern explicitly is the Greek historian Herodotus (ca. 484–425 BCE), who had set himself the task of reconstructing the Greco-Persian Wars (490, 480–479 BCE) in his *History*. There he makes the pattern of rise, prosperity, and decline history's basic structure: "For many states that were once great have become insignificant, and those that are great in my time were once small."[51]

Thucydides (ca. 460–400 BC) also identifies in the rise and prosperity of Athens and its disintegration during the Peloponnesian Wars parallels with other historical periods. He tries to offer an explanation for this pattern and believes he has found one by analogy with human nature, which he supposes to be the cause: the same pattern of rise, prosperity, and decline can serve people and states as "an aid to interpreting the future."[52] This view of the future, present, and past as an eternal pattern can also be found among the Pythagoreans and in the Greek tragedies.[53]

There were also historians who searched for a principle that expressed a causal connection. We see this in Polybius, who in his *Histories* (ca. 200 BCE) personally coped with historical events, sometimes even literally, trekking across the Alps following Hannibal's journey. As a hostage of the Romans, he was deprived of access to Greek libraries for years and was dependent for his sources on either contact with others or his own experiences. But this did not prevent Polybius from identifying one of the most interesting patterns, along with an explanatory principle. According to Polybius, Rome seemed to be an exception to the pattern of rise, peak, and decline that had occurred so often in the history of Greek cities as a cycle of monarchy, aristocracy, democracy, and back to monarchy through tyranny. Unlike Athens, Rome was immune to this cycle—and thus also to decline—owing to its mixed constitution. Rome simultaneously had a monarchy (consuls), an aristocracy (senate), and a democracy (popular assembly). And according to Polybius, the cyclic pattern had been broken by this simultaneity of phenomena.

Although we now know that Rome would be subject to decline just like all other states, Polybius sought an explanatory principle for the two basic patterns he observed: the rise, peak, and decline of the Greek city-states, and the continued prosperity of Rome. His principle consisted of a mixed versus unmixed constitution, which led to prosperity in one case and to decline in the other. Of course, Polybius's principle applied to only these two patterns, one of which later turned out to be incorrect. But that does not detract from the value of Polybius's search. Again we see a discipline, in this case historiography, where demands are made on the principles to explain or even predict the observed patterns.

Later Roman historians mainly focused on the linear pattern of prosperity without decline. The age-old Roman tradition of keeping annals (*Annales maximi*) formed its basis. Livy produced his grandiose *Ab urbe condita* (late 1st century BCE), a chronological reconstruction of the city in which he incorporated the ideas of earlier historians such as Polybius into his view of history. Although the linear pattern of continuous prosperity is seldom mentioned in so many words, it remained a constant in Roman historiography. The best-known work of Publius Cornelius Tacitus (55–120 CE), the *Annales*, also starts with a summary of historical highlights, after which he continues where Livius left off. Ammianus Marcellinus (ca. 330–400) continued Tacitus's work until the year 378. With Ammianus, annalistic historiography of Rome remains linear, without decline. Of course, by the 4th century, Rome's decline could no longer be denied, but Ammianus interprets the later Roman Empire as a form of maturity. Its imminent end was unthinkable for him. Like so many classical historians, Ammianus attributed the problems of the city not to structural changes, only to failing individuals.

Antiquity's greatest historiography comes from China. Just as in Greece, here the pattern of rise and fall is central. The Chinese historian who first worked out this pattern was Sima Qian (ca. 145–86 BCE), the court historian of the Han dynasty. His 130-volume *Shiji* (Historical records) describes the entire history of all the peoples and regions known to him inside and outside of China. Sima discusses the time of the mythical Yellow Emperor in the 27th century BCE up through the 1st century of the Han dynasty (2nd century BCE). Never before had a historian covered such a vast period. In his historiography, Sima detects a pattern in the succession of dynasties. Each dynasty begins with a righteous ruler chosen by heaven, after which each successive ruler becomes less virtuous, until heaven loses patience and revokes the mandate of the last worthless ruler, after which everything starts over from the beginning, including the era. This pattern of the rise and fall of dynasties resembles the pattern found by Herodotus and Thucydides, but Sima's case is rooted in Taoism. According to the Taoist worldview, the cosmos consists in the constant mutual influence of opposing phases of the same energy, *yin* and *yang*. Everything that exists is subject to this ordering principle, called the "way" (*tao*, also spelled *dao*), which becomes visible in the changes caused by the advancing of the present. When *yin* peaks, *yang* starts to emerge, and the converse. Although the principle of *yin* and *yang* is not sufficient to predict when the rise will give way to decline, it will be used to explain the rise and fall.

Just as in Greco-Roman Europe with Polybius, there were also historians in China who promoted a pattern of continuous prosperity. For example, in his *Hanshu* (The book of the Han), Ban Gu (32–92 CE) pronounces the heavenly mandate given to the ruling family of the Western Han dynasty irreversible regardless of the virtue or corruption of the ruler's behavior. This dynasty was not subject to the pattern of rise and fall. Perhaps Ban Gu's position had something to do with the fact that he himself worked as a court historian under the Western Han,[54] which did not prevent his execution in 92 CE for his alliance with Empress Dou. However, his work was brilliantly completed by his younger sister Ban Zhao (45–116), who may be the first woman historian of note.[55] Ban Zhao is also the author of the influential *Nujie* (Lessons for women), in which she recommends that women submit in obedience to men, although she also advocates for that women be decently educated. Many Chinese women knew the book by heart. The male–female opposition, and the hardness–friendliness opposition that Ban Zhao thought corresponded to it, is explained by the principle of *yin* and *yang*.

Poetics: Failed Convergence between Principles and Patterns in Literature?

In the study of literature, in antiquity also known as poetics, we initially find a development similar to that in the study of music. But where the latter led to fairly stable principles and patterns—at least for the simple consonant intervals of octave, fifth, and fourth—this was not the case in poetics. For example, the Stoics tried to explain the beauty of sentences according to the principle of *natural word order*.[56] The following patterns were consequences of this principle: (1) nouns precede verbs, (2) verbs precede adverbs, and (3) past events are mentioned earlier than later events. But these patterns were hardly subjected to empirical verification by the Stoics; on the basis of mainly philosophical considerations, it was assumed that they were valid for artful poetic composition.

And in particular, these patterns were deemed valid for the text that no Greek doubted was pure and poetic: the Homeric epics. Yet, as far as we know, the Stoics never tested the patterns they recorded (let alone the underlying principle) against these texts.

The first to do so and immediately refute the patterns reported by the Stoics was Dionysius of Halicarnassus (ca. 60 BCE–ca. 7 CE), who dealt with both historiography and rhetoric as well as poetics.[57] Dionysius also adhered to the Stoic principle of natural word order, which he substantiated on the basis of deeper considerations: for example, nouns should be placed before verbs because a noun represents the substance, a verb represents the "accident," and the substance naturally precedes its accidents. However, Dionysius was not satisfied with purely theoretical considerations, and he subjected the Homeric texts to an experiment. When Dionysius tested the patterns following the principle of natural word order, he was surprised to discover that in many cases Homer did not adhere to these patterns.[58] In a sense, Dionysius is to literary studies what Aristoxenus was to musicology and Hipparchus to astronomy. But Dionysius's reaction was diametrically opposed to that of Aristoxenus and Hipparchus. Instead of looking for different and better principles for the patterns he observed, Dionysius simply rejected the patterns reported by the Stoics, along with the principle underlying them.

Using digital methods, today we can easily establish that while the "poetic" patterns drawn up by the Stoics are not valid in an absolute sense, they are not far from reality. For example, in Greek, verbs actually precede adverbs more often than not. We can therefore establish that certain statistical patterns hold in poetic language. But the Greeks were not interested in statistical patterns. They wanted to discover regularities that were always valid, so for them a pattern was all or nothing. We do come across exceptions and "exception rules" in other disciplines (see below), but in Greek poetics, convergence between principles and patterns failed.

In China, too, there was a search for underlying principles for literary works, but here too it is unclear whether we can speak of a convergence of principles and patterns. The Chinese writer Liu Xie (465–521 CE) produced one of the most impressive works in poetics: the *Wenxin diaolong* (The literary mind and the carving of dragons).[59] This work comprises 50 chapters and gives an overview of no fewer than 32 writing styles known in Liu's time: from the most aesthetic to the most practical, ranging from literary and philosophical works to exam papers and even declarations of war. Liu then searches for underlying principles for the various stylistic patterns that he believes to be present in the principles of vitality, musicality, and parallelism. These principles make loose generalizations about the patterns he found, but the patterns cannot be explicitly derived from the principles.

Art Theory: Principles for Patterns in Good versus Beautiful Art

The oldest documented search for patterns and their underlying principles in the visual arts can be found with the Roman Pliny the Elder (23–79 CE). In his *Naturalis historia* (Natural history), Pliny endeavored to collect all the knowledge of antiquity at that time. He refers to earlier Greek authors in the field of art theory, such as Xenocrates of Sicyon and Antigonus of Carystus (both from the 3rd century BCE), but their works have not survived. Pliny believed he had found a pattern in the development of art, which we could describe as an attempt to represent the world in as lifelike a way as possible, a pattern known as *illusionism.* According to Pliny, illusionist painting begins with drawing a line along a person's shadow and goes through a type of monochrome (single-color) painting to a polychrome representation of reality.

In addition, Pliny also looks for principles for good and beautiful art, but he succeeds only in the former. He concludes that there are no general principles for *beautiful* art, which can be achieved only with good luck or inspiration. Pliny tells us about the painter Protogenes, who after many unsuccessful attempts to portray a dog foaming at the mouth, threw his sponge at the panel in frustration, thus achieving the desired effect—it was coincidence that led to beauty.[60]

Although there are no principles for beautiful art, according to Pliny there are principles for "good" art: certain proportions need to be taken into account for good architecture, good sculpture, and good painting. These proportions follow a canon, that is, they function as a standard for other works. This notion was already known before Pliny and comes from a lost treatise of Polykleitos (5th century BCE). Pliny describes how Polykleitos with his sculpture the *Spear Bearer* creates a canon, which he used to "create the art itself through a work of art."[61] Polykleitos's canon consists in exact proportions between the size of the head and body height, between head width and shoulder width, between the palm and the fingers, and so forth. Harmony and balance can be achieved as long as these pure mathematical relationships are observed. In his overview of architecture, Pliny also points to the importance of mathematical proportions, which he, like the architecture theorist Vitruvius (ca. 85–20 BCE), defines in terms of ideal proportions that he believes to be found in Greek architecture and that are based on Pythagorean relationships.

What we see here is a process from misalignment to alignment for achieving harmonic proportions in art: initially Pliny sought out principles for beautiful art, but in the end he discovered principles for "good" art to meet correct harmonic proportions. Although this is a step backward—good art is a necessary but insufficient condition for beautiful art—Pliny found the principles that would dominate Western art theory for centuries. His principles were initially descriptive (found in existing works of art) but soon came to be prescriptive: good art consisted primarily in following the correct mathematical relationships. Pliny is thus at the beginning of a process from descriptive to prescriptive—we have observed similar processes in earlier disciplines, such as in Babylonian jurisprudence (see chapter 2.4).

India and China also attempted to establish principles for good and beautiful art. In India the oldest known text is a theoretical tract about Buddhist painting, the *Sadanga* or *Six Limbs*. As with Pliny, this text, which according to tradition was written around the 1st century BCE,[62] contains only principles for good art, especially for representing proportions.

A few centuries later, an attempt to comprehend not only good but beautiful art using principles was made by the Chinese art historian and critic Xie He in his *Gu huapin lu* (Classification of painters, ca. 5th century CE).[63] Xie gives the following somewhat cryptic principles:

(1) Spiritual resonance or vitality: the energy transferred from the artist at work (according to Xie, without mental resonance it makes no sense to continue looking at a work of art)
(2) "Bone method": brush strokes that express self-confidence, strength, and elasticity (just like bones do)
(3) Similarity to the subject depicted
(4) Suitability: the adequate use of color and tonality
(5) Distribution: creating a composition, space, and depth.
(6) Transmission by imitation: the pictorial representation of models as an imitation of earlier works

Some of these principles are about good art, while others go one step further, such as the principle of spiritual resonance, "the energy transferred from the artist to the work." Yet this principle offers no recipe or procedure for achieving such art. It is merely a definition of a certain perception of beauty. The same applies to the second principle, which recommends self-confidence, strength, and elasticity in the brush strokes. In contrast to Pliny's principle of mathematical proportions, Xie's principles are not specified in such a way that they can be applied unambiguously.

Philology: Principles of Analogy and Anomaly

Philology emerged as an academic activity after the Library of Alexandria was established by Ptolemy II around 300 BCE. Bringing together hundreds of thousands of manuscripts from all over the Hellenistic world led to one of the greatest inconsistencies in the history of knowledge: among the often dozens or even hundreds of copies of a given text, no two copies were ever the same. The differences were sometimes minor, caused by a copying error, but at other times they could be substantial and consist of entire words or sentences. How could the original text, the archetype, be derived from all this material?

One of the biggest challenges was the problem of corrupted words: how to determine whether an unknown word was archaic or corrupted. Corrupted words can be very frequent and even form a stable pattern. This happens, for example, when an incorrectly copied word is repeated in the same way throughout the manuscript. For this reason, a stable pattern does not constitute proof of an unknown, archaic word. Aristophanes of Byzantium (ca. 257–180 BCE) was the first to address this problem systematically. He understood that the variations that words could exhibit were based on a number of criteria that could determine whether a word was real or a corruption. Aristophanes summarized these criteria with the notion of *analogy*.[64] If he could establish that an unknown word was formed and inflected in a similar, analogous way to a known word, then that word was an existing, archaic word. But these criteria of word formation and inflection were not sufficient. Aristophanes eventually came up with a total of five criteria that word forms had to meet to be considered analogous: the word forms had to match in (1) *gender*, (2) *case*, (3) *ending*, (4) *number of syllables*, and (5) *accent*. But even these criteria proved insufficient. His student Aristarchus of Samothrace (ca. 216–144 BCE) added an additional criterion: when comparing two word forms, both had to be either *compound* (complex) or *noncompound* (simplex).[65] It is on the basis of these six criteria that some of the best reconstructions of Homer, Hesiod, Pindar, Archilochus, and Anacreon came about. Thus, systematic philology begins with Aristophanes and Aristarchus. We will refer to their underlying principle of similarity in inflections and word formation as the *principle of analogy*.

In this analogical school we thus find a convergence between the principle of analogy and the patterns responsible for it. The number of criteria was increased until the patterns could be accounted for and corruptions could be detected with great accuracy.

Despite this success, not all philologists embraced the Alexandrian approach. There was also a contrasting Stoic school that was founded in Pergamon. The philologists of this school worked on the basis of the *principle of anomaly* attributed to Chrysippus of Soli (ca. 280–ca. 207 BCE).[66] According to this principle, text transmissions did not consist in regularities, as the Alexandrians believed, but rather in exceptions. According to the anomalists, the primacy of the exception was for language in general, since an exception in language always takes precedence over the general rule, such as with irregular verb conjugation, as well as with expressions and colloquialisms that are not rule based (see above). The observation that the special case takes precedence over the general rule can also be found in other disciplines, such as law, where this observation would be elevated to a sort of meta-law, the *lex specialis* (see below). And long before this we find the distinction between the rules and exceptions on Babylonian linguistic clay tablets, though without noting which takes precedence (see chapter 2.1).

The most ardent supporter of the anomalistic approach was Crates of Mallos (died ca. 150 BCE).[67] According to Crates, all the efforts of the Alexandrian analogists were vain and superficial. The only way to get to the original text was not to look for regularities but to choose the surviving document that comes as close as possible to the author's intentions and to follow it to the greatest extent possible. No matter how unempirical this approach may seem, it has produced extremely original works. In contrast to the mostly formal work of the Alexandrians, the anomalists created the most erudite commentaries.

Later Roman philologists from the 1st century BCE, such as Marcus Terentius Varro and Marcus Verrius Flaccus, seem to have been influenced by both schools, considering the attention they give to both the regular and the exceptional in their textual analysis. Yet the two approaches stood in opposition from roughly 300 to 50 BCE, after which the controversy appears to have been settled or at least disappeared. Even Julius Caesar, in 55 BCE, at the time of his northern conquests, wrote a book about the controversy, *De analogia*, in which he follows Cicero in assuming that language and law determine a people's identity.[68] His pursuit of absolute laws for language served to safeguard the Latin identity and heritage. The fact that Caesar, alongside all his clashing of arms in

Gaul, also found time to write a treatise on the analogy issue demonstrates the importance of this subject.

And Back to Linguistics Again

Although I discussed linguistics at the beginning of this chapter ("A Tale of Two Principles"), thanks to our notion of convergence between principles and patterns, we can now see linguistics in a new light. And then we are mainly talking about Panini's Indian linguistics, where recursive grammar (the principle of recursion) can be used to determine for each row of words whether it constitutes a grammatical or ungrammatical sentence in the language (see above). Although we know nothing about a process from misalignment to alignment between the principle of recursion and the language patterns of Sanskrit—simply because we know almost nothing about pre-Paninian linguistics—there is clearly a convergence of principles and patterns in Panini's own linguistics.

3.4 The Ultimate Convergence of Principles and Patterns? Mathematics and Logic

Nowhere were the relations between principles and patterns as close as in mathematics and logic. The Greeks established axiomatic principles with which they could deductively prove mathematical patterns (theorems). Here the convergence between principles and patterns was brought about by a formal inference in the form of a logical proof. This approach became extremely influential in other disciplines, such as in Ptolemy's Hellenistic astronomy (above) and in mechanics (below). Formal inferences in mathematics were also developed in China, but instead of a deductive approach, Chinese mathematicians used an algorithmic method. A mathematical problem was linked to its solution through a step-by-step procedure, demonstrating that the procedure was correct.

The Mythical Origin of the Proof: From Thales to Pythagoras

According to tradition, we find the oldest Greek mathematics with Thales of Miletus, around 600 BCE. In addition to the primordial principle of water (see above), several mathematical propositions have been attributed to him. The most famous concerns a triangle inscribed within a circle: if one side is the circle's diameter, it always forms a right triangle.[69] This is an extremely stable pattern that calls for an explanation.

In all likelihood, the pattern had been known for much longer, but Thales was the first to provide a proof that gave the proposition absolute validity. This is why he is called the father of deduction. In later centuries, Thales's proposition became the model of beauty and perfection, and in the Middle Ages it was even given a place in paradise in Dante's *Divine Comedy*.[70] Although nothing of Thales's proof is known, after him we see a growing concern with the notion of the proof, and within a few centuries almost all Greek mathematics would consist in a logical structure of definitions, propositions, and proofs.

According to tradition, Pythagoras, the second legendary figure, was active a generation after Thales. Nothing has survived from Pythagoras either, but that did not stop his followers from attributing many discoveries to him. As we have already seen in astronomy and musicology above, Pythagoras assumed that the entire world consisted in whole-number ratios and pure figures. The statement attributed to him that in any right triangle the square of the hypotenuse is equal to the sum of the squares of its other two sides—in modern notation $a^2 + b^2 = c^2$—was already known to the Babylonians, the Chinese, and the Egyptians (see chapter 2.2). However, according to tradition, Pythagoras was the first to prove it, although we do not know what sort of proof he used.

Like the Babylonians before them, the Pythagoreans were also active in collecting Pythagorean triplets, that is, three integers that satisfy the formula just mentioned, such as (3, 4, 5), (5, 12, 13), and (8, 15, 17). According to the Pythagoreans, whole numbers were the building blocks of the cosmos and enjoyed an almost sacrosanct status. The Pythagoreans therefore assumed that the length of the hypotenuse of a right triangle could always be expressed in terms of integers, possibly in the form of a fraction. However, this assumption led to a problem for right triangles with short sides of the same length. In the simplest case, where the short sides are equal to 1, according to the Pythagorean theorem, the long side is equal to $\sqrt{2}$. The story goes that the Pythagoreans were unable to find a ratio of two integers equal to the square root of 2,[71] that is, until a student of Pythagoras, named Hippasus, proved that no such relationship existed. In other words, he proved that $\sqrt{2}$ could not be expressed as a fraction of whole numbers, what is known as an irrational number.[72] Suddenly, a complete worldview—one based on the principle of whole numbers and pure figures—was turned upside down by a mathematical proof. Although Hippasus's proof has not survived, it was apparently so convincing that it was a threat to the Pythagoreans, who tried with all their might to keep it secret. According to the biographer Diogenes Laertius, Hippasus was thrown overboard into the Mediterranean Sea when he disclosed it. One cannot imagine a greater "misalignment" between principle and pattern.

Plato's Role

Although Plato is best known as a philosopher, he was also the most important patron of Greek mathematics. As evidenced by his beautiful quotation about patterns and principles above, Plato saw mathematics as a way to penetrate the "real world"—the world of ideas—which can be comprehended only using reason. At the Academy founded by Plato in 387 BCE, the first years of the curriculum consisted largely of mathematics.[73] He demanded that his students be able to draw up precise definitions and unambiguously formulated propositions and that they prove them through logical inferences. In addition, he held that geometric proofs needed to be carried out with a compass and ruler. For many proofs this is possible, but Plato also presented three (previously known) Classic Problems that mathematicians wracked their brains about for centuries: squaring the circle (that is, constructing a square with the same area as a given circle), doubling the cube, and splitting an angle into three. It wasn't until the 19th century that it was demonstrated that these problems could not be solved with a ruler and compass.

The idea of the proof as a logical inference probably stems from before Plato, but it is thanks to him that this form of proof was established as the mathematical method: mathematical patterns had to be reduced to a small number of basic principles. Although Plato did not succeed in finding these general basic principles, we do encounter such principles two generations later with Euclid. In mathematics, too, there has been a process from misalignment to alignment between principles and patterns.

A Proof as a Connection between Principles and Patterns: Euclid

In Euclid's (ca. 325–265 BCE) *Elements*, we gain insight into what a Greek proof looked like.[74] Like Plato, Euclid defines a proof as a logical inference: the reduction of a theorem to "self-evident" principles known as axioms. If the logical inference is valid, then the theorem is true. Euclid's great contribution was that he brought together all of the many mathematical theorems in geometry that were circulating in his time. Just how many personal contributions Euclid made to mathematics is not known, but his greatest contribution is that he managed to base all assembled theorems on a total of 10 basic principles: five geometric axioms and five so-called universal truths.

The impact of the *Elements* was tremendous: no scientific or scholarly work has ever been in circulation for so long and used as intensively as the *Elements.* Well into the 20th century, everyone in the Western world and beyond was introduced to mathematics with Euclid. The *Elements* was also one of the first Greek works to be translated into Arabic. And the 16th-century Jesuits took it to China to showcase what European science could do. Euclid's tremendous fame stands in stark contrast to what we know about his life, which amounts to almost nothing, apart from a few apocryphal anecdotes, such as the story that King Ptolemy I purportedly asked Euclid whether there were no shorter route to mastery of the subject than through the *Elements,* to which Euclid replied that there was no royal road to geometry.[75]

In contrast to the earlier Babylonians and Egyptians, Euclid expressed almost all mathematics in terms of geometric concepts. Numbers were lengths of line segments, squared values were represented with squares, and prime numbers were also defined in terms of a line segment. Expressing mathematical problems through geometry is not always practical, but that did not concern the Greeks. Their goal was to reduce all mathematical statements to a small number of geometric and general principles.[76] Euclid presents these principles in two groups: (1) axioms that apply specifically to geometry and (2) general truths that apply to all knowledge about quantities (and that are therefore also axioms). This dichotomy corresponds to the Aristotelian distinction between "special insights" for specific domains of knowledge and "general insights" that underlie all thinking (see below). In addition, Euclid also gives a number of definitions, such as for point, line, plane, and circle.

Euclid applies these five axioms specifically to geometry:

1. Any two points can be connected with a straight line.
2. Any straight line can be endlessly extended as a straight line.
3. Any line segment can be a radius emanating from the center of a circle.
4. All right angles are congruent.
5. If two lines intersect a third line in such a way that the sum of the inner angles on one side is smaller than two right angles, these two lines must inevitably intersect each other if they are extended sufficiently.

The first four axioms indeed look like self-evident principles. However, number 5 looks less like a principle and more like a proposition that has yet to be proven. Has Euclid made a mistake here? For centuries and up through modern times, mathematicians have made admirable attempts to prove axiom 5 using the other four axioms so that the total number could be reduced to four (see chapter 4.3) but without success: it appears that at least five axioms are needed for the geometry of flat surfaces.

Euclid's second group of principles applies to all human knowledge:

1. Things that are equal to some other thing are also equal to each other.
2. If equals are added to equals, the sums are equal.
3. If equals are subtracted from equals, the remainders (differences) are equal.
4. Things that coincide with one another are equal to one another.
5. The whole is greater than the part.

We do not know whether it was Euclid himself who formulated the five axioms and the five general truths. Today it is assumed that they were already in circulation in some form before him. Either way, these 10 basic principles constitute the basis for Euclid's argumentation. As an example, let's take proposition 17 from book 1 with the corresponding proof:

For any triangle, the sum of two angles is always smaller than that of two right angles.

Given a triangle ABC.

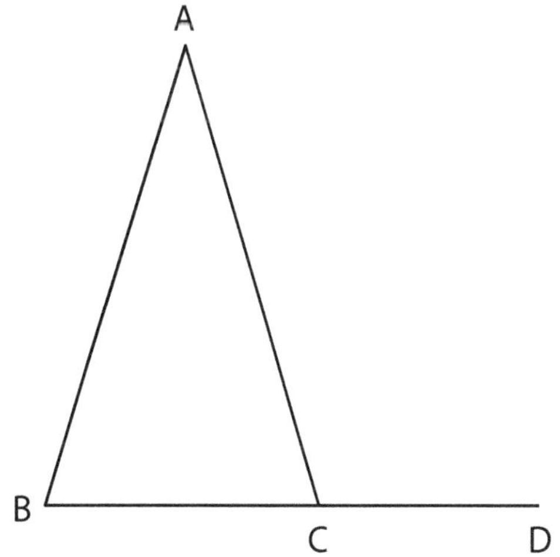

I say that the sum of two angles of the triangle ABC is smaller than two right angles.

> Extend BC to D.
> Since the angle ACD is an outer angle of the triangle ABC, it is larger than the inner angle and the opposite angle ABC. Add the ACB angle to both. Then the sum of the angles ACD and ACB is greater than the sum of the angles ABC and BCA. But the sum of the angles ACD and ACB is equal to two right angles. So the sum of the angles ABC and BCA is smaller than two right angles.
> Similarly, we can prove that the sum of the angles BAC and ACB is also less than two right angles, and therefore also the sum of the angles CAB and ABC.
> Therefore, in any triangle the sum of two angles is always smaller than that of two right angles.

The way this and other Euclidean proofs were constructed set the standard for Western mathematics. In the above proof, we see the use of axioms, such as axiom 2, when BC is extended to D. We also see the use of general truths, such as general truth 5, where adding angles ACB to ACD and ABC is used to deduce the fact that the sum of the angles ACD and ACB is greater than the sum of the angles ABC and BCA. This leads to an inference built step by step from the axioms and general truths until one arrives at the proposition to be proven. Use is also made of propositions that have already been proven in the book and that can be used in the same way as an axiom to construct the proof.

However, Euclid regularly omits part of a proof, as introduced above with the sentence "Similarly, we can prove that . . ." According to Euclid, once something has been proven for a particular case, the proof for similar cases can be considered trivial. Meanwhile, some of Euclid's proofs are believed to contain nontrivial gaps that were filled only later.[77] Furthermore, Euclid also fails to state which logical rules he has used to proceed from one step to the next in the course of a proof. He may also have found this trivial. Indeed, the quest for logical rules is a characteristic of logic rather than of mathematics. Logic can in fact be viewed as the study of the validity of inferences. The question that was not asked by Euclid is whether there is a procedure for verifying that a proof is valid. In other words, are there underlying principles for mathematical reasoning?

Even Further Convergence in Logic

Aristotle did for logic what Euclid did not do for mathematics: develop principles for constructing valid reasoning. Aristotle believed that all correct argumentation could be constructed from *syllogisms*, in which certain statements are accepted as true and from which other statements necessarily follow. Although Aristotle lived about half a century before Euclid, he also wanted his logic to be valid for mathematical reasoning, such as the proofs that would be assembled by Euclid. Aristotle took a surprising step: he made his system—*syllogistic logic*—independent of whether the axioms were true. For him, the focus was entirely on the validity of correct reasoning or on the validity of the relationship between axioms and the conclusion, regardless of whether the axioms were true. What exactly was that relationship?

Aristotle tried to make the step from axioms to what had to be proven (the conclusion) watertight. In doing so, he hoped to get a grip on both logical and mathematical reasoning. While he succeeded in the former, he did not in the latter. Although Aristotle found a number of powerful reasoning schemes, ultimately they could not be used for mathematical proofs. In a series of works referred to as the *Organon*, Aristotle analyzes the structure of reasoning, in which syllogisms play a central role.[78] To be more precise, a syllogism is a logical argumentation in which a proposition (the conclusion) is derived from two other propositions (the premises). A syllogism consists of three parts: a major premise, a minor premise, and a conclusion, as exemplified here:

(1) *Major premise*: All humans are mortal.
(2) *Minor premise*: All Greeks are humans.
(3) *Conclusion*: All Greeks are mortal.

According to Aristotle, this syllogism can be generalized as follows:

(1) *Major premise*: All A are B.
(2) *Minor premise*: All C are A.
(3) *Conclusion*: All C are B.

Whatever words we plug in for A, B. and C, the inference is always valid.

In addition to the term "all," syllogisms can also include the terms "some," "none," and "not":

(1) *Major premise*: All informative things are useful.
(2) *Minor premise*: Some books are not useful.
(3) *Conclusion*: Some books are not informative.

This syllogism is of the type "All A are B, some C are not B, some C are not A."

A syllogism comprises three propositions, the first two of which—the premises—have exactly one term in common, and the third of which, the conclusion, contains the two non-common terms of the premises. Although there are an infinite number of possible syllogisms, the four predicates "all," "some," "none," and "not" make a total of 256 distinguishable types, of which no more than 16 are valid. So, for example, the following syllogism is *not* a valid argument:

(1) *Major premise*: All humans are mortal.
(2) *Minor premise*: Euripides is mortal.
(3) *Conclusion*: Euripides is a human.

While both premises and the conclusion are correct in this case, the reasoning is invalid. From the premises that all people are mortal and that a certain Euripides is also mortal, it does not necessarily follow that Euripides is a human. For example, if Euripides refers to a dog by that name, and who is indeed mortal, the conclusion is incorrect. In short, the syllogism of the type "All A are B, C is a B, so C is an A" constitutes invalid reasoning, even if the premises and conclusion are correct.

With the 16 valid types of syllogisms, Aristotle was able to construct new valid arguments through repeated application of syllogisms, thereby establishing the underlying basic system of syllogistics. As nice as that may be, Aristotle's logic was of limited use outside of the system of syllogisms: it could not describe even the most elementary reasoning in Euclid's *Elements*. At the most, syllogisms are useful for reasoning of a more everyday nature, such as in dialogues and speeches.

In his *Metaphysics*, Aristotle gives some basic principles of reasoning, which are similar to Euclid's "general truths."[79] The first of Aristotle's basic principles is the *law of noncontradiction*, which contends that a statement and its negation can never be true at the same time. The second is the *law of the excluded middle*, which says that any statement is either true or false. These two laws can be seen as the criteria for sound thinking. While these laws cannot prove that an argument is either valid or not (this is only possible in a number of cases, with the limited syllogistic arguments), they do constitute the conditions to which all valid arguments and correct proofs are subject, including the mathematical proof of proposition 17 above.

Although the system of syllogisms has limited application, it constitutes the most far-reaching convergence between principles and patterns in ancient times. In contrast to Euclidean geometry, where the step from axioms to theorems still contains various implicit assumptions, the syllogistic schemes in Aristotelian logic, or combinations of such schemes, can be applied directly to new situations. This maximizes convergence. Aristotelian logic is both the most attainable form of convergence and the most trivial.

After Aristotle: Propositional Logic

Aristotle was not the only person looking for principles for constructing valid reasoning. Within his own generation, the Stoics developed another form of logic, a system that has become known as *propositional* logic. In this branch of logic, the truth or untruth of combinations of statements (propositions) is derived from the truths or untruths of the individual or partial statements. Statements can be combined by *connectives* such as *conjunction* ("and"), *disjunction* ("or"), and *implication* ("if . . . then . . ."), as well as by *negation* ("not").

The statement "John is smart and Peter is stupid" consists of the conjunction of the statement "John is smart" and the statement "Peter is stupid." This conjunction is true when the two statements "John is smart" and "Peter is stupid" are both true. In contrast, if only one of the two coupled statements is false, the entire statement is false. Things are different with disjunction: the statement "John is smart or Peter is stupid" is only untrue if both "John is smart" and "Peter is stupid" are false. And with implication, the statement "If John is smart, then Peter is stupid" is only false when "Peter is stupid" is false and "John is smart" is true. In this way, what are known as truth tables can be constructed indicating the truth value of a complex statement based on the possible combinations of the truth values of the individual statements. The oldest truth table is attributed to Philo of Megara, from around 300 BCE, for the logical implication *if A then B*.[80] And a generation later, Chrysippus of Soli, whom we already encountered in philology above, established propositional logic on an axiomatic basis.

Mathematics after Euclid: From Archimedes to Hypatia

While Euclid provided an overview of Greek geometry, Archimedes (ca. 287–212 BCE) focused on both arithmetic and geometry.[81] It is to him that we are indebted for the relationship between the surface and the volume of the sphere, as well as the most accurate calculation of the number π to date. In addition, Archimedes developed an ingenious system for expressing very large numbers—in his case for estimating the number of grains of sand that would fit into the universe. We will return to Archimedes in more detail in the section on generating patterns. One generation later, Apollonius of Perga (ca. 262–190 BCE) provided an overview of the conic sections, to which we owe the curves of the hyperbola, parabola, and ellipse.

It is curious that, just as in astronomy, scant new developments took place in mathematics between the 1st century BCE and 3rd century CE. But in later antiquity (between 250 and 400 CE) mathematics flourishes again in the Roman Empire. For example, in his *Arithmetica*, Diophantus (3rd century CE) not only provides solutions to 150 algebraic problems; he also casts these solutions in the form of symbolic mathematics.[82] However, no one built upon this *algebraic* mathematics until the Persian al-Khwarizmi elaborated algebra as an independent branch of mathematics in the 9th century (see chapter 4.3).

The Greco-Roman tradition concludes with the first great woman mathematician, Hypatia of Alexandria (ca. 350–415). Although none of her manuscripts has survived, comments on the works of Diophantus, Apollonius, Ptolemy, and Euclid are attributed to her.[83] She also simplified the complex calculations in Ptolemy's *Almagest* and, as far as is known, wrote the first and only commentary on Diophantus's *Arithmetica* (which survives thanks to Arabic translations).[84] She was brutally murdered in 415 in Alexandria. After Hypatia, Western mathematics produced nothing new for centuries, and most of what had been achieved would soon no longer be read or understood.

China: Algorithmic Proof and Mohist Reasoning Principles

Whereas in the West mathematics was one of the seven liberal arts, mathematics in China was part of the six arts, the *liu yi*, attributed to Confucius. The other five arts in China were music, ritual, archery, chariot racing, and calligraphy. The function of these arts was similar to that in the Greco-Roman world: they constituted a pedagogical program that every well-educated man was expected to have mastered; the intention was not to produce new research. But the Confucian school was just one of many in China. The Mohist school, founded by the followers of the philosopher Mozi (ca. 470–391 BCE), was especially successful in mathematics and logic. The Mohist Canon, dating from ca. 330 BCE, contains the oldest work of Chinese geometry, the *Mo jing*.[85] In addition to problems from mechanics (see below), the *Mo jing* also provides definitions of the point, line, parallelism, circumference, diameter, and volume.

The work that put Chinese mathematics on the map was the *Nine Chapters of Mathematical Art*, or the *Jiuzhang suanshu*.[86] This work, which dates back to before the great book burning of 213 BCE, discusses 246 problems of various kinds. Among the solutions, those of Liu Hui (3rd century CE) are extraordinarily impressive. Unlike the Greeks, Liu expresses his calculations in decimal fractions. He also provides a method for calculating π, which he approximated to five correct decimal points, three decimal points beyond Archimedes, which he was not aware of. With Liu we also encounter a concept of the proof: he gives a diagram for the Pythagorean theorem, which appears as one of the problems in the *Nine Chapters*. Another problem that Liu Hui tackled was that of linear equations. The way he approached this problem bears an uncanny resemblance to today's matrix calculus. He provides a step-by-step procedure that most closely resembles an algorithm, just as the Chinese astronomer Liu Hong had previously done for calculating lunar motion (see above). But Liu Hui goes even further: on the basis of the steps in his algorithm, he argues that the procedure is correct and shows again, step by step, that this procedure must lead to the correct solution. Liu's argumentation is nothing less than a proof, although diametrically opposed to the Greek notion of the proof, which is constructed axiomatically.[87] We can best identify Liu's argumentation with the notion of the *algorithmic proof*. Such an algorithm leaves nothing to chance and always specifies what the next step will be. The Greek approach, on the contrary, is the *geometric* or *axiomatic proof*, where a logical inference is constructed between axioms and a theorem. One should keep in mind that although the Greek notion of the axiomatic proof was fruitful in the deductive disciplines, such as mathematics, it was much less successful in the more inductive disciplines, such as the natural sciences.

Later Chinese mathematics is also often characterized by algorithmic procedures for solving mathematical problems. The mathematician Zu Chongzhi (429–500 CE) is one prominent example, especially because he succeeded in calculating π to one further decimal place—the world's most accurate value of π until the 15th century.[88]

The interaction between mathematics and logic was less strong in China than in Greece. Logic is mainly found in the Mohist Canon (330 BCE), where it is presented as the basis of all other disciplines. According to the Mohists, correct reasoning and argumentation require general principles, the first of which is that of two contradictory statements, one must be false. The second principle suggests that conflicting statements cannot both be false.[89] These principles are surprisingly similar to Aristotle's general reasoning principles, the law of noncontradiction and the law of the excluded middle, but come about 200 years earlier.[90] Thus, it appears that the Mohists were the first to discover these fundamental laws of logic. There was no intellectual contact between the Greeks and the Chinese in antiquity, although this knowledge may have been exchanged through trade contacts. Logic in ancient China did not end well: during the Qin dynasty (221–206 BCE), under strict Legalist rule, Mohism was banned. It was not until the 7th century CE that work on logic was taken up again in China.

India: Mathematical Innovations and Their Linguistic Origins

Panini occupies a prominent position not only in the history of linguistics (see above) but in that of mathematics as well.[91] This is due to his notion of recursion, which had a major impact on later mathematics. Yet Panini worked exclusively with linguistic notions such as phonemes and morphemes, rather than with numbers or geometric objects. The integration of language and mathematics goes a step further with the linguist Pingala (ca. 3rd century BCE), especially in his work on prosody, the branch of linguistics that studies the metric aspects of language. Pingala distinguished between metric patterns of short and long syllables in Sanskrit. He described these patterns in a way that follows the successive numbers in the binary number system (see chapter 2.2 for number systems).[92]

The main mathematical works of Indian antiquity are the *Sulba Sutras*, which I mentioned in the previous chapter; given their date between the 8th and the 5th century BCE, they might as appropriately be placed in early antiquity.[93] The sutras contain approximations of irrational numbers, such as $\sqrt{2}$, and calculations of Pythagorean triplets. Besides these sutras, the most important Indian mathematical insights can be found in astronomical texts, especially in the *Siddhantas* from the Gupta dynasty (3rd–6th century CE). These texts contain the oldest examples of the well-known trigonometric notions of *sine* and *cosine*. These terms are corruptions of the words *jiya* and *kojiya* in Sanskrit.[94]

Indian logic is completely different from its Greek counterpart: instead of being deductive, the most important logical tradition in India is inductive, where generalizations are derived from observations. This logic is known as *nyaya* (Sanskrit for "inference") and goes back to texts by Aksapada Gautama from around 200 CE. Four sources of knowledge are distinguished in *nyaya* logic: *observation*, *inference*, *comparison*, and *testimony*. For logic, inference, or *anumana*, is particularly important and comes about in the form of inductive reasoning.[95] The *nyaya* inference consists of five steps, as in the following example:

(1) There is fire on the hill (*pratijna*, the thing that needs to be proved).
(2) Because there is smoke (*hetu*, the reason or cause).
(3) Where there is smoke, there is fire, as in a kitchen hearth (*udaharana*, the example).
(4) Just like on the hill (*upanaya*, application of the example to the case).
(5) Therefore, there is fire on the hill (*nigamana*, the conclusion).

The characteristic feature of the *nyaya* inference is in the emphasis on the example and its application to a new situation. Unlike Aristotelian logic, this form of reasoning is not deductive, making it less strict but more broadly applicable in practice. It was used in rhetoric, or the art of persuasion, in ancient India, where inductive proofs usually predominated and to which Aristotelian syllogisms were unsuited.

3.5 Coexisting Principles and Patterns: Medicine

The pursuit of convergence between principles and patterns was not successful in all disciplines. In medicine we find influential principles, but there is no convergence between these principles and clinical patterns, despite the many attempts.

The Principles of Hippocrates: Theory Is Central

The greatest physician of Greek antiquity is Hippocrates of Kos (ca. 460–370 BCE), although we know nothing about his life. The works attributed to him were probably assembled by his students, and even on that point we are uncertain. The *Hippocratic Corpus* consists of texts on various topics, such as epidemics, prognoses, fractures, and the famous Hippocratic Oath. The most influential Hippocratic text is undoubtedly *On Human Nature*, in which *humorism*, or the doctrine of the humors, is explained. This theory describes a general theory of illness and health based on four humors, or temperaments.[96]

The concept of four humors predates Hippocrates and may be of Egyptian or Mesopotamian origin.[97] Although there were several versions of the doctrine of humors, roughly speaking, it consisted in four personality types: sanguine, phlegmatic, choleric, and melancholic. These temperaments were associated with four bodily fluids: blood, phlegm, yellow bile, and black bile. With the sanguine type, blood dominates; this type of person is cheerful and in good spirits and has many interests. In a phlegmatic person, phlegm predominates, and this type of person is peaceful, calm, and dreamy. In the choleric type, yellow bile predominates; this type is busy and active. And finally, with the melancholic person it is black gall that dominates; this type is melancholic and serious. A person can exhibit several personality types according to the humor theory, but when there is a surplus or deficiency of one of the four bodily fluids, the personality falls out of balance. It was a small step to apply this doctrine to human health, and that is exactly what Hippocrates did:

> The human body contains blood, phlegm, yellow bile and black bile. These are the things that make up its constitution and cause its pains and health. Health is primarily that state in which these constituent substances are in the correct proportion to each other, both in strength and quantity, and are well mixed. Pain occurs when one of the substances presents either a deficiency or an excess, or is separated in the body and not mixed with others.[98]

Thus, according to Hippocrates, diseases were caused by an excess or shortage of these fluids in a certain part of the body. For example, gout was caused by too much of the bodily fluids dropping to the feet. And a displacement of phlegm from the head to the lungs was thought to cause coughing and lung disease. When the balance between bodily fluids was disrupted, the physician's job was to recognize and restore it by removing certain fluids, such as by bloodletting. Moreover, humorism was highly individualized: each patient had a unique humorous composition. This made the doctor's job an exercise in interpretation, for which a holistic approach was indispensable. The relationship between mental and physical processes thus formed the core of the theory of the humors.

With this theory Hippocrates's followers thought they could explain all symptoms of disease. For example, the Hippocratic text *On the Sacred Disease* argues that epilepsy is caused by phlegm blocking the airways, leading to convulsions of the body struggling to free itself. But could humorism actually cure illnesses in addition to providing these kinds of explanations? That was rarely the case: most disorders were handled mainly by preexisting treatments whose efficacy had withstood the test of time. The *Hippocratic Corpus* describes, for example, how fractured bones should be set, splinted, and bound and how bladder stones could be catheterized. The medical observations were sometimes presented in the form of the well-known *if–then* patterns, like in Egyptian and

Babylonian medicine (see chapter 2.5), such as "If sleep puts an end to delirium, then that's a good sign." But an explicit relationship between the principles of humor theory and long-standing successful treatments was usually impossible to establish. When one did assume the doctrine of the humors, the treatment method derived from it, such as bloodletting, was generally harmful rather than beneficial. Nonetheless, humorism did occasionally lead to effective treatments, such as the suppuration of a wound where, according to Hippocrates, one needed to allow the pus to leave the body freely. However, a similar line of reasoning could not be followed for blood, which always had to leave the body in a controlled manner. In addition, humorism gave no indication of how much blood to let—in practice, bloodletting was almost always harmful. So, what we see is that although several successful clinical treatments did exist, they could rarely be associated with theoretical principles.

Here we find a fascinating discrepancy: while in most disciplines the Greeks sought ever greater convergence between principles and patterns, in medicine the Hippocratic doctors apparently satisfied themselves with principles that were largely independent of patterns (that is, patterns that were thought to be successful). Perhaps this discrepancy can be explained from the classical Greek worldview, according to which the microcosm reflected the macrocosm. The four human bodily fluids—blood, phlegm, black bile, and yellow bile—corresponded to the four cosmic elements—fire, water, earth, and air—and even to the four seasons and the four stages of human life—childhood, adolescence, adulthood, and old age. The doctrine of the humors thus provided an all-encompassing view of the world that was not easy to deviate from (something similar would happen later with the Aristotelian view of the world in medieval Europe; see chapter 4). In addition, medical principles and patterns diverged because the medical practitioner was primarily a scholar, while surgical practices were left to specialized artisans. The Hippocratic Oath even prohibited doctors from cutting into the body. As a result, there was less attention to empirical patterns than was the case in other disciplines.

Alexandria: Empiricism Is Central

Concrete discoveries were also made, especially in Hellenistic Alexandria, where empiricism reigned supreme in medicine. Not only did the study of anatomy boom, but hernia and even eye and windpipe operations were performed for the first time. Herophilus of Chalcedon (ca. 330–260 BCE) distinguished arteries from veins, finding that arteries alone have a pulse.[99] Herophilus was also the first to perform anatomical dissections publicly. His greatest achievement is probably the dissection of the nerves, which he discovered were connected to the brain. Herophilus also linked the nervous system to movement and experience, and he situated intelligence in the brain. His contemporary Erasistratus of Chios (ca. 330–255 BCE) also devoted himself to brain research, and he conducted experiments not only on animals but also on humans.[100] Among his greatest insights is the distinction between the cerebrum and the cerebellum. Moreover, he distinguished motor nerves from sensory nerves. The Alexandrian discoveries and insights are certainly impressive, but they had much less influence on medicine than one would expect. Theory and philosophy continued to dominate medical practice.

Research into medications was of a different nature. As a student of Aristotle, Theophrastus (ca. 371–287 BCE) was the first in the West to systematically examine plants, shrubs, and herbs for their medicinal effects (see also below). But it was Pedanius Dioscorides (ca. 40–90 CE), a Greek surgeon in Nero's army, who, in his lavishly illustrated *De materia medica*, arranged plants and herbs according to their pharmacological properties.[101] Many of his recipes consisted of ordinary herbs, such as cinnamon, which was considered beneficial for inflammation and snake bites. Some of his recommendations were downright bizarre: for example, one formula prescribed for malaria fever consisted of mashed bedbugs mixed with meat and beans. It is hard to imagine that this remedy came into existence as the result of experimentation.

Galen's Attempt to Integrate Theory and Empiricism

The Greco-Roman physician Galen (129–216 CE) had an exceptional goal: he wanted to combine Hippocrates's philosophy with the empirical-surgical approach of the Alexandrines. Galen was one of the last doctors to perform brain and eye surgery, something that would not be done again for centuries after him. And since dissections on human bodies were taboo at the time, Galen dissected countless animals, from which he subsequently drew conclusions about the human body. On a theoretical level, Galen largely embraced the Hippocratic humors, presenting his own work as the ultimate perfection of Hippocrates's legacy. With more than 350 titles to his name, he was particularly prolific, and he penetrated the highest echelons of the Roman Empire, reaching even the emperor, which certainly contributed to the dissemination of his ideas. Galen's highest goal was to unify the theoretical Hippocratic principles with the more empirical clinical treatments, including the *materia medica*, but whether he succeeded in that endeavor is questionable. With Galen, too, speculative principles ultimately win out over clinical patterns. He states that fever is caused by a surplus of yellow bile, black bile, or phlegm. To restore the balance of the humors, it was necessary to apply energetic bloodletting, which could be as frequent as twice a day, where patients had to bleed until they lost consciousness. The motivation for this treatment was purely theoretical, and it did not lead to a convergence with existing clinical practices.

For Galen, blood was not something pumped through the heart but was continuously produced by the liver and then transported to all parts of the body, including the heart. Blood could be venous or arterial—these two types followed different paths and had different functions related to three body centers: the liver, which was responsible for nutrition and growth; the heart, which was responsible for vitality; and the brain, which was in charge of feeling and reason. Nutrition and growth were brought about by venous blood from the liver, while vitality was transmitted by arterial blood. According to Galen, once distributed to the various parts of the body, all the blood was used up and was not sent back to the heart. Although Galen could never support his theory empirically, his authority was such that his ideas were considered sacrosanct for almost 1,500 years, in both European and Arab medicine.

Ayurvedic Medicine: Humors in India

The Greek theory of humors has a number of surprising similarities with Indian and Chinese medicine. In India the most important medical system was *Ayurveda*: the knowledge (*veda*) of long life (*ayus*).[102] The oldest surviving Ayurvedic texts are the *Caraka samhita* and *Susruta samhita*, both from around the 1st century CE. According to Ayurveda, there are three bodily humors—air, bile, and phlegm—which correspond to the macrocosmic forces of wind, sun, and moon. Ayurvedic teachings cover all aspects of life, such as rules for washing, exercise, and diet. These aspects are part of Hindu thinking about rebirth, abstinence, and the balance of the soul. Treatments usually consist of prescribing herbs and ointments, as well as enemas and massages, right up to surgery.

Just as in classical Greece, most of the surgery in India was left to artisans, but this more practical sort of medicine was highly developed. For example, plastic surgery probably dates from Indian antiquity and remained in use there for centuries. It was described in detail by 18th-century English doctors.[103] And in this way, the technique for reconstruction of a cut nose became known in Europe as the "Hindu method."

The Principles Qi, Yin *and* Yang, *and the Five Phases in China*

Even more than in Europe and India, harmony between microcosm and macrocosm was central to Chinese medicine. The number of surviving Chinese medical works is staggering: more than 10,000 texts from the 2nd century BCE, making it even more impossible to do justice to this wealth of literature than for other Chinese disciplines discussed in this book. The work that put Chinese medicine on the map is the *Huangdi neijing* (Inner canon of the Yellow Emperor).[104] It is generally assumed that this work dates back to the 2nd century BCE, even though it is only first mentioned in *The Book of Han*, completed in 111 CE by Ban Zhao (see above). The *Huangdi neijing* contains descriptions of the structure of the body, including the circulation of *qi*, the life force or vital energy, that is part of everything that exists. The *Huangdi neijing* also discusses the origin and course of diseases, and their treatment using needles. The body is in harmony with heaven and earth, and all body parts and organs align with the natural cycles of the seasons and stages of life, as the following example shows:

> The location of the spleen is central to the body, just as the earth forms the center. In the distribution of the four seasons, the time of the spleen consists of the last eighteen days of each season. So the spleen does not really belong to a particular season. Its function is to transform and transport the essence of food and fluids from the stomach. Just as the earth symbolizes the nurturing of all things in nature, so the spleen is responsible for nurturing each individual body part. That is why the spleen does not correspond to a particular season, but rather has control over each of the seasons.[105]

The quest for harmony between microcosm and macrocosm in Chinese medicine is even stronger than in Greece. The art of healing consists in the knowledge of how this harmony can be consolidated or restored, to which end the doctor needs to be both a philosopher and a craftsperson. To maintain good health, people must care for and cherish the *qi* life force. However, *qi* can also have a disruptive effect and cause illnesses. *Qi* is governed by the *yin/yang* dichotomy that we have already encountered in other disciplines in this chapter. Like all natural processes, a disease goes through an active *yang* phase and a passive *yin* phase. When a disease has reached a point of crisis, it changes to another phase. There are also the Five Phases (*wu xing*): wood, fire, earth, metal, and water, which correspond to the heart, liver, spleen, lungs, and kidneys. The body is a microcosm whose condition—sick or healthy—is determined by the universal principles of *qi*, of *yin* and *yang*, and of the *wu xing*.

Thus, the worldview according to which the macrocosm is reflected in the microcosm is found in all three regions. In later medical work, such as the *Shanghan lun* (On cold damage, 3rd century CE), for each condition is described in detail how the principles *qi*, *yin/yang*, and Five Phases determine the best method of treatment.[106] However, it is unclear whether these treatments actually had a curative effect. For this reason, it remains difficult to determine whether there was a convergence between theoretical principles and successful treatment patterns in Chinese medicine. But there was certainly a desire to bring the two together.

3.6 The Generation of Patterns by Experiment: Statics, Mechanics, Zoology, Botany, and Geography

In addition to the search for principles, the search for patterns continued in ancient times. Many of the patterns discovered were not passively observed but were actively generated. However, historical sources rarely state whether a pattern was determined with the help of an "experiment."[107] Reports of empirically established facts were more often questionable than reliable. A famous example is that a magnet would lose its magnetic power when rubbed with garlic. This pattern was presented as an empirical fact by Plutarch in the 2nd century CE, and it held this status until the 17th century.[108] But Plutarch obviously never carried out the experiment and relied instead on conventional wisdom.

Pythagoras and Empedocles

The oldest experiment is attributed to Pythagoras, who in the 6th century BCE used a single-string instrument (the monochord) to generate the patterns of consonant and dissonant intervals on the basis of which he established his famous principle of ratios between simple numbers (see above). However, it is by no means clear whether Pythagoras himself carried out these experiments. The patterns were probably already known before him. The story that he discovered them when walking past a blacksmith's shop where striking anvils produced harmonies that were at times consonant and at other times dissonant seems unlikely.[109]

But it was not only in music that patterns were created: it is told that in the 5th century BCE, Empedocles discerned that air, albeit invisible, was a physical substance and could block water. He demonstrated this with a clepsydra: a barrel with a hole in both the top and bottom. When he pushed the barrel into the water, he noticed that water flowed in through the bottom. But when he stopped up the hole at the top with his finger, the water stopped flowing in until he removed his finger. From this, Empedocles deduced that it was the air in the vessel that was stopping the water and that air was therefore not "nothing" but must be a material substance. However, it is again far from certain whether Empedocles himself carried out this experiment. But the result led him to include air—alongside water, fire, and earth—as one of the elements that made up the world.[110]

Aristotle as an Experimentalist: Zoology

We have encountered Aristotle on several occasions in this chapter, in the disciplines of natural philosophy, astronomy, musicology, mathematics, and logic. Aristotle was a theorist in all these fields. But in other areas, such as in zoology, Aristotle was also active as an experimenter. Although Aristotle professed reluctance to intervene in phenomena, he was certainly not a passive observer. In the *Historia animalium* (4th century BCE), he noted that the growth of an avian embryo is identical for all the birds but that there is a connection between the size of birds and the speed of their development. He discovered this pattern by opening fertilized eggs:

> The growth of the egg occurs identically in all birds, although the time to fullness varies, as we have said. In the case of the chicken, the first signs of the embryo can be observed after three days and nights; this takes longer for larger birds, a shorter time for smaller ones. During this time the yolk moves upwards . . . and the heart is no larger than a small blood stain in the albumen. This spot beats and moves as if it is alive; and from there, as it grows, two vein-like vessels with blood take a winding path to two adjacent membranes. . . . A bit later the body can be distinguished, initially very small and pale. The head is clear and the eyes very swollen; this takes a long time, and it is later that they contract and become smaller.[111]

In addition to this relationship between the size of the bird and the speed of development, Aristotle also provided a number of principles. According to him, there was a steering dynamic force of nature that he called *entelechy*, a kind of purposefulness. All nature was based on a plan that could be understood through the principles of necessity, possibility, and impossibility. In *De partibus animalium* (ca. 350 BCE), Aristotle explains the position of the mouth and the presence or absence of a neck. For example, according to him, a given animal had a neck only if it was needed for breathing. That was why animals without lungs, such as fish, have no neck. It may sound ad hoc to attribute the fact that fish are neckless because of their gills, but Aristotle's functional explanations proved extremely inspiring for later biologists.

Botany and Geography: From Theophrastus to Eratosthenes

Aristotle's successor Theophrastus (ca. 371–287 BCE) is considered the "father of botany." In his 10-part *Historia plantarum* (Inquiry into plants), Theophrastus ranks plants based on their size, reproduction, location, method of sowing, and practical applications as food, herbs, and juices. Much of the knowledge presented by Theophrastus is based on remote observation rather than active intervention. But the texts about sowing and soil preparation, including fertilization, can only have come about through empirical practices—although he may have simply reported an existing agricultural tradition.

In the Hellenistic period, empirical research was being conducted in many areas—from mechanics, the study of light, and geodesy to poetics and philology.[112] Without attempting to give a complete account, I will at least mention Eratosthenes (ca. 276–195 BCE), the librarian of Alexandria, who estimated the earth's circumference on the basis of shadow lengths in two identical wells, one in Aswan and one in Alexandria. In Aswan, the sun did not cast any shadow at its highest point on June 21, while the shadow in Alexandria measured more than 7 degrees, or 1/50 of a circle. Assuming that Aswan was exactly south of Alexandria and that the earth was perfectly spherical, its circumference had to be 50 times the distance between Aswan and Alexandria (which was approximately known), resulting in 250,000 stadia, corresponding to roughly 44,000 kilometers or about 27,000 miles. Today we know that the earth is oblate, that is, flattened at the poles, rather than a perfect sphere, so the perimeter is not the same everywhere. The average circumference of the earth is now held to be around 40,000 kilometers, so Eratosthenes was only about 10% off.

Eratosthenes was also a historian, philologist, astronomer, mathematician, and music theorist, but above all a geographer.[113] One of his greatest contributions to the history of systematic knowledge was the mapping of the then-known world. In Alexandria, Eratosthenes had access to travel books about many places and countries, and he set himself the goal of forging this information into a whole. Based on his knowledge of the shape and circumference of the earth, Eratosthenes in the *Geographika* divided the world into five climate zones (as Aristotle had previously suggested): two cold zones around the poles, two temperate zones, and one zone straddling the equator and the tropics. To show all information about the locations of countries, cities, seas, and lakes, he introduced a way of arranging the globe that is still in use today and that would be adopted by Hipparchus a generation later for the arrangement of the celestial globe (see above). Eratosthenes placed a grid of intersecting lines across the surface of the earth, which he called parallels and meridians. This allowed him to determine the distance between each pair of points. He also placed the locations on this grid for more than 400 cities. In this way Eratosthenes managed to unify the many geographical fragments using a simple underlying method: the grid or coordinate system. Unfortunately, the *Geographika* has not survived, although several fragments have been handed down through Pliny, Strabo, and above all through the geographic work of Claudius Ptolemy (see above), whose atlas with improved map projection remained the standard until the Renaissance.

Statics and the Relationship with Mathematics: Archimedes and the Aristotelians

Archimedes (ca. 287–212 BCE) is seen as the most important of all Hellenistic scientists. In addition to being a physicist, he was a mathematician, astronomer, engineer, and, above all, an inventor. Archimedes was living and active in Syracuse (Sicily), but his fame was so great that stories about his life and work appeared throughout the Greco-Roman world. According to tradition, he carried out his best-known experiment with the upward force of water. The story goes that Archimedes was prompted to do this by King Hiero II of Syracuse, who had commissioned a golden crown but suspected that the gold was mixed with silver.[114] The king asked Archimedes to determine whether the crown was made of pure gold without melting it down. While taking a bath, Archimedes realized that he himself became lighter in the water. Together with the fact that gold is on average heavier (has a higher density) than silver, Archimedes deduced that a pure golden crown should have a smaller volume than an equally heavy crown made with silver or a mixture of lighter metals. By submerging the crown in a bowl of water and measuring the displacement of the water—by measuring how much the water rose—he could calculate the volume of the crown, and by combining that information with the weight, he could determine the crown's purity. According to the anecdote, Archimedes then ran into the street shouting, "Eureka!" On the basis of this experiment, not only was he able to determine that the crown was an alloy, but he also came to his famous "law," which states that a body immersed in a liquid is subject to an upward force equal to the weight of the liquid displaced (*On Floating Bodies*, ca. 250 BCE). In our terminology this law is a pattern: it is a stable regularity, rather than an underlying principle from which this law can be explained or derived.

The famous law of the lever is also attributed to Archimedes. However, the lever had been in use for centuries before a "law" or rule for it was formulated. The simplest lever is a balance with arms of equal length where weights are placed at the same distance from the pivot point. This sort of equilibrium balance can already be found in Mesopotamia at the end of the 3rd millennium BCE (the Ur-III period). In the course of the 1st millennium BCE. we also come across balances with uneven arms. With this sort of unequal balance, the weight at the end of the longer arm needs to be smaller than that on the shorter arm to achieve equilibrium. The greater the distance from the pivot point, the smaller the weight needed to achieve the same effect. This pattern is described in Aristophanes's *Peace* in 421 BCE,[115] but we do not find a rule that describes the precise ratio between distance and weight. The first time we encounter such a rule, along with an underlying principle, is in the Aristotelian text *Mechanical Problems*. It is not known whether this text was written by Aristotle himself, but it is likely that large parts come from the later Peripatetic school (ca. 300 BCE).[116] The text asks how a small force can move a large weight with the help of a lever. The pattern is primarily discussed qualitatively, followed by a quantitative formulation:

> The ratio of the moved weight to the moving weight is inversely proportional to the distances to the pivot point.[117]

This formulation means that the weight multiplied by the distance to the pivot point is the same on both sides. That is, if the distance from a weight to the pivot on one side of the lever is made twice as long, two times less weight is needed to achieve the same effect. The age-old qualitative pattern has become quantitative![118] The Aristotelian text then also attempts to explain the pattern using a more general principle:

> If moved by the same force, the part of the radius of the circle farthest from the center moves faster than the smaller radius that is closer to the center.[119]

Can this principle make generalizations about the law of the lever and explain it? The statement merely indicates that a part of the radius farther from the center of a circle moves faster than a part closer to the center. It is not made clear how this depends on different weights. In other words, the quantitative pattern cannot be deduced from this principle, either formally or informally. Instead of the principle and the pattern coming together, it would seem a discrepancy emerges between the two. But barely a generation later, Archimedes gives a mathematical derivation in his text *On the Equilibrium of Planes.* He does something that has never been seen before: he presents a reduction of the law of the lever to Euclid's theory of proportion, and then to the Euclidean axioms (see above). With Archimedes, the underlying principles of the law of the lever have become the same as the principles of Euclidean geometry. In this way he succeeds in "proving" this law as if it were a mathematical theorem reduced to the axioms of geometry. Although contemporary physicists and historians of science are of the opinion that Archimedes's proof is not entirely conclusive, Archimedes's major innovation is that the theory of equilibrium, that is, statics, became part of mathematics.[120] We can therefore also perceive a process of convergence between principle and pattern in statics.

But to what extent did Archimedes actually experiment with the lever? His texts seem above all to bear witness to theoretical reflection on existing practical knowledge. But since Archimedes built a tremendous number of technological artifacts in addition to writing his theoretical work—from the slingshot, turntable, pulley, and wedge to the anchor winch—it is likely that he was also actively engaged in experimentation. In addition, Archimedes explained his method in a letter to Eratosthenes. This letter, discovered as a palimpsest in 1906,[121] gives a surprising picture of Archimedes's working method: with his principles he could not only prove the aforementioned technological applications but also implement them.

With his successful mathematical approach to mechanics, we would expect Archimedes to have gathered a large following, but that is not the case. During his lifetime he made his work known to Alexandrian scholars, but his research was not carried forth after his death, and many of his writings were not preserved. This is partly due to the complexity of his approach, which was understood by only a few, but it also has to do with Roman disinterest in Archimedes's theoretical work. Although his name and fame remained undiminished, it took more than 1,500 years for his work to be taken up again, when he was rediscovered in the early modern period by humanists such as Petrarch and Bracciolini (see chapter 5.1), who mainly saw him as an inventor.[122] Only with Galileo do we see a continuation of Archimedes's mathematical approach to mechanics (see chapter 5.3).

Generation of Patterns in China and India

In China, the oldest known text that mentions empirically generated patterns is the Mohist Canon (see also above), fragments of which have survived. There must have been many more texts in circulation, but a large part of them were lost in the great book burning of 213 BCE. The Mohist Canon contains works on politics, logic, mathematics, ethics, economics, and knowledge of nature. The texts also deal with experiments, especially in the field of statics, hydrodynamics, optics, thermodynamics, magnetism, and acoustics. In the *Mo jing* (which is part of the Mohist Canon) we read, "With regard to the unequal-arm balance, let a quantity of material and a weight be balanced, and let the distance from the pivot point to the point where the material is attached be shorter than the distance from the pivot point to the point where the weight is attached. . . . Now, if the same mass is added to both arms, the weight goes down."[123] This fragment describes the result of an experiment with an unequal-arm balance: if the masses on both arms are increased, the longer arm goes down. This is indeed a pattern, but it describes only balances with both sides weighted with the same masses. We are still quite far from the pattern that describes the inverse proportionality of weight and distance of a balance and lever.

The canon also provides explanatory principles, or something of the sort. The oldest principle seems to come from the *Huainanzi* that we encountered previously in musicology and astronomy. The balance or lever is explained as follows:

> Therefore, if one has the advantage of position, a very small weight can support something very large. That which is small but essential can dominate something that is wide and broad. So a beam only 10 *wei* long can support a house weighing 1000 *jun*; a hinge of only 5 inches in length can handle the opening and closing of a large gate. It does not matter whether the material is large or small. What matters is the exact position.[124]

It is primarily a part of the first sentence that indicates an underlying principle: "The advantage of position." We could summarize this by saying that the position of a weight is the essence of its force. Can we also use this position principle to derive the pattern given above? The text fragment says only that the position (of a weight) is important, without further specifying this position. So here we are dealing with an informal principle: it may be correct, but we cannot use it to derive a testable pattern.[125]

We find something similar in Chinese texts that refer to the upward-moving force of water, such as the problem of weighing an elephant, which no one could solve until, according to tradition, a six-year-old boy suggested placing the elephant on a boat and marking the water level. Afterward, a large number of heavy objects needed to be weighed and placed on the boat until it dropped to the same level.[126] However, this description is not so much about the upward force of water as it is about the insight that a given degree to which a boat sinks corresponds to an equal amount of weight.

These two examples resemble what we can call "qualitative physics": on the basis of established principles it was possible to reason what would happen without further quantification. It was known that the long arm of a balance goes down, but the question of the exact position of the weight to be suspended was left unanswered. And it was known that the loaded boat would displace the water, but the precise relationship of the weight to the amount of displaced water was not made explicit. Qualitative physics has a long history and is still of value for reasoning about physical systems without making calculations.[127]

So while the Hellenistic Greeks took a quantitative direction in statics, the Chinese opted for a more qualitative approach.[128] This qualitative approach did not apply to all Chinese disciplines: think of Liu Hong's astronomical calculations or the design of the 12-tone system in Chinese music theory discussed earlier in the chapter, both of which were highly quantitative. Moreover, we must realize that Mohism was only one of many Chinese schools, some of whose texts have not survived.

The unequal arm balance is described in India as well, such as in the Hindu text *Arthashastra* from the 3rd century BCE, but without specifying a clear pattern or principle. The study of the balance was in fact an almost global activity—from Asia and Africa to Europe and even pre-Columbian America.[129] Although the balance could be considered a (nearly) worldwide phenomenon, such was not the case for another invention from classical antiquity: the compass. We find descriptions of magnetic spoons that always point south in the Chinese writings of Wang Xu from the 4th century BCE. Subsequently, in the 1st century BCE, this extremely stable pattern was used in China to develop the oldest compass, while in Europe the first instance of a compass doesn't appear until the 13th century and elsewhere not until even later.

Mechanics: Aristotle versus the Mohists

This section concludes with the area of mechanics known as dynamics, not because it is less important but because it is unclear whether patterns were actively generated or were simply perceived passively. In either case, there is no doubt that reflection was going on in both Greece and China on perceived patterns of motion, such as free fall and projectiles. But the difference in interpretation between these two regions can hardly be greater. Furthermore, the term "motion" is far from unambiguous: Aristotle's theory of motion, as set out in his *Physica*,[130] focuses on the general notion of change; namely, *change of location*, and is therefore broader than what was later considered by Galileo and thereafter as motion (see chapter 5.3). It is to this meaning that we will limit ourselves here, for both Aristotle and the Mohists. As we saw above, according to Aristotle all natural motions in the sky are circular, while on earth all natural motions are oriented to the center of the earth (the "natural place"). When an object is kept moving by an external force, Aristotle speaks of unnatural or forced motion. Much of what Aristotle describes seems to correspond to direct perception: when one releases an object, it falls to the earth and comes to a standstill, and if one no longer pushes an object, it also comes to a standstill. So the absence of a force leads to rest. Aristotle further states that the speed at which an object moves due to a certain force is proportional to its weight. This also seems to correspond to direct observation, because if we look at falling objects, an iron bullet falls faster than a feather. Centuries later, Galileo will argue on the basis of experiments he conducted himself that the absence of a force, including frictional force, leads not so much to rest as to motion with a constant speed. That is, without an effective force, an object remains either in motion or at rest.

According to Aristotle, a constant force is needed to keep a body moving. And since in practice there is always friction, a force is always needed (in practice) to keep an object moving; otherwise, the object comes to a standstill and tends toward the center of the earth, and thus toward a state of rest. So, it seems obvious that a person at the time of Aristotle would have to interpret the motion patterns in the way he did. This was the prevailing view among many historians of science for a long time, until it became clear that in China the Mohist contemporaries of Aristotle had come to completely different conclusions when looking at patterns of motion. They also saw that an object ceases to move if no pushing or pulling force is exerted on it, but the principles they derived from this observation were different. In the *Mo jing*, chapter 64, on motion, we read:

> Motion is the result of a certain looseness [that is, of an absence of an opposing force].
> There is motion [if a force] can work on the edge.
> The cessation of motion is the result of [the opposite force of] a "supporting pillar."
> If there is no [opposing force] from a "supporting pillar," the motion will never stop. This is as true as the fact that an ox is not a horse.[131]

The above is almost the opposite of what Aristotle argued: it is not the presence of a force but the absence of one that keeps a moving object from coming to a standstill. How can two observers arrive at such different interpretations? That we don't know, but both interpretations are plausible. The Mohistic observation that if no opposing force is applied, a moving object will not stop may seem like an incontrovertible truth. The reason Aristotle did not ascertain this is because he was assuming a different worldview (see above), according to which all motion on earth tends to come to a standstill, while all motion in the sky tends toward pure circular movements that are maintained by an unmoved mover. According to Aristotle, this is the way the cosmos was constructed, from which it followed that there could be no earthly perpetual motion.

Did the Mohists engage in experimentation to come to their surprising insights? Unfortunately, we cannot learn anything about this from the fragmented texts. But we do know that the Mohists did not use their principles of motion to predict concrete phenomena. However, many Mohist texts were lost during the great book burning of 213 BCE; if any concerned other motions as well, from falling and spinning bodies to clashing bodies, they are unknown.

3.7 Principles for Legal Patterns: Jurisprudence

It is often claimed that the Romans played a relatively insignificant role in the history of science and the humanities. But in one discipline they were superior: jurisprudence. The great Roman jurists bore names like Gaius, Papinian, Ulpian, Paulus, and Modestinus. While nowadays they are known only to legal historians, these figures were some of the most influential scholars in the history of knowledge. With Roman legal scholars, the relation between the general and the exceptional arises again. They thought that many of the principles that applied to legal rules could also be modeled on principles from linguistics, considering that language, just like law, consists in both regularities and exceptions. The result was unparalleled: astronomers like Ptolemy drew inspiration from principles taken from legal scholarship. And many of the legal principles formulated by the Romans are still applied to this very day. Jurisprudence in Greece, India, and China also flourished, but what we do not see there is an endeavor to find a formal relation between legal rules and underlying principles.

Greece: Legal Practice versus Natural Law

It is striking how unrefined Greek jurisprudence is, compared to other Greek domains of knowledge. We already see this in Athens's oldest known legal system, issued by Draco in 621 BCE. These infamous Draconian laws were known for their severity but were largely based on the Babylonian law of retaliation: an eye for an eye and a tooth for a tooth (see chapter 2.4).[132] Other legal notions, such as administrative law, were largely absent. This last fact also applies to the laws in the Greek colonies of the Magna Graecia in the 7th and 6th centuries BCE. It is only with the emergence of the first Athenian democracy under Solon (6th century BCE) that we encounter a codification of the law with rules for family and inheritance law, criminal and procedural law, and economic and social law, but this codification did not lead to a specialized legal literature. This dead end has everything to do with the fact that the 500 men chosen to represent the public assembly and the people's court were bound by their oath to a literal interpretation of the written laws (the *nomoi*). This form of legal positivism led to a situation where many concrete cases fell outside the scope of the legal system, in which case a person's innocence or guilt was determined on the basis of a roll-call vote. For this reason, Attic orators were more interested in the art of persuasion than in analyzing the legal system. The democratic ideal was apparently something they felt so strongly about that the Greeks assumed that the principle of roll-call votes was always just, including in jurisprudence.

Although the Greeks had only a moderately developed legal system, they were all the more active in the philosophy of law. They were the inventors of *natural law*: the idea that certain laws of nature are given, making them universal, whereas other laws are time and location dependent.[133] Although we encounter notions of natural law in Plato, it is usually Aristotle who is viewed as the father of these ideas. According to him, on the one hand there are special laws that each people has created for itself, while on the other hand there are general laws that follow nature. Aristotle collected more than 100 constitutions of various (city-)states, but we do not find a developed system for jurisprudence with him either. Through the Stoics, the notion of natural law found its way to the Roman philosopher and statesman Cicero (106–43 BCE). According to him, natural law consists in two general principles: a prohibition on disturbing order in the community and an instruction to actively participate in the general good of the community.[134] According to Cicero, the purpose of law is to safeguard citizens' safety and happiness. But these principles of natural law do not yet lead to any specific legal rules, let alone the possibility of deriving a legal system from them. For this reason we cannot speak of an endeavor to bring about a convergence between principles and patterns, as we have seen in other disciplines, but such an attempt at convergence was made by the later Roman legal scholars.[135]

Rome: Principles for Legal Rules

Roman law includes a body of principles, rules, laws, and judgments that are unparalleled in antiquity. Legal scholars introduced general principles with the same ease as procedural principles, and both types applied to all legal rules. Many of these principles, which were drawn up more than 2,000 years ago, are still relevant today. Roman law was not a closed corpus of legal rules but remained in constant motion and continued to grow after the fall of the Western Roman Empire. What we now refer to as "Roman law" is actually the revision and codification of this law under the Byzantine emperor Justinian (482–565 CE), known as the *Corpus iuris civilis* (see chapter 4.6). To find the first legislation in Rome, we need to go back to the Twelve Tables (ca. 450 BCE). They were created after a long struggle between the lower-class plebeians and the upper-class patricians and stipulated that patricians could no longer change laws pertaining to plebeians at will.

Laws could be created in a variety of ways. Earlier I mentioned the process of development from customary law to written law, which I termed a *process from descriptive to prescriptive*. As early as the 4th century BCE, Roman law was formulated per magistrate, per region, city, and ethnicity. Furthermore, judicial authority was temporary: a magistrate held his position for a maximum of a single year, after which he was no longer eligible for election and could still be held to account. This control mechanism was lacking in Greek law. Tribunes, consuls, praetors, and—after the fall of the Republic—emperors could also issue laws and edicts. Starting in 400 CE, the right to issue laws was reserved for emperors.

Roman judges arrived at a judgment not by voting, as with the Greeks, but by weighing the pros and cons. To gain knowledge of the law, the judge sought the advice of the praetor, who in turn sought the advice of legal scholars. This practice resulted in a specialized legal literature that laid the basis for the Roman legal discipline. The collected recommendations (*responsae*) became just as important as the laws themselves. This caused the number of legal rules to increase so greatly over time that there was no longer any consistent system. There were exceptional cases, special cases, and contradictory rules that cut across the various legal domains. Many Roman legal scholars took up the task of providing a rationale for these rules and organizing them, searching for valid principles for all legal rules and procedures.[136]

What continues to impress us two millennia later is that the Roman legal principles look so different from their Babylonian and Greek counterparts. Perhaps the best-known principle originating from Roman law is, "A person shall be considered innocent until proven guilty." This is a loose translation of the original principle posed by the jurist Paulus Prudentissimus (ca. 200 CE), which states that the burden of the proof lies with the plaintiff rather than with the defendant: *ei incumbit probatio qui dicit, non qui negat*. Although we know hardly anything about the life of this Paulus, his influence was so great that he is listed in the Law of Citations of 426 CE as one of the five jurists whom every judge must consult before coming to a decision. Paulus's principle was included in the *Digest* of Emperor Justinian I, setting the norm for how a prosecuting authority was supposed to behave—it applied to all rules of criminal law. The principle was of tremendous import: it remained a leading principle in the Byzantine Empire, was adopted by Islamic civilization, and in the late 11th century it was rediscovered in western Europe in a manuscript by Irnerius of Bologna (whose School of Glossators is considered the beginning of the University of Bologna—see chapter 4.6). The principle of presumed innocence is just as valid today as it was more than 1,800 years ago. But in Paulus's time it was rather unique. In the Germanic legal system, for example, a totally different sort of principle prevailed. There it was defendants who had to prove their innocence after being accused. They could demonstrate that by having 12 men swear that they could not have committed the act of which they were accused. As far as is known, only Indian Hindu law appears to resemble the Roman principle (see below).

Another Roman legal principle that has survived the test of time is the principle of listening to both sides of the story, or in Latin, *audi et alteram partem* (listen to the other party as well). This principle enshrines the right of the accused to present their side of the story and counterevidence to refute the evidence used against them, regardless of how strong it is. This principle is important for ensuring not only that the process of establishing the truth is complete but also that the verdict is balanced. Although the principle is ascribed to the Romans, it may actually predate the codification of Roman law: it closely resembles certain proverbs in the Old Testament.[137] The principle was grossly violated by Cicero when he had the Catilinarian conspirators executed without a fair trial. Other principles that come from the time of the great legal scholars (ca. 200 CE) include *unus testis nullus testis* (one witness is no witness), *ne bis in idem* (not twice for the same thing), *nemo plus iuris ad alium transferre potest quam ipse haberet* (one cannot transfer more rights than one has), and *impossibilium nulla obligatio est* (there is no obligation to do the impossible).

While these principles have a primarily normative character, there are also procedural principles that regulate the legal rules themselves. Because over the course of the Roman Empire many legal rules were created—more than 3 million—it was inevitable that rules would contradict each other. Sometimes a particular case was covered by more than one rule. Which should be followed? And what was the relation between the exceptional rules and the general rules? It is not clear what legal scholar drew up the first procedural principles, but we encounter them in the *Codex Hermogenianus*, which was compiled by Hermogenes between 291–295 CE and issued by Emperor Diocletian. One of the most important procedural principles stipulates that a more specific law takes priority over more general laws: *lex specialis derogat legi generali*. For instance, when one legal rule requires that "mistreatment shall be punished" and another legal rule states that "a person is not punishable who commits an act dictated by the necessary defense of one's own person or that of another," the latter rule prevails if it is better tailored to the case at hand. So, the *lex specialis* is a procedural principle that makes these two contradictory laws consistent by stating that the more specific law (for the exceptional case) takes priority over the general law. We can view this principle as a sort of meta-rule that unites the general and the exceptional.

We already encountered the contrast between rule and exception in linguistics, especially in the discussion between two Greek schools of thought that were all the rage in the 1st century BCE: the analogistic (rule-seeking) school and the anomalistic (exception-seeking) school.[138] One of the Romans' challenges was finding a way to bring these two schools into harmony not only for language but also for law and politics. Cicero argued that language, law, and institutions define a people's identity, lending considerable legitimacy to the endeavor to harmonize rules and exceptional cases. As we saw in the section on philology above, Julius Caesar himself dedicated a book to the issue (*De analogia*). The *lex specialis* indeed seems to bring about a unification of the general and the special. Furthermore, this issue was not restricted to the Roman Empire; we also encountered the problem of exceptions versus rules in another part of the world; namely, in Panini's Indian linguistics in the 6th century BCE, where we saw at the beginning of this chapter that special cases (such as rules for the conjugation of irregular verbs) had priority over the more general case (such as rules for regular verbs), just as is the case with the *lex specialis*.

Legal inconsistencies were still not completely solved with this *lex specialis*. The centuries-long accumulation of laws had developed into a situation in which two legal rules could conflict with each other without one being more specific or general than the other. To handle these situations, a procedural principle was introduced that stated that the later law had priority over the earlier one: *lex posterior derogat legi priori*. The earlier law could still be applicable in certain well-suited cases, but in situations where they were in conflict, the more recent law prevailed. Despite its ad hoc character, the *lex posterior* principle is quite significant: it shows the nature of jurisprudence as a historical accumulation of rules that may be mutually inconsistent but that can be made into a consistent whole by means of a general principle.[139] And, once again, this principle is not unique to jurisprudence but can also be found in linguistics, this time rather literally: "If two rules contradict each other, the latter rule prevails" is how Panini puts it in his grammar (meta-rule 1.4.2 in the *Ashtadhyayi*, see the discussion at the beginning of the chapter). This would suggest a deep similarity between law and language that cuts across geographic boundaries, because as far as we know the Roman legal scholars had no knowledge of Indian linguistics.

Roman Jurisprudence as Compared with Other Disciplines: Astronomy

It is perhaps not surprising that in systems that develop over time, such as law and language, sooner or later inconsistent rules arise. But what is fascinating is that in different disciplines—and even in different regions—the same solutions have developed to tackle these inconsistencies. Identical procedural principles were created for both legal rules and rules of grammar to regulate and resolve contradictions.

Did a convergence between principles and rules also come about in jurisprudence, as in most other domains of knowledge in antiquity? This appears to indeed be the case. For example, the principles that had originally been formulated (such as by Paulus Prudentissimus) were simply the first step toward the creation of a consistent system for the numerous legal rules. However, these principles could not regulate cases in which legal rules lead to contradiction. It was only with the introduction of the later procedural principles, especially that of *lex specialis* and *lex posterior*, that the entire body of legal principles was gradually brought into harmony with existing legal rules, allowing the creation of a consistent whole.

So, in the area of law, the Romans had a boldness and originality comparable to that of the Greeks in astronomy. Just as in Greek astronomy (see above), the Romans continued to refine their legal principles until they could satisfactorily account for their patterns. And just as in astronomy, where extra notions were introduced for the exceptional planetary motions (from eccentric and epicycle to equant), principles were devised to "rescue" the exceptions. And in both astronomy and jurisprudence, principles proceeded from predetermined ideas about the world—whether they concerned perfect circular movements or the assumed unity of language, law, and institutions. But in the area of the formal relation between patterns and principles there was a big difference between these two disciplines. While astronomical patterns could be derived mathematically from the principles formulated by Ptolemy (and could thus be predicted), legal rules could not be derived from the legal principles. The rules needed to satisfy the principles but could not be predicted by them. For example, the amount of a fine for a violation could not be derived from the underlying legal principles, except that "there is no obligation to do the impossible." Legal principles were formulated in general terms, indicating the conditions that the rules were required to meet but nothing more. This is in accordance with what I termed a *declarative system of principles* in chapter 2.4. There is no logical inference between rules and principles, but the principles constitute the logical restrictions that the rules need to meet. Conversely, in astronomy, as well as in mechanics, mathematics, and (Pythagorean) musicology, there is a *procedural system of principles*: once you know the principles, the patterns can be derived from them by logical or mathematical inference.

Parallels between jurisprudence and astronomy were also drawn in antiquity, by the astronomer Ptolemy himself. In his text *On the Criterion*, numbering a mere 24 pages, Ptolemy makes a systematic comparison between investigating a legal case and investigating the natural world.[140] For example, he argues that the revelation of evidence in jurisprudence is analogous to the discovery of a phenomenon in nature based on sensory perception. He also argues that in both disciplines, the notion of judgment is a central issue: just as a judge passes judgment on a given matter, the researcher studying nature arrives at a judgment on a given phenomenon. In both cases, the legitimate goal is to determine the truth.

India: Dharma *and Refinement of the Burden of Proof*

In India, we also see a preoccupation with legal principles in Hindu law. The most important concept is that of *dharma*, which in the context of law is seen as the comprehensive duty to do what is right at every moment in one's life. This concept is described in *Dharmasastra*, which probably dates to around 600 BCE and which comprises both religious and legal prescriptions.[141] The legal parts of the *Dharmasastra* include the *Vyavahara*, which describes rules for legal proceedings, and the *Prayascitta*, which lays out the rules concerning punishments for violations of the *dharma* rules. Both parts are quite detailed. For example, the *Vyavahara* gives an overview of the entire litigation process, beginning with the role of the court, the king, and the judges. It then turns to the nature of the charge and the various answers that can be given by the defendant (from confession to denial). Unlike in Roman law, the burden of proof is on both the plaintiff and the defendant. When the defendant denies the charge, the burden of proof is on the plaintiff, but if the defendant appeals to an exception or an earlier judgment, then the burden of proof is on the defendant. There is no burden of proof if the defendant admits to the charge (although that can entail other problems of its own). The way the concept of the burden of proof is elaborated is an excellent specimen of legal discernment, especially for a text written in 600 BCE. In the parts concerning fines and penance for the various violations, the *Prayascitta*, the lines between religious and secular matters become blurred. Although mention is made of violations that fall outside the rules, no procedure is indicated for how to deal with them—all we find is a remark to the effect that Brahmans should impose the most appropriate fine or punishment possible.

Like all Indian scripture, the *Dharmasastra* is an impressive book, but it is not the sort of work in which we encounter a process of convergence between legal principles and legal rules. While there may have been such a process, we could only find it by reconstructing the history of the *Dharmasastra*. But the *Dharmasastra* looks like a finished work, an end point, giving rise to the question of how inconsistencies between rules of law were dealt with in practice. Sooner or later, cases arise that "fit" multiple legal rules, specific or general. How was this sort of case dealt with in Indian antiquity? Did jurisprudence have procedural principles like the Roman *lex specialis*? This is something we don't know.

China: Confucianism versus Legalism

The *Shujing*, or *Book of Documents*, contains the oldest surviving descriptions of law in China. The work is ascribed to Confucius (551–479 BCE) and concerns the relation between the ruler and his ministers and between the ruler and the populace. The work consists largely of reports from consultations, instructions, explanations, and orders. It does not contain any criminal laws. The Confucian worldview was based on a strict distinction between social classes, where punishments were relevant only for those engaging in activities outside the boundaries of civilized behavior. The six Confucian virtues were central to this worldview: benevolence, obedience, justice, decency, loyalty, and reciprocity. Those lacking these virtues were social outcasts. For this reason, criminal law was long seen as something primitive, relegated to barbarians.

Under the short-lived Qin dynasty (221–206 BCE), China was united for the first time, and a need arose for a new system of explicit (criminal) laws, known as Legalism. This stringent dynasty was firmly opposed to Confucianism. According to the legal scholar Han Fei (ca. 280–233 BCE), the imposition of severe corporal punishment with great ostentation was the most important instrument of government. Offenders were deemed incorrigible and needed to be eliminated. Everything was in the service of achieving a new social order in which differences between social classes were rigorously excluded.[142] Strict as they were, the Legalists recognized equality before the law, just as the Romans did, and in this they were opposed to Confucianism, which was based on class distinctions.

In the Han dynasty that followed (206 BCE–220 CE), the ruler's strict legal rules were also at the forefront, although Confucian values were restored and other legal domains, including administrative law, were valued. Nevertheless, until the Tang dynasty, Chinese legal scholarship remained highly underdetermined: when inconsistencies occurred, it was left to the emperor to judge.

3.8 Conclusion: Relations between Principles and Patterns in Classical Antiquity

While early antiquity is characterized by the search for patterns (see chapter 2), a shift occurs in classical antiquity toward the search for principles. Principles occupy a central position in practically all disciplines and regions, and the greatest challenge is getting to the root of the relation between principles and patterns.

Predictive Principles versus Restrictive Principles

If we try to get a picture of classical antiquity as a whole, we can make out a certain dichotomy. On the one hand, there are domains of knowledge where principles are laid out that can derive and predict patterns through logical steps. We see this in astronomy, mathematics, mechanics, (Indian) linguistics, portions of art theory, and musicology. On the other hand, there are other domains where the principles discovered generalize over the patterns but cannot derive or predict them. These principles are merely preconditions or restrictions that indicate what patterns are possible or how patterns are arranged. We see this in disciplines such as jurisprudence, poetics, zoology, medicine, and history. So, this dichotomy between predictive and restrictive principles does not fall along the line dividing modern natural sciences and humanities. Furthermore, the dichotomy is not absolute for practically any discipline. In Aristoxenian musicology, for example, patterns can be derived from principles, but not in Pythagorean musicology. And in Greek statics (especially Archimedes's study of balance), the patterns can be derived from principles, whereas in China that is much less true of the study of balance, and in India it is not at all the case. And in medicine, it is only the Chinese tradition that comes close to deriving medical patterns from principles.

The quest for the *nature* of the relation between principles and patterns is most explicit in Greece, where it consists principally in logical deduction. The most far-reaching type of deduction—in Aristotelian logic—is also the most trivial. In contrast, in China, algorithmic relations were introduced between principles and patterns, allowing patterns in mathematics and astronomy to be derived from principles using a step-by-step procedure. In Roman jurisprudence the principles are in accord with preconditions that are neither deductive nor prescriptive but merely declarative (see also chapter 2.4). So, the relation between principles and patterns can take forms as varied as the disciplines themselves.

The Search for Patterns Continues

The search for mere patterns continues, all the more so because patterns are in many cases the shortest way to useful knowledge. Ptolemy's *Handy Tables* are the prototypical example of patterns used to summarize a complex mathematical model. Making astronomical predictions based on Ptolemy's principles and concepts was extremely time consuming, but they could be carried out almost as accurately with tables and a few rules of thumb. The only limitation was that the rules of thumb did not give insight into the properties of the planetary system. In addition to pattern-based summaries of complex models, there were also cases where the search for principles simply failed—as in poetics—so that the center of gravity shifted back to patterns.

Descriptive versus Prescriptive

The process from descriptive to prescriptive, which I described in the previous chapter, is something that we also encounter in classical antiquity. This process is not limited to art theory and jurisprudence—where a prescriptive approach appears to be the norm—but can also be found in most other disciplines. Even for Greek astronomy the initially observed and thus descriptive circular motions of the planets were quickly taken to be a norm. Planets were shoehorned into (an at times complex combination of) circular motions. It was practically impossible for the Greeks to deviate from the idea of perfect circles. Once certain choices are made regarding principles, it can take centuries for them to be broken out of. In this regard, studying physical phenomena does not differ essentially from studying expressions of culture.

Unique versus General

Not every phenomenon constitutes a pattern. Unique occurrences arise everywhere. But it wasn't everywhere that exceptional phenomena were considered interesting. The Greeks showed little interest in deviant astronomical phenomena such as sun spots, whereas the Chinese noted and studied such peculiarities alongside astronomical patterns. We also find attention paid to the exceptional or unique in historiography, in Indian linguistics, and in Roman jurisprudence. In linguistics and jurisprudence, exceptions and regularities were even fused into a coherent whole using overarching principles such as the *lex specialis*. These are the oldest known cases where the regular and the exceptional are brought together.

Exchange of Ideas

The exchange of ideas between the various knowledge activities occurs so often that limiting oneself to either the history of natural science or that of the humanities distorts the picture of the history of knowledge as a whole. For example, there was an exchange of the Greek principle of numerical proportions between musicology, astronomy, and art theory. And we have also seen how the Chinese notion of *yin* and *yang* cut across the many diverse disciplines, from historiography and medicine to the knowledge of nature. The same applies to the notion of an arithmetical procedure or algorithm. There is an exchange of concepts, principles, and methods that does not seem to be concerned with specific disciplines. What works in one discipline can work just as well in another. However, it remains a mystery how some principles, especially the Chinese and Greek logical laws of noncontradiction and the excluded middle, were discovered in completely different regions that had no contact.

Many of the similarities we have found between disciplines were not noticed previously, possibly because the idea had not occurred to scholars to compare such disparate disciplines like philology and astronomy. This is in spite of the fact they become obvious if we look at sources from that period. Then we see that the historical actors themselves had no problem connecting astronomy with jurisprudence, or linguistics with mathematics. The entire constellation of disciplines seems to be much more tightly knit than was long thought possible, cutting across the modern notions of humanities and natural sciences.

Myths about the History of Science

Few are the myths in the history of science that escape an eventual debunking in our history of knowledge. One of these persistent myths is that the Romans played no significant role in the advancement of knowledge. As we have seen in this chapter, Roman legal scholars, philologists, and linguists successfully searched for both patterns and principles. However, there is a difference between Greek and Roman scholarship: Roman legal scholars and philologists were looking for principles *only* where the large number of patterns led to a contradiction and where principles were needed to eliminate this contradiction.

In some domains, such as Lucretius's atomism, we see a Roman quest for knowledge that is barely distinguishable from that of the Greeks. Lucretius, just like Pliny, is usually considered an exception. But they are only exceptions if we limit ourselves to the natural sciences. If we instead take all knowledge activities into consideration, the Romans turn out to be extraordinarily multifaceted, exceptionally active in jurisprudence, rhetoric, art theory, linguistics, medicine, and knowledge of nature.

The oft-heard supposition that there was no notion of mathematical proof outside of Greece also turns out to be based on prejudice or myth. A proof does not necessarily need to be deductive; it can alternatively consist of an algorithmic procedure that connects a pattern to a principle, as was the norm in Chinese mathematics and astronomy.

CHAPTER FOUR

The Reduction of Principles

Postclassical Period

500–1500: The Islamic World, China, India, Africa, Europe, America, Oceania

While the sciences and the humanities in classical antiquity differed greatly from those in early antiquity—as seen, for example, in the use of principles—such a distinction is harder to detect between classical antiquity and the period that followed. The search for principles, patterns, and inferences continued in almost all regions. All indications are that few new notions were introduced other than an endeavor to *reduce* the number of principles, as we shall see. Thus, we could just as well extend the "classical era" until new concepts start to be introduced. After all, it would actually be quite a coincidence if sociopolitical history were completely in sync with intellectual history. Yet more often than not this appears to be the case, as we already saw in the transition from the Neolithic to early antiquity and subsequently to classical antiquity. But above all, the political and social changes around 500 CE are so drastic that we cannot get around acknowledging a new era. This era is usually referred to as the Middle Ages, but I prefer the term "postclassical period," unless referring explicitly to medieval Europe.

The year 500 CE roughly corresponds to the fall of two major empires: the Western Roman Empire in 476 and the Gupta Empire in 550. After the fall of the Sassanid Empire in 651, the empires of the classical world are gone. This also applies to China, where the Han dynasty fell in 220, but Chinese science and scholarship flourished again, with the Tang dynasty (618–907) as its new high point. A new culture also emerged: Islamic civilization, where an impressive reconstruction of classical learning took place starting in the 7th century, followed by many new insights. So the year 500 is less significant as a turning point for China and the Near East than it is for Europe and India. This also applies to Mayan civilization, whose classical period dates from 250 to 900 CE and whose postclassical era runs from 900 to 1537; there is no "middle ages" there either.

4.1 History: Historical Source Criticism as the Basis for All Disciplines

All disciplines use sources, be they astronomical reports, medical diagnoses, philological texts, musical scores, lab journals, or historical testimonials. In addition to textual sources, there are also material, visual, and oral sources. So, determining the reliability of a source is important for all disciplines but is of supreme importance in history, where contradictory testimonies are the rule rather than the exception.

Already in the 5th century BCE, the Greek historian Herodotus (see chapter 3.3) was acutely aware of contradictory testimonies; in his *Histories*, he strove to get as close as possible to his sources about his main subject, the Greco-Persian Wars (490, 480–479 BCE). Herodotus held that the many conflicting witness reports had to be compared, after which one needed to select the *most probable source*. But Herodotus was silent as to how this choice should be made, and more often than not he put forth several contradictory reports without making any choice. The subjectivity of Herodotus's method led his contemporaries to accuse him of embellishing sources or even distorting them. That is why he has the dubious honor of being called both the "father of history" and the "father of lies."[1] A generation later, Thucydides (see chapter 3.3) believed that only *firsthand eyewitness reports* were reliable. Of course, Thucydides had a point, but unlike Herodotus he was writing a history of the Peloponnesian Wars (431–404 BCE), eyewitnesses to which were still living, including him. In the vast majority of cases, however, we do not have access to eyewitness testimony and are obliged to make do with transmitted reports.

Many patterns can be discovered in the degree of reliability of sources. One is that the more hands a source passes through, the more unreliable it tends to become. Another is that the more a given source is in agreement with other independent sources, the more reliable it appears to be. The question that soon arises is whether we can identify general principles to establish the reliability of a source. Above all, can the degree of reliability be quantified? It was at the time of Islamic civilization that these questions were asked for the first time, especially in the field of history. The solutions devised by Islamic historians have been used in almost all other disciplines, from linguistics to astronomy and from mathematics to poetics. It is thanks to the historical reconstruction of primarily Greek sources that the ancient disciplines were successfully revived and that Islamic science and humanities came to full maturity. While much has been written about the Islamic natural sciences, the Islamic humanities are often neglected in overview works.[2]

Islamic Transmission Theory and the Principle of the Isnad

One of the greatest challenges Islamic scholarship faced was to reconstruct the life of the prophet Muhammad (ca. 570–632). This task was not taken up until about a century after his death, when everyone who had known the prophet personally had already died. To reconstruct Muhammad's deeds and sayings as accurately as possible, all historical information had to be traced back to the prophet himself, and Islamic scholars believed this had to be done on the basis of precise chains of transmission. This transmission theory is known as the science of *Hadith* (report, account, narrative).

Since there were many stories about Muhammad in circulation, it was important to determine the relative reliability of the different sources. Various methods were devised, of which the historical method of the *isnad*, or "chain of transmission," became the most common. Each Hadith was accompanied by an isnad, such as "A heard it from B, who heard it from the mouth of C, who heard it from Muhammad's companion D," followed by a *matn* (the actual utterance). Isnads were carefully examined to determine whether the chain of transmission was actually possible, such as by making sure that all transmitters had actually existed and that they had lived in the same area at the time of transmission. In the first centuries following Muhammad's death in 632, Islamic theologians and legal scholars discussed the question of which of the traditions handed down were authentic and which had been fabricated at a later time. The biggest problem that arose was the multiple, inconsistent, and partially overlapping chains of transmission. Which chain is the most reliable? Since the most important elements of a transmission chain are the informants or transmitters, it is they who need to be evaluated, and Islamic scholars established the following criteria for this:

- Could the transmitters have possibly met, considering where they were in time and space?
- Is there any report of their meeting or of collaboration or common interests?
- Do the transmitters have sound morals, untinged by political motivation?
- Is the information that has been transmitted logically consistent? Is it coherent?
- Is the chain of transmission free of hidden defects?

Based on the answers to these questions, each source transmitted is assigned to one of the following four categories:

- *Sahih*, "authentic"
- *Hasan*, "good"
- *Da'if*, "weak"
- *Mawdu'*, "fabricated"

These categories, treated systematically by the Hadith scholar Ibn al-Madini (778–849), assign a degree of likelihood to each source.[3] They can therefore be seen as a formalization of source criticism based on fundamental criteria. In our terminology, an isnad traces statements to the original source on the basis of general criteria (principles). It didn't take long until no text or source was considered credible unless it was accompanied by an isnad. The many inconsistent transmission chains were evaluated, triaged, and forged into a coherent whole, leading to the first complete biography (*sira*) of the prophet by Ibn Ishaq (704–767).[4]

A few generations later, the isnad method fell prey to controversy. The reconstruction of Muhammad's life on the basis of the isnad took place in a highly politicized context at the time of the Abbasid caliphate, just after the overthrow of the Umayyad dynasty. The groups from which the Sunnis and Shiites would eventually emerge came out with rival versions of the history of Islam. The isnad method, however precisely and impartially defined, was seen as a politically motivated instrument by both groups.

Spreading the Isnad

And then something extraordinary happened. Instead of being set aside, the isnad came to be used outside of the Hadith, first by historians and subsequently by jurists and other scholars (see below). They recognized that the method was quite useful for determining the reliability of a source in general. Historians stripped the method of its religious context and applied it in a new, nonreligious context. This led to an unprecedented degree of historical precision, for which Arabic history is renowned.[5] The great isnad historians include al-Dinawari (815–896), who was also active as an astronomer and botanist; al-Baladhuri (died 892); al-Tabari (838–923); and al-Masudi (896–956). For example, in his *History of the Prophets and Kings* (*Tarikh ar-rusul wa l-muluk*), al-Tabari wrote an overview of history from Creation to 915.[6] In this 40-volume magnum opus, the first 4 volumes roughly follow the Hebrew Bible. For these volumes, an isnad is unnecessary and even impossible. After a volume on the pre-Islamic empires, the remaining 36 cover more than two centuries of Islamic history in great detail, constructing chains of informants in the best isnad tradition. Al-Tabari describes the many caliphates, conquests, crises, collapses, revivals, uprisings, and the establishment of the Abbasids in Baghdad. His work is often seen as one of the most accurate histories of early Islam, thanks to his use of the isnad method.

The isnad method was the first empirical theory in Islamic civilization and thus became the model for many other scientific and scholarly activities, the first of which were the *adab* disciplines that had begun to constitute a flourishing curriculum over the course of the 9th century. *Adab*, today also known as *studia adabiya*,[7] consisted of the study of history, grammar, poetics, rhetoric, and moral philosophy. These were the same disciplines that later in Europe, around 1400, would form the *studia humanitatis* (see chapter 5.1). The historically documented isnad method soon became the foundation of all source-based disciplines, from medicine, law, astronomy, music theory, and mechanics to mathematics. These disciplines were largely reconstructed from Greek sources, but also from Persian and Sanskrit sources. The historian al-Masudi (896–956), known as the "Herodotus of the Arabs" (see below) expressed the importance of historical reconstruction of disciplines: "If scholars had not recorded their thoughts over the centuries, the foundations of knowledge would have collapsed and their conclusions would have been lost. For any branch of knowledge to exist, it must be derived from history."[8] The revival or even rebirth of the classical disciplines using historical means has been one of the Islamic scholars' greatest achievements.

These scholars also made new discoveries, but they succeeded in doing so only through the massive efforts made to translate inherited works into Arabic in places such as the House of Wisdom (Bayt al-Hikma) in Baghdad. This desire for historical reconstruction may explain why Islamic scholars were more interested in maintaining continuity with ancient insights than breaking with them. Arab astronomers, for example, took Ptolemy as a starting point and improved upon his work. The Islamic works in the field of poetics were a continuation of Aristotle (see below). Even al-Khwarizmi, who developed a new form of mathematics—algebra—relied on the Greeks for his method of constructing a proof, particularly on the geometric method of Euclid. The importance attached to building upon the ancient disciplines may partially account for the fact that a revolution in the natural sciences was unthinkable in the Islamic world.[9] But in the field of history, Islamic scholars were on their own, and they developed a theory of transmission that was nowhere to be found in classical antiquity and that was unique in postclassical times.

Reducing Principles with the Isnad Method

It is instructive to compare the isnad method with Herodotus's principle of the most probable source and Thucydides's principle of eyewitness accounts. The four categories of the isnad assign a degree of likelihood to each source and can therefore be seen as a formalization of Herodotus's approach. At the same time, the isnad methodology offers a solution to Thucydides's problem of second- or thirdhand eyewitness reports, at least if the transmission can be demonstrated to be possible, consistent, and without defects. In this way, isnad methodology formalized and unified Herodotus's and Thucydides's principles.[10] While Herodotus seemed to apply his principle randomly without any method, isnad-based historiography exhibits a well-defined set of criteria. And while Thucydides foresaw problems with second- or thirdhand eyewitness testimony, isnad-based historiography provides a method for incorporating eyewitness accounts of this kind. In short, the isnad not only provides a better method for source criticism but also reduces the number of principles for this from two to one: the chain of transmission. In this way it aspires to rationally reconstruct the past.

The isnad method certainly had its limitations. To start off with, the method provided only a way to find the most reliable source transmitted, not a way to reconstruct an original text. The latter was a different exercise previously undertaken by the Alexandrian philologists using the principle of analogy and the corresponding six criteria (see chapter 3.3). It is not until the early modern period that we find a method that unites historical source criticism and philological reconstruction (see chapter 5.1). Another, perhaps even greater, shortcoming of the isnad method is that an authentic transmission is not necessarily a true source, as the historian, astronomer, and mathematician al-Biruni (973–1048) pointed out in his criticism of the isnad.[11] After all, a lie that is transmitted flawlessly and with the greatest accuracy is still a lie.

These shortcomings led al-Masudi and al-Biruni to abandon the isnad method, although not completely. For example, in his historiography, al-Masudi made extensive use of personal experiences that he, like Herodotus before him, had had in the Persian provinces Armenia, India, East Africa, Sri Lanka, China, and probably Russia.[12] In addition to his unique reports on the Tang dynasty, the Khazars, and the Russians, his approach was strikingly hybrid. On the one hand, he made use of historically documented facts derived by the isnad method using a chain of verifiable sources and stories. On the other hand, he alternated these facts with less reliable anecdotes, poems, and even jokes, without any investigation into the chain of transmission.

With his description of India, the *Kitab al-Hind*, al-Biruni went on an anthropological journey. Instead of relying on a principle based on transmitted reports, he let the Indians speak for themselves. Al-Biruni was the author of another historical work from the year 1000, with the beautiful title *The Remaining Signs of Past Centuries* (*Kitab al-athar al-baqiya ʿan al-qurun al-khaliya*).[13] This book contains a comparative study of calendar systems from different civilizations interspersed with historical and astronomical information, discussing the customs of the various peoples and their religions (ranging from Manichaeism and Buddhism to Christianity). All traces of the isnad method have disappeared in this work. Indeed, it was not even applicable to *The Remaining Signs*, since al-Biruni used existing historical works.

And with the greatest Islamic historian, Ibn Khaldun (1332–1406), every "vain superstition" and "uncritical acceptance" of historical sources is criticized.[14] Ibn Khaldun was born in Tunis but was active for a long time at the University of Fez, founded by the prosperous merchant's daughter Fatima al-Fihri in 859.[15] Ibn Khaldun provides a sociological elaboration of the famous cyclical pattern of the rise, peak, and decline of civilizations (see chapters 2.7 and 3.3). When a given civilization becomes a dominant culture in a region, a period of decline usually follows its height. This means that the next coherent group conquering this civilization will be, by comparison, barbarians. When these barbarians have established their dominion over the conquered civilization, they are attracted by its more refined aspects such as its arts, letters, and sciences, which they then appropriate. A subsequent group of barbarians repeats this process, so that the pattern of peak and decline leads to an accumulation of knowledge and culture rather than to an absolute decline. This pattern can indeed be found with the Romans, who conquered Greece, or the Arabs, who conquered the Persian Empire.

With Ibn Khaldun we are actually dealing with a "modern" historian who could just as appropriately be discussed in the next chapter, which covers the (early) modern period. Nevertheless, like the other Islamic historians, al-Biruni and Ibn Khaldun would not have amounted to anything without the isnad method. This method brought three things: (1) a critical view of sources and their transmission, which was important for all disciplines; (2) an investigation into the origin of a source; and (3) the most accurate citation possible of this origin. Whereas with scholars elsewhere in the postclassical world we can only guess at the origin of their sources, with Islamic scholars this is almost always clear, thanks to the isnad tradition.

Chinese History: Continuity with Antiquity

Whereas Islamic civilization developed a new approach to history, in China there was a high degree of continuity with antiquity. In most of the postclassical period, Sima Qian represents the most important historical tradition with his *Shiji* (see chapter 3.3). His historiographical division into annals, tables, treatises, genealogies, and exemplary traditions served as the model for the dynastic chronicles drawn up since the Tang period (618–907). The Tang dynasty is considered one of the peaks of Chinese civilization, with printing as its most important invention. The first printed book in China, the *Diamond Sutra*, dates from 868 (587 years before the Gutenberg Bible), and the Confucian classics have been in print since 932. Production of official court writings was tasked to officials who used documents from the imperial archive. The main goal was often to legitimize the takeover of power. These court chronicles, collectively also referred to as the *Twenty-Four Histories*, followed both the form and content of *Shiji* (which is also part of it).[16] The *Twenty-Four Histories* cover a period of 1,832 years, consisting of 3,212 volumes and some 40 million words. With court historiography, Chinese historical production became a state affair and got its own Historiographic Office (Shi Guan).[17] The dynastic histories eventually come to resemble a sort of source publication that is very thorough but as dry as dust.

Liu Zhiji's Historical Criticism

Despite China's heavy historiographic bureaucracy (or perhaps thanks to it), there is also historical criticism, in particular by Liu Zhiji (661–721), who objected to the mechanical nature of the dynastic histories. In 710 he wrote his brilliant *Generality of Historiography* (*Shitong*), the first Chinese work devoted entirely to the subject.[18] After a critical analysis of the historical works prior to the Tang dynasty, Liu discusses the various historiographical methods, such as appropriate styles, documentation issues, and how criticism should be applied in research.

According to Liu, historians should first and foremost be as objective as possible. They should not base their assessments on moral points of view or other value judgments. Moreover, they should approach any given theory with skepticism. All that matters is evidence, and when describing an event historians need to give an overall picture obtained from *all possible sources*. In Liu's view, all factors—cultural, social, economic, and geographic—must be taken into account, and their presentation should be detached and unbiased.

Liu's criticism is to some extent similar to that of Ibn Khaldun, but the way he goes about his work is completely different: not based on the sociological analysis promoted by Ibn Khaldun, it is based instead on as many sources and factors as possible. Although we find Liu's emphasis on objectivity and socioeconomic factors almost seven centuries later with Ibn Khaldun, it would appear that the latter could not possibly have been familiar with the work of the former.

At the time of the later Song dynasty (960–1279), a total of approximately 1,300 historical works were compiled, a number surpassed only during the later Qing dynasty with its dazzling 5,478 works of history. It is beyond the scope of this book to examine all these works for patterns and principles, but the following picture can be drawn from the works of some important historians of that time, such as Xue Juzheng (912–981), Ouyang Xiu (1007–1072), and Sima Guang (1019–1086). First and foremost, all these historians emphasized the practical utility of history. Historiography was expected to show facets of human behavior in conjunction with the natural environment and social changes. This allows history to be used as a guide for citizens. Historiography is a basic virtue for every intellectual, whose main purpose is to serve the state.

Christian Historiography in North Africa

The first Christian historiography originated in Africa and built upon older Roman annalistic methodology, which focused on a linear history starting from the foundation of the city (*Ab urbe condita*; see chapter 3.3). The Christian chroniclers went a step further, however, constructing a history that spanned from Creation to their own time or even to the "end of time." Sextus Julius Africanus and Eusebius gave the impetus to such chronicles, which are the first works of universal history (also called *salvation history*). Central to these histories were the person of Jesus and a preoccupation with periodization. For example, in his *De civitate Dei*, the Berber church father Augustine (354–430), working in Hippo Regius (present-day Algeria), proposed a periodization of six eras.[19] This division, which had already been made by Sextus Julius Africanus in the 3rd century, is analogous to the six days of Creation: (1) from Adam to the flood, (2) from the flood to Abraham, (3) from Abraham to David, (4) from David to the Babylonian exile, (5) from the Babylonian exile to the birth of Christ, and (6) from the birth of Christ to the end of the world. Augustine compared these six periods to stages in human life: childhood, puberty, adolescence, adulthood, middle age, and old age.

Augustine's periodization essentially overlaid biblical history onto world history, leaving only a very modest place for contemporary history. It was thought that the end of time would take place during the Roman Empire. Although officially, Christian historiography continued to be based on the classical method of written sources, eyewitness reports, or personal experience, the sources were either hardly tested or not at all for reliability or accuracy, let alone for possible chains of transmission. What counted was the authority of the source and especially the extent to which the source was in accordance with biblical history. Among the more fanatical Christians, there was even a conviction that all other classical sources and texts should be rejected. It was to Augustine's great merit that he convinced these radical Christians that pagan insights should not be distrusted but appropriated: "All good and true Christians should understand that truth, wherever they may find it, belongs to their Lord."[20] Thanks to Augustine's authority, studying the classics was legitimized time and time again, and we can rightly speak of a postclassical era in Christian Europe as well.

World history was reinterpreted by Christian historians on the basis of what I will call the *principle of biblical coherence*: According to Justin Martyr, Moses had a decisive influence on Homer; following Melito of Sardis, the Roman Empire was created to facilitate the spread of Christianity; and according to Clement of Alexandria, Plato and Aristotle were not completely wrong but were instead not fully informed.[21] A problem for all Christian historians was the interpretation of the final era: from the birth of Christ to the end of the world. While there seems to be a clear pattern in history before Christ (at least in Augustine's periodization), there seems to be little divine structure in the post-Christian period until the end of time. Miracles and prophecies were considered of great importance because they bore witness to the omnipresence of God.

The pattern of cyclicity was abandoned. Classical Roman historians had already replaced this pattern with a linear historiography, but for Christian historians acyclicity was given a theological foundation: universal world history follows a *linear pattern* from a *unique* beginning (Creation) to an *ultimate* goal (the final Day of Judgment). With a bit of effort, one can also detect this linear pattern in the history of Rome itself: Rome continued to exist despite the raids and looting of the city. For Augustine, the notion that the Roman Empire would not be around for the end of time was unthinkable. Instead, he spoke of the Old Rome and the New Rome, with the latter reaching perfection that the former had not achieved.

From North Africa to Europe

While up to the 5th century, North Africa was the intellectual center of early Christian science and scholarship—Sextus Julius Africanus, Tertullian, Origen, Augustine, and Orosius all worked in Africa—this center quickly moved to Europe. One reason for this was undoubtedly the invasion of North Africa by the Vandals in the 5th century, during which Augustine died. The conquest of North Africa by the Arabs at the end of the 7th century was the final blow for Christian science and scholarship in this region. The principle of biblical coherence, required to explain all patterns, remained the most important generalization. For example, Gregory of Tours (ca. 539–594) penned a historical overview from Creation to the death of his predecessor, Martin of Tours, after which he concentrated on the history of Gaul.[22] Gregory chronicled the contemporary history of the Franks firsthand. But when he explains historical events, he does so on the basis of prophecies and allegorical interpretations of the Bible. For example, he tells of a Frankish princess's marauding entourage and then argues that this event is prophesied in the Bible, in Joel 1:4. If we would like to summarize this type of explanation in the form of a principle, then it is at worst a principle of allegory and at best the above-mentioned principle of biblical coherence: in practice, any rumor, miracle, omen, or prediction can be considered acceptable as long as it does not contradict ecclesiastical doctrine.

The work of the Venerable Bede (ca. 673–735), who was active in the Anglo-Saxon kingdom of Northumbria, started the spread of the influential anno Domini (AD) dating convention. Apart from this innovation, we do not find any new principles in Bede's historiography. This, of course, does not detract from the incalculable value of his history of the Anglo-Saxon people from the time of Caesar up to the year 731 (*Historia ecclesiastica gentis Anglorum*). Like Gregory, Bede mainly describes the wonders of his time on the basis of hearsay.[23] Although his reports may seem unlikely to us, historians like Bede saw miracles as pointing to a divine plan and as such constituting an essential component of universal history. Two handbooks of great importance for historiography were Bede's *Liber de temporibus* and *De temporum ratione*. These works provided a foundation for the discipline of timekeeping, or chronology, and they also gave a decisive boost to the anno Domini dating system. This system was not Bede's own invention but that of the Scythian monk Dionysius Exiguus (ca. 470–544),[24] but it was Bede who first dated historical events using anno Domini in his *Historia*. Bede was also the first to use the Latin equivalent of "before Christ," as well as the custom among historians not to use zero to indicate a year (the number zero was unknown in Christian Europe).

Whereas Bede wrote about the Anglo-Saxons, in the 7th century Paul the Deacon wrote his *History of the Lombards*. In fact, each region was graced with its own historian: in the 9th century, the Celtic monk Nennius wrote his *History of the Britons*, which also contained the entire Arthurian epic. Widukind wrote his *Saxon History* in the 10th century; in his *Chronicle of the Slavs*, Helmold focused on the 12th century; and chronicles of Latvia, Estonia, Bohemia, Poland, Denmark, West Francia, and Normandy were written in a similar fashion. These chronicles have little or nothing to do with universal history and have everything to do with legitimizing new states or dynasties through a mostly imposed collective genealogy.

Was historiography the foundation of the other disciplines in Europe as well? This is not as clear as in the Islamic world. In Christian Europe, sources were not subjected to the rigorous investigation of reliability that they were in the Islamic world. The decisive factors were authority and biblical coherence. But starting in the 12th century, many Greek and Arabic writings were translated into Latin during what is known as the *12th-century Renaissance*. On the borders of Islamic and Christian civilization, such as in Toledo, hundreds of translations were produced in unified cooperation among Jewish, Christian, and Islamic scholars, the greatest wave of translation that Europe would ever know. For example, the Italian monk-translator Gerard of Cremona spent 40 years in Spain, where he translated 87 works from Arabic, which themselves had been translated from Greek.[25] In Christian Europe, too, the continuation (or revival) of the once-thriving ancient disciplines could take place only through historical reconstruction of scholarly texts, which had to be reread and understood. But it would not be until the 15th century that a real source criticism was developed in the Latin world (see chapter 5.1).

India

It was in the middle of the 12th century that India's first historical work (to our knowledge) was written: *Rajatarangini*, or *River of Kings*, by Kalhana. This tract gives an overview of the history of the kings of Kashmir from the beginning of time. Until the 12th century there was no work in India that would fall within the notion of history of the type we find in abundance in China, Africa, the Arab world, and Europe, and it appears that the *River of Kings* is the only history written in Sanskrit.[26] The list of kings in Kalhana's work dates back to 1900 BCE, and although some of these kings can be identified on the basis of inscriptions in archaeological finds, Kalhana's seemingly precise chronology has been called into serious question. Because of the virtually complete absence of previous sources, the *River of Kings* is seen primarily as a narrative that is accurate with respect to the way contemporaries understood the knowledge of their past.

Ethiopia

According to tradition, the first Ethiopian rulers were descendants of the biblical king Solomon, who had an affair with Makeda, the legendary Queen of Sheba from Yemen. From this relationship a son was born, Menelik, who became ruler of the Kingdom of Aksum. Ethiopia was Christianized in the 4th century CE by the Greek-Syrian monk Frumentius, and biographies of saints in Ethiopic (Ge'ez) started appearing in the 5th century. Yet it is not until the 14th century that the first surviving Ethiopian chronicle appears on the scene: the *Kebra nagast*, or *Glory of the Kings*, written in the golden age of Ethiopian literature. This anonymous work combines historiography with allegory and symbolism. The *Glory of the Kings* begins with Adam and Eve and extends to about the 4th century CE with the Christianization of the Ethiopians, who were converted from the pagan worship of the sun, moon, and stars to the Christian worship of the "Lord God of Israel." A glorious narrative describes the arrival of the Ark of the Covenant in Ethiopia, the abdication of Makeda in favor of Menelik, and the Christianization of the Kingdom of Aksum.[27] Although no clear principle can be distilled from the text, it at least seems to adhere to the principle of biblical coherence: all events are interpreted in biblical terms and brought into agreement with Holy Scripture in the most coherent way possible. The similarity between the structure of the *Kebra nagast* (from Adam to the present) and that of the Christian universal histories in North Africa and Europe is striking. The *Kebra nagast* may have been created under the influence of the Greek-Syrian monks who Christianized Ethiopia.

4.2 Astronomy: Seeking Models with Fewer Principles

The Reduction of Principles by al-Tusi and Ibn al-Shatir

Just like Islamic historiography, Islamic astronomy practically had to be built from the ground up. While there was a pre-Islamic astronomy, it was elementary and untheoretical. The Islamic expansion during the first century after the death of Muhammad in 632 brought Muslims into contact with astronomical theories from Greece, India, Persia, and elsewhere. The superior accuracy of the Ptolemaic system was soon recognized, followed by a revival of the Hellenistic tradition. The importance of centuries-long work by Islamic scholars in the field of astronomy can hardly be overestimated: the classical texts were translated, explored, and annotated and in this way were brought back to life. The historian al-Masudi was therefore not far off when he stated that in Islamic civilization "every field of study needs to be derived from history" (see above). In addition, astronomy had an almost self-evident legitimacy in Islam: astronomers were able to determine the time and direction of prayer and the beginning of Ramadan, the month of fasting, with greater accuracy than ever before. At the time of the Abbasid caliphate, astronomy at the House of Wisdom achieved an unparalleled level of prestige.

The first important astronomical work from Islamic civilization is by al-Khwarizmi (ca. 780–850), who was also the founder of algebra. In his *Zij al-Sindhind* (*Astronomical tables of Sindh and Hind*) from 830, he introduced Ptolemaic astronomy on the basis of tables and calculation rules for the motions of the sun, the moon, and the planets.[28] It was the first in a long tradition of Arab *zijs* that build partly on Ptolemy's *Handy Tables* (see chapter 3.2) and partly on Indian models but do not analyze them critically. Yet al-Khwarizmi's work was a turning point: until then, Muslim astronomers had only translated and studied the writings of others, but al-Khwarizmi managed to expand the Ptolemaic model with new calculation methods. The astronomer al-Farghani in 850 proposed further enhancements with improved values for the slope of the ecliptic (the apparent orbit of the sun on the firmament) and for the circumference of the earth.

The heyday of Islamic astronomy begins in the 11th century; this is when the Ptolemaic system was not only closely studied but also criticized and extended. Although we cannot speak of a revolution—Islamic astronomers operated within the geocentric worldview—attempts were made to reduce Ptolemy's system to a smaller number of principles. The equant in particular was a thorn in the side of Islamic astronomers. They wanted a model that relied solely on circles and epicycles without additional concepts such as the equant. This is similar to the aspiration of the earlier classical Indian astronomers we encountered in chapter 3.2, although it is not entirely clear whether they were consciously trying to eliminate the equant. We do know this for Ibn al-Haytham (965–1040), also known in the West as Alhazen, who criticizes several parts of the Ptolemaic model in his *Al-shukuk 'ala Batlamyus* (*Doubts on Ptolemy*).[29] He argues that certain mathematical concepts, the equant in particular, do not correspond to the requirement of uniform circular motions proposed by Ptolemy himself. Ibn al-Haytham also states that real movements should not be explained by imaginary mathematical points, lines, and circles. Many astronomers took up Ibn al-Haytham's challenge, and the equant problem became the greatest undertaking in astronomy, first in the Islamic world and then, after the major wave of translations in the 12th century, in Europe as well, until the time of Copernicus and Kepler.

An elegant solution was suggested by the Persian al-Tusi (1201–1274) in his *Tadhkira* (Memorandum).[30] He developed a geometric technique that bears his name—the Tusi couple—which he used to generate linear motion by "adding" two circular motions together (see figure 9). The black dot on the diameter of the large circle makes a linear motion when the smaller circle, which is exactly half its size, rotates within the larger.

Al-Tusi showed that placing his couples on the site of the Ptolemaic epicycles rendered the equant superfluous. This applied to all planets except Mercury. About a century later, the astronomer Ibn al-Shatir (1304–1375) would extend

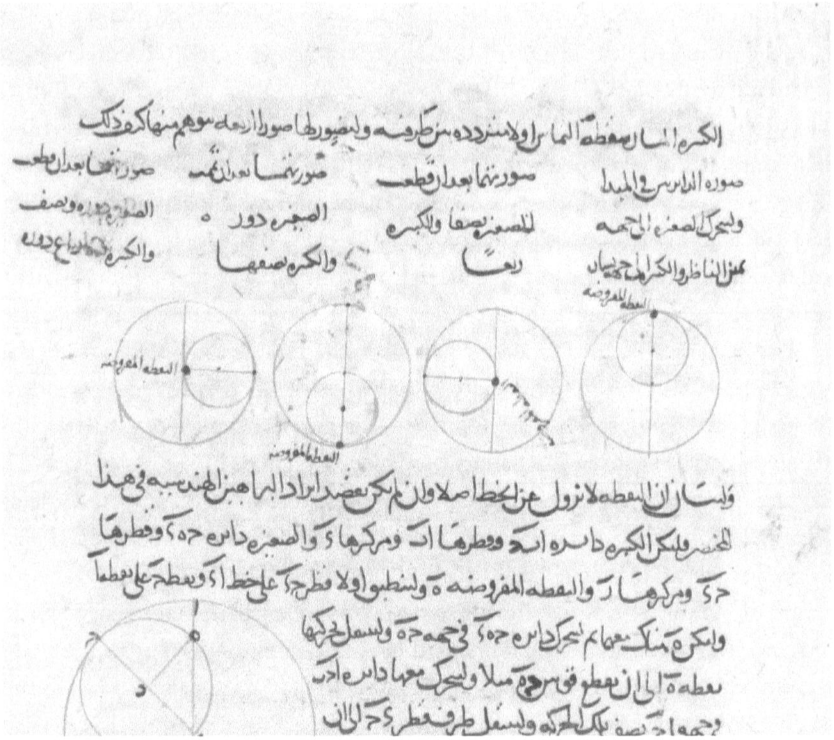

Figure 9. The Tusi couple. Notice how the black dot moves in a linear fashion on the diameter as the smaller circle rotates within the larger circle. Vat. Arabic ms 319; https://commons.wikimedia.org/wiki/File:Tusi_couple.jpg.

the Tusi couple to also account for the movement of Mercury, as well as the other planets. In this work, with the brilliant title *The Final Quest concerning the Rectification of Principles* (*Kitab nihayat as-sul fi tashih al-usul*),[31] al-Shatir introduces a triple epicycle based on the Tusi couple. This system of couples renders not only the equant superfluous but the eccentric circular orbit, as well. Now all planetary motions could be predicted with a simpler geometric model that was based not on three principles (eccentric, epicycle, and equant) but on a mere two: the epicycle and the couple. Moreover, with these two principles, al-Shatir's system achieved the same accuracy as Ptolemy's: the celestial patterns were deduced from the principles in an entirely mathematical way. This was an unparalleled result, and Islamic astronomers finally surpassed Greco-Roman astronomy. A few centuries later the Tusi couple seems to make a comeback with Copernicus, this time in a heliocentric model (see chapter 5.2).

Although almost all Muslim astronomers were convinced that the geocentric model was correct, there were also some unorthodox astronomers, such as al-Biruni. In addition to being a historian (see above), al-Biruni was also a mathematician, anthropologist, and astronomer. In his book on India (the *Kitab ta'rikh al-Hind*), he speculates widely about the possible movement of the earth and concludes that it can be neither proven nor refuted.[32] He also criticizes Aristotle's steadfast assumption of the immutability of the spheres. In al-Biruni's greatest surviving work on astronomy, the *Mas'ud Canon*, he places the earth at the center of the universe but states that the astronomical facts can be explained just as well by assuming that the earth revolves around the sun.

Europe: An Intermediate Model and the Influence of Islamic Civilization

In the Latin West there is hardly any astronomical research to speak of until the 12th century, but we do encounter an interest in conceptual models, such as in Martianus Capella's 5th-century work *De nuptiis philologiae et Mercurii* (*On the marriage of philology and Mercury*). Although this book has largely been forgotten, until the 13th century it was the standard introduction to the *artes liberales*. These arts—or skills—were taught at cathedral schools and early universities and consisted in a *trivium* (place where three roads meet) and a *quadrivium* (place where four roads meet). The trivium included the linguistic disciplines of grammar, logic, and rhetoric, while the quadrivium comprised the mathematical disciplines geometry, arithmetic, astronomy, and music. Once students became proficient in these two domains, they could be admitted to one of the three main subjects: medicine, law, or—the highest attainable field—theology. For centuries the astronomical curriculum in the Latin West was essentially Martianus Capella's textbook, along with a few fragments from Plato (*Timaeus*), Pliny, and Macrobius.

Although Martianus's textbook reflected the predominant—and thus orthodox—view, it came with a surprise in the field of astronomy. Instead of assuming the standard geocentric system in which all celestial bodies, whether or not equipped with epicycles, orbit the earth, Martianus discusses a system in which the earth is circled by the sun, moon, Mars, Jupiter, and Saturn, while Mercury and Venus rotate around the sun. Martianus asserts that this accounts for why Mercury and Venus can sometimes be seen above the sun and sometimes below it.[33] We encountered this "intermediate model"—in which the outer planets revolve around the earth while the inner ones orbit the sun—earlier with Heraclides Ponticus from the 4th century BCE (see chapter 3.2).

But given the dominance of the geocentric model—which was adhered to by all astronomers starting with Plato and Aristotle and which gained absolute dominance with Ptolemy—it is remarkable that Martianus is the only person to discuss Heraclides's deviant system (though without mentioning him by name).[34] Apparently, even after Ptolemy, several conceptual systems remained in circulation. In terms of predictive power, however, this intermediate model was much less accurate than Ptolemaic and Islamic astronomy. Martianus's description is sketchy and conceptual, while the Ptolemaic and Islamic models were also worked out mathematically so that the planetary positions could be calculated at any time. Although Martianus was concerned about the relations between planetary patterns and underlying principles, he did not provide enough details to make calculations using his model. So with Martianus we find no search for deductions from principles to patterns.

Although we find no historical reconstruction of the ancient disciplines in the Latin West until the 12th century, new ideas do arise. We see this in the work of the 7th-century Venerable Bede, whom we encountered above as a historian. Using the anno Domini dating system, he calculated the dates of all Easter Sundays up through the year 1595.[35] Easter falls on the first Sunday after the first full moon of spring. The resulting cycle therefore comprises both 19-year lunar cycles and 28-year solar cycles and consequently has a period of 532 years. With his Easter cycle, Bede showed his mastery of all the knowledge of his time, from history and mathematics to astronomy and theology. In contrast to Martianus, with Bede we find a mathematical inference of patterns (from the solar, lunar, and Easter cycles) to an underlying principle-based calculation model. And thanks to the many Carolingian copies of his work, Bede's ideas spread throughout western Europe.

Starting in the 10th century, the first Christian scholars began studying Islamic astronomy. For example, Gerbert of Aurillac, who would later become Pope Silvester II, traveled to Sicily and Spain, which were occupied by the Arabs and Berbers, to find any truth to rumors of the flourishing astronomical science in the Islamic world.[36] He would return with little more than an astrolabe, but his journey took place before the major astronomical innovations in the Islamic world in the 11th century. After Gerbert, numerous Christian scholars went to Spain, where at the border of the Christian and Islamic civilizations they translated many Greek and Arabic texts into Latin, including the works of Aristotle and Ptolemy's *Almagest*.[37] Thus finally, the Latin West joined the historical reconstruction of classical science and scholarship. The Aristotelian worldview in particular enjoyed attention, with its theory of the spheres. Ptolemy's *Almagest* was studied less and was probably less understood. Most astronomers used tables, such as the Arabic *zijs*, which were based on the *Almagest* and could be used to calculate the positions of the planets, sun, and moon using rules of thumb (see chapter 3.2). New astronomical tables with updated observations were soon drawn up that could make better predictions. The 13th-century *Alfonsine Tables*, dedicated to the monarch Alfonso X of Castile, enjoyed extensive circulation: for three centuries, these were the most widely used astronomical tables in Europe—even Copernicus had a copy in the 16th century (see chapter 5.2).[38]

It is tempting to describe the history of postclassical astronomy as consisting in the development of planetary models, but for several centuries the best European astronomers—from Johannes de Sacrobosco to Richard of Wallingford—focused on calculation models rather than principle-based planetary models. In this way, they did not greatly differ from their earlier Chinese colleagues (see chapter 3.2). However, the consequence was that the text of the *Almagest* was largely ignored in the Latin West. If there was a dominant conceptual planetary model in the later postclassical era, it was the Aristotelian worldview. While the predictive power of Aristotle's model lagged far behind that of the astronomical tables and the Islamic models, it was able to "explain" quite a few things. With Aristotle, everything came together: the natural, circular motions in the sky; the center-oriented motions on earth; and above all, the ordered spheres, which were interpreted as a heavenly hierarchy. This arrangement was worked out by the 5th-century Byzantine mystic Pseudo-Dionysius the Areopagite in his *De coelesti hierarchia* (*On the Celestial Hierarchy*), in which he brought the theory of the spheres in line with Christian doctrine.[39] Each sphere was propelled by a class of angels, with the exception of the outermost *primum mobile* sphere, where all movement originated. Dionysius subdivided heaven and earth into a hierarchy of ranks populated by beings including seraphim, cherubim, archangels, and angels for the heavenly spheres, and bishops, priests, penitents, and those possessed by demons for the earthly spheres. His work became known in western Europe when the Byzantine emperor Michael sent a copy to Louis the Pious in 827. Various translations into Latin soon followed, including one by the Irish philosopher Johannes Scotus Eriugena. It took several centuries for *De coelesti hierarchia* to become the dominant theory, but after the new 12th-century translations of Aristotle had reached the Latin West, Dionysius's integration of the theory of the spheres and the Christian worldview was embraced by theologians such as Thomas Aquinas, who quoted

Dionysius no fewer than 1,700 times. Aristotle's theory of the heavenly spheres subsequently pops up in all the arts—from painting and music to literature—with Dante's *Divina Commedia* being a notable example, and the theory becomes inextricably linked to the Christian-European worldview.

India: The Double Epicycle Model and Mathematical Astronomy

Compared to the Arab world and Europe, in India there was greater continuity with antiquity. In the first centuries of the era, a double epicycle model was developed that, like the Tusi couple, made the Ptolemaic equant superfluous (see chapter 3.2). This tradition is continued in the work of the mathematician and astronomer Aryabhata (476–550), who completed his most important work, the *Aryabhatiya*, at the age of 23.[40] In it he presents a model for the planetary patterns where the sun and the moon are carried by single epicycles, while the movements of the planets are supported by a complex of double epicycles: a small, slow (*manda*) one and a larger, fast (*sighra*) one. In addition, Aryabhata provides a list of "constants" for the periods of the planets and for solar and lunar eclipses. In contrast to the Tusi couple, Aryabhata's model is not a simplification of Ptolemy's: while the double epicycle model makes the equant superfluous, it does not obviate the need for the eccentric. So we are still dealing with three underlying principles here. But in India we can speak of a continuous search for astronomical principles, as well as for inferences from the observed (planetary) patterns to principles.

Aryabhata's work was known to Muslim astronomers. Al-Khwarizmi translated it into Arabic in 820 but eventually opted for the Ptolemaic model, which he enhanced with improved calculation procedures. It is nonetheless because of al-Khwarizmi's translation that Indian mathematics, together with the number zero, entered the Arab world and found its way to Europe in the 12th century. Aryabhata had several followers, including Varahamihira (505–587), who used the double epicycle model for astrology and also wrote an impressive overview of Greek, Egyptian, Roman, and Indian astronomy.[41] His work shows that the Indians had a thorough knowledge of existing astronomical models.

A few generations later, Brahmagupta (598–668) also improved the calculation rules for the double epicycle model, using the number zero (see more about this below). In the centuries after that, Indian astronomy seems more and more like a mathematical affair, although empirical measurements also became more precise: for example, the orbital period of the earth was measured to the ninth decimal place. The Indian dyad of mathematics and astronomy was further institutionalized in the Kerala school, which flourished from the 14th through the 16th century. The works of this school may have found their way to Europe through Jesuit missionaries, although there are no indications that European astronomers actually used the Indian results.[42]

The First Geometric Model for Planetary Motion in China

Although interest in mathematical astronomy visibly waned in China starting in the 4th century, great attention was paid to star catalogs. The most impressive of these, the *Dunhuang* catalog, dates from around 700 and covers 1,585 stars grouped into 257 constellations, most of which are hard to see with the naked eye.[43] This star catalog is more than one and a half times as extensive as the one Ptolemy included in the *Almagest* (1,022 stars). The absence of mathematical astronomy in China is offset by a number of impressive astronomical artifacts, including Su Song's astronomical water clock dating from 1095. This clock, which is about eight meters high, was topped with an enormous armillary sphere (a celestial globe).

Shen Kuo (1031–1095) was undoubtedly the most important astronomer in postclassical China. As a mathematician he occupied himself with the mathematical study of music and harmony. He then set his sights on astronomy. While previous Chinese astronomers had focused on arithmetical, algorithmic descriptions of planetary movements, Shen was the first Chinese astronomer to present a geometric model. He assumed that every planet followed a circular orbit until it reached a "willow leaf" section of the orbit. This willow leaf could lie on either the outside or inside of the track, but when the planet in question arrived at this part of the orbit, it first followed this detour before continuing on its original path.[44]

Shen's model illustrates just how different the models for planetary patterns can be: from calculation rules to pure spheres and from triple epicycles to willow leaves combined with circular orbits. So, the epicycle was only one of the many possible solutions for explaining backward planetary motions. A willow leaf worked just as well—even better, in fact—because it could explain the variation in speed during the retrograde motion. But a willow leaf shape would have been unacceptable to Greek and Muslim astronomers, who would not have been willing to give up the notion of "pure figures."

As varied as the history of astronomy may be, the reduction of planetary motions to an underlying principle-based model was also pursued in China. It is not clear how accurate the predictions of Shen's model were as compared to the models of his Chinese predecessors, such as Liu Hong's calculation rules (see chapter 3.2). But given the very sketchy way Shen's model is worked out, we must categorize it as a conceptual model, just like that of Martianus Capella in Europe. The "circle plus willow leaf" orbit is perhaps the most fanciful model in the history of astronomy, but it could get even stranger: a few centuries later, a European would propose that the planets do not move in any sort of circle, but in the form of what no one until then had even considered a possibility: an ellipse (see chapter 5.2).

Pre-Columbian Astronomy and the Oldest Book in the Americas

Pre-Columbian astronomy enjoyed its greatest flourishing in the Mayan civilization. Of the many hundreds of Mayan books, however, only 15 have been preserved, the most important of which is currently in Dresden. The *Codex Dresdensis* from around the 12th century is the oldest written book from the Americas, even though it is probably a copy of an earlier manuscript from around the 8th century.[45] It is not entirely clear how the book ended up in Europe, but it may have been presented as a gift by Hernán Cortés to Charles V, emperor of the Habsburg Empire, and later fallen into German hands. All Mayan codices are written on amatl paper, which was made from tree bark that was boiled, flattened, and stretched. The codices contain detailed tables for calculating the phases of the moon and solar and lunar eclipses, but above all the motions of Venus. This planet was central to the Mayan religion and was, among other things, the guardian of war. Battles took place at the time of Venus's highest position in the sky.[46] The precision of the Venus tables still arouses the imagination: the almanacs have a margin of error of less than one day every 6,000 years. The Mayan calendar system is also based on Venus, alongside the sun, moon, Jupiter, Saturn, and Mars. This unique calendar system consists of 18 "months" of 20 days each, with five leap days.

The 15 surviving Mayan codices have now largely been deciphered and consist mainly of tables and patterns; principles or underlying models seem to be lacking. That does not necessarily mean that the Mayans had no underlying principles for the solar, lunar, and planetary motions. Indeed, such principles seem to have been incorporated into Mayan architecture. For example, the impressive temple complex of Uaxactun in Guatemala includes a platform on top of a pyramid.[47] Using this platform as a vantage point, a number of celestial phenomena, including the position of the sun at the solstice, can be predicted by linking the edges of three other buildings. This temple complex is therefore a de facto underlying (geometric) model—but in architectural form—and this model forms the basis for the patterns of the celestial bodies.

However, using the Uaxactun complex we can predict only the highest and lowest positions of a celestial body, such as the sun; we cannot derive the complete Mayan tables. This means that although there are inferences from certain patterns to the principles incorporated in the architecture, this is not true of all patterns known to the Maya. In any case, Mayan astronomy shows that when searching for principles we should examine not only texts but buildings and other material objects as well, just as we saw for Neolithic astronomy in Europe (such as Stonehenge; see chapter 1.2). We also found the importance of material objects in a completely different discipline: art theory from classical antiquity (see chapter 3.3). Here too the underlying principles were incorporated into buildings (with proportions in the Doric, Ionic, and Corinthian orders of Greek temples) and into sculptures (the proportions in Polykleitos's *Spear Bearer*).

Polynesia, Oceania: Solar Motions and the Galaxy

The oldest astronomical sources from Oceania date from the early 13th century and are architectural in form. In the languages of this region, which are all closely related, there is no discrete term for astronomer; instead, the local word for "navigator" is used. This is no coincidence, considering that the people of Oceania were excellent seafarers, who had emigrated from Asia to the groups of islands in Oceania long before our era, and knowledge of the starry sky is indispensable for orienting oneself correctly at sea. For this reason, most knowledge practices of these navigators fall under the general notion of astronomy. Moreover, this knowledge turns out to be specialized if we are to believe the later sources, such as those of 18th-century British visitors, who found that astronomical knowledge was the domain of specialists with a religious function.[48] The oldest observatories in Oceania appear to have been religious and served as temple complexes for recording the direction of the sunrise at the spring and autumn equinoxes, as well as at the summer and winter solstices.[49] The most impressive of these observatories can be found on the island of Tonga and is said to have been erected by Tu'itatui, the 11th Tonga leader, who ruled around the year 1200. We also find observatories where the solar movement could be studied in finer gradations.[50] It is quite possible that these temple complexes incorporated implicit principles, but unlike Mayan astronomy, there are no tables of celestial patterns in Oceania that would allow us to discover any underlying principles.

Astronomical patterns and possibly principles were also incorporated into Polynesian meeting houses, the *marae*. These houses can be found in large parts of Oceania, including the Cook Islands, Tahiti, Easter Island (Rapa Nui), and in New Zealand. They were the community's central gathering place and were (and still are) richly decorated. For example, the precolonial meeting houses of the Maori on the east coast of the North Island of New Zealand are usually embellished with an image of the Milky Way, which traditionally takes the form of a "big fish."[51] At the start of the new year, which on the Maori calendar starts with the winter solstice, the house faces the rising sun and is therefore in line with this big fish. These meeting houses are also embellished with decorations resembling constellations, although we also find images of food and other objects. It is not known whether the images express deeper principles in addition to the pattern of the solstice. The Maori had no written culture until the 19th century, and oral traditions do not provide any answers concerning the nature of any principles. However, written sources have been found on Easter Island in the rongorongo script. This writing system has yet to be deciphered, apart from a few short fragments on wooden tablets showing that they include astronomical calendars, probably with cosmogonic descriptions (see also chapter 5.1).[52] We can only hope that the script will someday be deciphered and provide a definite answer to the question as to whether Polynesian astronomy made use of principles.

4.3 Mathematics (and Mechanics): The Attempted Reduction in the Number of Axioms

Islamic Civilization: The Parallel Postulate and Non-Euclidean Geometry

Just as Islamic historians managed to reduce the methods of Herodotus and Thucydides to a single underlying principle (the isnad), and just as Islamic astronomers reduced the number of principles of the Ptolemaic model from three to two, so too did the Islamic mathematicians seek to reduce the number of principles in Euclidean geometry. They were not alone in this; Proclus (410–485) had preceded them, but without success. Of the five Euclidean axioms, the fifth, the so-called parallel postulate, was a thorn in the mathematician's side. The axiom goes as follows: if two lines intersect a third line in such a way that the sum of the inner angles on one side is smaller than two right angles, these two lines must inevitably intersect each other if they are extended sufficiently.

As noted in chapter 3.4, this parallel postulate looks more like a theorem that still needs to be proven than like an axiom. An axiom needs to have a certain self-evident character so it can serve as a basis for proving other propositions. For centuries, Islamic mathematicians tried either to derive the parallel postulate from the other four Euclidean axioms or to find a fifth axiom that was self-evident. So the pursuit to reduce the number of principles appears to be a constant in the Islamic world. We already saw this with Ibn al-Haytham (965–1040), who argued that the number of concepts in Ptolemaic astronomy needed to be reduced, especially the equant (see above).[53] The parallel postulate became to mathematics what the equant had been to astronomy. However, Ibn al-Haytham did not manage to derive the postulate from the other axioms, nor was he able to prove it by contradiction (i.e., by first accepting the denial of the postulate and then demonstrating that this leads to a contradiction). The versatile Persian poet, astronomer, and mathematician Omar Khayyam (1050–1123) was more successful in proving the parallel postulate with the help of another, more natural and therefore more axiom-like postulate: "Two converging straight lines intersect, and it is impossible for two converging straight lines to diverge in the direction in which they converge."[54] But this postulate was later proven to be equivalent to the parallel postulate. As an added benefit, Khayyam discovered that the parallel postulate was valid only for plane geometry (Euclidean geometry). If we give up the fifth postulate, we get some different form of geometry, such as spherical geometry, elliptical geometry, or hyperbolic geometry. For example, on a spherical surface, such as a globe, two straight lines that make a right angle with the equator (so-called "meridians") will converge and intersect at the north and south poles in contradiction of the parallel postulate. In this way Khayyam invented non-Euclidean geometry.

The astronomer and mathematician al-Tusi (1201–1274) wrote an extensive critique of Khayyam's attempt but also failed to prove the parallel postulate. The same was true of his son Sadr al-Din (also known as Pseudo-Tusi), who wrote an influential work on the postulate in 1298, which, although not providing a proof, did present the first non-Euclidean hypothesis equal to the parallel postulate, building on Khayyam.[55] His book was published in Europe in 1594 (in Rome) and served as the starting point for the study of this postulate there. Although we now know that it is not possible to prove the parallel postulate in Euclidean geometry on the basis of the other four axioms, this centuries-long research led to a new geometry that would play a decisive role in the development of modern physics. Sometimes the incidental discoveries resulting from a study end up being more important than the original research goal.

The Origin of Algebra: Al-Khwarizmi

The Islamic mathematicians did not limit their activities to geometry. Already in the 9th century the astronomer and mathematician al-Khwarizmi (ca. 780–850; see also above) wrote about the so-called Hindu numbers and methods for solving equations. His *Book on Addition and Subtraction in Indian Arithmetic* from 825 led to the spread of the Indian numeral system, first in the Islamic world and subsequently in Europe. Unfortunately, the Arabic original of this work has not survived, but there are several Latin translations from the 12th century. The book contains extensive calculation procedures for adding, subtracting, multiplying, dividing, halving, and doubling numbers, as well as for determining square roots. Such procedures were called *algorismi* in the Latin translations, a corruption of al-Khwarizmi's name, which is how we get the English word "algorithm." Incidentally, al-Khwarizmi's calculation procedures were not the first algorithms in mathematics: step-by-step procedures for solving mathematical problems had already been formulated by the Babylonians and Egyptians (see chapter 2.2).

Al-Khwarizmi's most important mathematical work is *The Compendious Book on Calculation by Completion and Balancing* (*Al-Kitab al-mukhtasar fi hisab al-jabr wa l-muqabala*) from around 830.[56] Here the algebraic approach introduced by Diophantus (see chapter 3.4) is elaborated into a full-fledged branch of mathematics: algebra. Although today the word "algebra" is also used for Greek algebra, which was only short lived, the word was coined in the 12th century as a corruption of the word *al-jabr* that appears in the Arabic title of al-Khwarizmi's book. *Al-jabr* means "completing" or "restoring" and refers to the operation of moving a subtracted number on one side of an equation to the other. For example, the conversion from equation (1) to equation (2), in contemporary notation, is an example of *al-jabr*:

(1) $5x - 3 = 27 - 10x$
(2) $15x - 3 = 27$ [$10x$ has been added to both sides]

Equation (2) can be further "completed":

(3) $15x = 30$ [3 has been added to both sides]

And from here, by dividing both sides by 15, it can be deduced that x equals 2.

In the introduction to his work, al-Khwarizmi states that this is primarily a practical handbook, not a theoretical work, but in reality he provides the first systematic study of many types of equations. He also devised a number of rules to solve these equations, which boil down to the fact that if the same operation is applied on both sides of the equation (such as addition, subtraction, or multiplication), the equation remains valid. An equation can be solved algorithmically by repeated application of these rules. Al-Khwarizmi expresses his method in words, without using any symbols.

In addition to rules for linear equations, Al-Khwarizmi also provides a method for solving quadratic equations, such as $6x^2 - 3 = 2x^2 - 1$. But when al-Khwarizmi presents a proof, he first defines his problem in geometric terms (for example, a squared value is represented with a square), just as Euclid had before him. The originality of al-Khwarizmi's algebra lies not in his axiomatic method of proof, however, but in the algorithmic approach with which he transforms the study of equations into an autonomous branch of mathematics that is independent of its applications.[57] Equations came to be studied for their own sake, and in this new mathematics one discovery was made after another. In Egypt, Abu Kamil (ca. 850–930) expanded algebra to include irrational numbers, developing methods to solve increasingly complex equations. And al-Karaji (953–1029) expanded algebra into what would later become known as "Newton's binomial theorem." He also solved equations to the fourth degree (quartic, or biquadratic, equations). Calculation with decimals was introduced, and algebra split into several subfields, with al-Kindi's cryptography as one of the first (ca. 801–873).

Europe: Period of Stagnation Followed by Study of Hindu-Arabic Mathematics

Until the 12th century, knowledge of mathematics in the Latin West did not amount to much more than Martianus Capella's curriculum (see above). In Martianus's textbook, mathematics was stripped down to the most elementary knowledge of arithmetic and geometry. Thanks to the late-antique Roman philosopher Boethius (480–525), some parts of Euclid's *Elements* were also handed down in Latin (see also chapter 3.3). Gerbert of Aurillac (ca. 956–1003; see also above) is perhaps the only mathematician of importance in early medieval Europe, yet no new discoveries are known to have been made by him either. After his visit to Islamic Spain, he brought the Hindu-Arabic numeral system and the abacus to the Latin West, but they would not disseminate until much later.

It wasn't until the 12th century that Arabic and Greek mathematics become known in Europe owing to the great translation efforts taking place on the edges of Islamic civilization, such as in Spain and Sicily. Al-Khwarizmi's algebraic work was translated into Latin by Robert of Chester,[58] and the rest of Euclid's *Elements* also became available. The famous mathematician Leonardo of Pisa, better known as Fibonacci (ca. 1170–1250), became acquainted with the Hindu-Arabic numeral system in Algeria, where his father held a diplomatic post as a merchant. Fibonacci realized that the Indian positional numeral system was far more efficient than the additive system on which the Roman numeral system was based, which was still in use in Europe.[59] In the Indian system, the value of a number is determined by the position of its digits, such as with the number 731, where the 7 has the value of 700, 3 has the value of 30, and 1 has the value of 1. In contrast, in the Roman system, the value of a number was determined by the sum of the constituent symbols, which needed to be added or subtracted, such as MXL, where the M has the value of 1,000, L has the value of 50, and X has the value of 10 but—because it appears to the left of the L—needs to be subtracted, yielding the total value of 1,040. The Hindu-Arabic system of calculation greatly facilitated calculation, which was of considerable importance for trade and accounting. In 1202, Fibonacci wrote his *Liber abaci* (*The Book of Calculation*), in which he introduced the technology and corresponding algorithms to Europe, and it was now widely disseminated, unlike Gerbert of Aurillac's attempt a few centuries earlier.[60] Fibonacci's name is also linked to the fascinating sequence of numbers where each number is the sum of the two previous numbers. The sequence starts with 1 and 1, after which the rest follows automatically: 1, 1, 2, 3, 5, 8, 13, 21, 34, 55, 89, 144, 233, 377 . . . This series of numbers is said to have been inspired by the way rabbits multiply, which is why it is sometimes called the rabbit series. Although Fibonacci was the first to introduce this sequence to Europe, it had been known in India since the 6th century.[61]

European mathematics from the 13th and 14th centuries is strongly linked to the study of kinetics. For example, Thomas Bradwardine (ca. 1290–1349) showed how the speed of a body increases with the ratio between force and resistance. We also see calculations by Nicole Oresme (1323–1382) and Giovanni di Casali (ca. 1320–1374), who independently calculated the area under a curve expressing a body's motion with constant acceleration. In mechanics, this surface corresponds to the total distance traveled.[62] In addition, in a commentary on Euclid, Oresme shows how a body accelerating at a constant rate attains an increase in distance traveled that corresponds to the odd numbers. This may sound somewhat cryptic, but since Euclid had proved that the series of the sum of odd numbers corresponds to the series of the ascending squares, the total distance traveled increases with the time squared. This result is usually attributed to Galileo, who would become all the rage a few centuries later (see chapter 5.3), but Galileo was probably not familiar with Oresme's work. With Oresme we are back to deductive Greek geometry, including the formal inference from motion patterns to principles, now enhanced with Arabic algebra.

After nearly a thousand years of stagnation, European mathematics had made great strides in catching up. This is further highlighted by Levi ben Gershon (1288–1344), also known as Gersonides, who was one of Europe's most original mathematicians. Gersonides worked out a mathematical notion of *inductive proof*.[63] This notion had already appeared in preliminary form in the Euclidean proof that there is no largest prime, but Gersonides applied his method to a large number of other problems that had not been addressed before, especially in the area of combinatorics. However, his work would remain unread for 200 years.

India: Zero, Negative Numbers, and Infinite Series

As we saw in the previous chapter, around 400 CE, Indian mathematics had introduced the trigonometric notions of sine and cosine that would prove so important for celestial mechanics. India's greatest mathematician was Brahmagupta (598–668), who is considered to be the inventor of zero (see also above). In his *Brahmasphuta-siddhanta* (Correctly established doctrine of Brahma), zero is used in counting for the first time, rather than as a symbol expressing a "lack of quantity," in the way that Ptolemy and the Romans had used it.[64] We find a symbol for "nothing" much earlier with the Babylonians, but only with Brahmagupta can arithmetical operations be performed with zero. In addition, he shows how zero can be combined with the notion of negative numbers in the *Brahmasphuta-siddhanta*:

> The sum of two positive numbers is positive, the sum of two negative numbers is negative; the sum of a positive and a negative number is their difference; if they are equal, their sum is zero. The sum of a negative number and zero is negative, that of a positive number and zero is positive, and the sum of two zeros is zero.[65]

And also:

> The product of a negative and a positive number is a negative number, the product of two negative numbers is positive, and the product of two positive numbers is positive.[66]

Brahmagupta also found negative solutions to quadratic equations—something that al-Khwarizmi would avoid two centuries later in his study of equations. Indian knowledge of zero and negative numbers reached Europe through Latin translations of al-Khwarizmi's adaptation of Indian mathematics. But this knowledge met with great resistance until deep into the 17th century. An exception was Fibonacci, who in the 13th century allowed for negative solutions in accounting, where he interpreted negative numbers as debts or losses.

Among the many currents in Indian mathematics, the Kerala school from the 14th to the 16th century was the most impressive (we already encountered this school in astronomy; see above). The Kerala school produced the oldest mathematical series for trigonometric functions, such as sine, cosine, and arctangent (the inverse of tangent).[67] The founder of the school, Madhava of Sangamagrama (ca. 1340–1425), is probably the person who discovered these series.[68] In a text from more than a century later, the *Yuktibhasa* (ca. 1530), this discovery is indeed attributed to Madhava, and geometric proofs utilizing circles and triangles are provided.[69] The Kerala series for sine, cosine, and arctangent are all presented in Sanskrit rhyme. With these series, the values of the sine, cosine, and arctangent can be calculated for any angle without having to measure them. The famous formula for approximating π can also be derived, simply by entering the number 1 in the arctangent sequence (since the arctangent of 1 equals $\pi/4$). In contemporary mathematical notation, this formula for π looks like this:

$$\frac{\pi}{4} = 1 - \frac{1}{3} + \frac{1}{5} - \frac{1}{7} + \cdots$$

However, this beautiful series is known not as the Madhava's formula but as Leibniz's formula. The same applies to many other series that were first discovered by Indian mathematicians but were later attributed exclusively to European mathematicians, as we already saw with the Fibonacci series. With current knowledge of the history of mathematics, it would be appropriate to attribute this series to the first mathematician known to have discovered it, or, in the case of independent discoveries, to at least use a hyphenated name. Occasionally such a term is actually used, as in "the Madhava-Leibniz series."

The Origin of Algebra in China: Zhu Shijie

Independently of Islamic civilization, a form of algebra was also developed in China. One of the founders of the Chinese algebra is Qin Jiushao (1202–1261), who in his *Mathematical Treatise in Nine Sections* (*Shushu jiuzhang*) developed a number of procedures for solving different types of equations, up to the fourth degree.[70] Using an iterative algorithm, Qin was able to approach the solutions as accurately as he wished. Here, as we previously encountered with Liu Hui (see chapter 3.4), we are dealing with a step-by-step procedure for how to advance from a problem to a solution; it's just that with Qin this procedure was not one single absolute procedure but an iterative, converging procedure.

The person who truly put Chinese algebra on the map is Zhu Shijie (1249–1314), one of the greatest mathematicians of all time. Zhu roamed China for 20 years as a teacher, after which he wrote his masterpiece, *The Jade Mirror of the Four Unknowns* (*Sijuan yujian*).[71] This is an incredibly rich and original book that deals with no fewer than 288 problems. For the first four problems the method of the four unknowns is explained. Zhu shows how a problem expressed in human language can be converted to a system of polynomials up to the 14th degree! A polynomial is a sum of exponents. So (in contemporary notation), $3 + 2x - 7y^2$ is a second-degree polynomial with two unknowns, x and y, and the polynomial $5 - 3x + 2y^2 - 7z^3 + 4x^4 - 23y^5$ is a fifth-degree polynomial with three unknowns. Zhu demonstrated how a system of polynomials with four unknowns—which he called *heaven*, *earth*, *human*, and *matter*—can be reduced to a single polynomial equation with only one unknown by means of successive substitution and elimination of the unknowns. Unfortunately, Zhu is extremely brief in his description of his method, and he skips all sorts of steps as he proceeds from the problem to the solution; these gaps may have been included for pedagogical purposes, intended to be filled in by students or teachers. Furthermore, Zhu quite often provides only the solution without any intermediate steps.

Despite these small gaps, *The Jade Mirror* contains a wealth of interesting mathematical problems. For example, Zhu show how systems of linear equations can be solved by putting their coefficients (the values before the variables) into a matrix and reducing them to a diagonal form. This now common method would be rediscovered by Blaise Pascal in the 17th century.

Zhu marks the end of a long tradition of great Chinese mathematicians. After the fall of the Mongolian Yuan dynasty in 1368, there is a clear decline in Chinese mathematics. The succeeding Ming dynasty associated mathematics with the previous dynasties and sidelined it as a suspicious activity.

Mathematics Elsewhere in the Postclassical World

Unlike many other disciplines, mathematics can be found all over the world. And just like certain astronomical knowledge, mathematical knowledge is not limited to texts but can take form in a wide variety of media. For example, knowledge of mathematics was represented using complex decorative figures found almost everywhere in the world. And in the Incan Empire, mathematical knowledge was recorded in the form of *quipu*, cords with knots, which could also store all sorts of other information. Mathematics also underpins intangible kinship structures (including the question of who could marry whom). These structures are sometimes so complex that in modern times they have led to new mathematical approaches, especially in unraveling the kinship relationships with the Malekula people in Melanesia and the Warlpiri in Australia.[72] Although these nontextual mathematical patterns are currently being studied within the discipline of ethnomathematics, it remains extremely difficult to find concrete principles used in these age-old patterns, let alone to determine whether there is a notion of inference or proof. It seems that some notion of principle can be found in all these forms of mathematical knowledge, even if they often remain implicit, as we saw with regard to the astronomical temple complexes in pre-Columbian America and Oceania (see above).

4.4 Reduction of Principles in Linguistics, Musicology, and Poetics

Not only in the fields of historiography, astronomy, and mathematics, but also in linguistics, musicology, and poetics, a quest was on to reduce the number of existing principles. But whereas in classical antiquity the disciplines of the humanities were ahead of astronomy, that is not the case in postclassical times.

Linguistics: Example-Based Grammar versus Rule-Based Grammar

Islamic Civilization: Sibawayh and Example-Based Linguistics

Arabic linguistics finds an early climax with the Persian scholar Sibawayh (ca. 760–793). As a non-Arab, he establishes the first grammar of Arabic in his *Al-kitab fi al-nahw* (The book of grammar), often simply called the *Kitab*.[73] Just as Dionysius Thrax's grammar was meant to acquaint foreign speakers with Greek (see chapter 3.1), Sibawayh's *Kitab* aimed at helping non-Arab Muslims understand the Quran. But the *Kitab* is many times more detailed than Dionysius's 30-page grammar textbook: in more than 900 pages, Sibawayh covers almost all facets of Arabic. His basic linguistic concepts seem to come directly from the Greek grammar tradition, such as the notions of word form, declination, and the distinction between two genders and three verb tenses. Although most Greek works were not translated into Arabic until after the 8th century, it is believed that Dionysius Thrax's grammar was known to Sibawayh, as it had been translated early into Syriac, a language understood and read in large parts of the Persian Empire.[74]

While the elementary categories in Sibawayh's grammar are thoroughly Greek, Sibawayh takes a decisive step in his *Kitab* toward an *example-based description* of language. Such a description had already been made in rudimentary form by the Greek linguist Apollonius Dyscolus (see chapter 3.1) and proceeds from the following idea: rules (such as for conjugations and inflections) are given wherever they can be identified, and where they cannot, the linguistic phenomenon is described as well as possible on the basis of concrete examples. Thus, Sibawayh uses a very large number of examples to show how Arabic works. However, one cannot construct or understand new sentences based on mere examples. To this end, Sibawayh introduces two linguistic principles: *analogical substitution* and *lexical dependence*. With the first, he shows that words and word combinations are interchangeable, provided that they are in similar, analogous contexts. We could illustrate this in English with the analogical substitution of nouns. Words preceded by the definite article "the" and followed by the verb "is" can be substituted with one another. So, we can take the sentences "The man is walking down the street" and "The woman is walking through the forest" and make the new sentences "The woman is walking down the street" and "The man is walking through the forest," because "man" and "woman" appear in similar, or analogous, contexts. This substitution rule sometimes leads to ungrammatical sentences, especially if the contexts are not entirely identical, but the more similar (analogous) the context, the better the substitution rule works. With the second principle, that of lexical dependence, Sibawayh shows how a given lexical element's form depends on the form of another lexical element (for example, in many languages the form of an adjective depends on the gender of a noun, such as in Spanish: *un muchacho listo*, "a smart boy," versus *una muchacha lista*, "a smart girl").[75]

So, Sibawayh's *Kitab* seems to be primarily a list of examples, and for that reason it has been described as "a collection of all the peculiarities and exceptions of the Arabic language."[76] But thanks to the principle of analogous substitution, the student of Arabic can construct and understand an unlimited number of new sentences using the set of examples from the *Kitab* by substituting words and phrases in the examples. Thus, in Sibawayh's view, the enormous amount of language data is based not on a consistent system of grammatical rules (as with Panini) but exclusively on analogous substitution and lexical dependence. This is somewhat reminiscent of the many cases in Roman law, which according to classical legal scholars were based not on a consistent rule-based system but solely on a few "ruling" principles, such as the *lex specialis* and the *lex posterior* (see chapter 3.7). In a sense, Sibawayh reduces all of linguistics to just two underlying language principles, but these principles can work only alongside a database of a great many examples (the linguistic data in the *Kitab*).

A shortcoming of the *Kitab* is that Sibawayh does not define analogical substitution but merely illustrates it, again, with examples. Sibawayh's aim seems to be for students of Arabic themselves to generalize over the examples provided, and he provides them with a tool for this in the form of the two principles. With his example-based grammar, Sibawayh begins a long tradition that would later be revived in a surprising way in contemporary linguistics (see the conclusion of this book).

Although Sibawayh's example-based grammar is diametrically opposed to Panini's rule-based grammar, the two are sometimes compared for their treatment of phonology, the study of sound systems, which both works discuss in staggering detail. The pronunciation of the Quranic verses is of utmost importance for Muslims because it concerns the pronunciation of the language of the Creator himself. Later Islamic linguists must have been familiar with Panini, albeit only superficially. In 1030, the astronomer and historian al-Biruni, in his description of India, the *Kitab al-Hind* (see above), devotes a chapter to Indian linguistics and examines the phonological aspects of Panini's grammar, but he immediately notes that "we Muslims cannot learn anything from this because it is a branch from a tribe that is not within our reach—I mean the language itself."[77] Al-Biruni considered Arabic so far removed from Sanskrit that Panini's rule-based method cannot be applied to it.

Europe: Universal Grammar and Hierarchical Sentence Structure

As with most other disciplines, it is difficult to determine at exactly what point we can speak of a "European linguistics." The early monastic schools mainly used Donatus's *Ars minor* for linguistic training, until Priscian's linguistic tract, the *Institutiones grammaticae*, was rediscovered in the Carolingian period (see also chapter 3.1). However, this rediscovery does not lead to new linguistic insights. This would only change when the European world was broken open by Islamic civilization through Sicily and especially Spain. Scholars finally learned about Aristotle, who at that time was largely unknown in Christian Europe apart from two works of logic from the *Organon*, but would become the champion of European scholastics.

It was not Aristotle's works on logic that would influence European linguistics but his metaphysics.[78] Aristotle divided knowledge into practical and theoretical knowledge, but only the latter leads to truth.[79] He maintained that only three disciplines were truly theoretical: natural philosophy, mathematics, and theology. The 13th-century linguists in Europe now questioned whether language could also be studied in a theoretical way, alongside the practical, descriptive way already known. These 13th-century linguists included scholars such as Roger Bacon, Boethius of Dacia, and Thomas of Erfurt, all from northern Europe.[80] Their linguistic movement, which culminated between 1270 and 1320, is called *speculative grammar*, where "speculative" should be understood as meaning "theoretical." These speculative grammarians were intensively focused on a search for the universal in language and its relationship with reality. Words could not be universal, of course, considering that they differed from language to language, but these grammarians held that grammatical categories were universal. The term they used for grammatical categories was *modi* (manners). And although almost every speculative grammarian had his own set of *modi*, there was some agreement about the basic classification. The *modi* were subdivided into (1) categories of "being" (*modi essendi*), (2) categories of "understanding" (*modi intelligendi*), and (3) categories of "meaning" (*modi significandi*). In fact, all categories were designated by *modi*: word classes, cases, genders, conjugations, and so on.[81]

According to speculative grammarians, or Modists as they were also called, grammatical categories embodied reality through these modes. For example, they held that each verb could be traced back to a "mode" independent of its specific meaning, leading them to postulate that every verb could be reduced to a copula (that is, a form of a verb like *to be*, *become*, *remain*, *seem*, or *appear*) and an adjective. For example, the sentence "John grows" can be paraphrased as "John becomes large." In this way, all sentences can be reduced to "simple" sentences with a smaller number of concepts and words consisting of only a copula and a small number of adjectives. In his treatise *De modis*, Boethius of Dacia proposes that the universal principles of language can be identified on the basis of such progressive linguistic reduction.[82] Linguistic reductionism was carried to the extreme by Roger Bacon, who argued that there was indeed a universal grammar that embodies the principles of all languages.[83]

Although the Modists did not construct any "practical grammars" for concrete languages, their theoretical reflections did lead to hypotheses, some of which were relatively easy to test, such as the proposition that complex sentences can be reduced to simpler sentences, resulting in a smaller number of concepts and words. The Modists assumed that an infinite number of (linguistic) phenomena could be accounted for on the basis of a finite number of concepts. Universal grammar has remained an attractive idea for centuries. It would return in the 16th and 17th centuries with scholars including Dalgarno, Wilkins, and Leibniz and would be revived in 20th-century linguistics with Noam Chomsky.

In addition to the notion of universal grammar, the Modists are also considered the founders of the theoretical notion of *hierarchical sentence structure*, dissecting a sentence into phrases that can be dissected further into individual words.[84] The Modist Thomas of Erfurt illustrates this notion with the sentence *homo albus currit bene* (the white man runs well), which he first dissects into two parts, the subject *homo albus* and the predicate *currit bene*, and he subsequently shows the dependencies between the words, whereby *albus* depends on *homo* and *bene* depends on *currit*. We encountered the notion of lexical dependence earlier in Sibawayh's grammar. Although we have no indication that the Modists were familiar with Sibawayh, it is possible that they had seen his work or that of his followers. Many Christian scholars had access to libraries in Islamic Spain, especially after parts of al-Andalus were taken during the Reconquista. However, there are also numerous differences between the Modists and Sibawayh: the

Modists were primarily theoretical, while Sibawayh provided a practical grammar of Arabic.

India and China

Panini's precise system of rules for Sanskit must have made an overwhelming impression, considering that for 22 centuries Indian linguists largely focused on writing commentaries on, interpretations of, and modifications to the work of the great master. The distinction between (late) antiquity, the postclassical period, and the early modern era is almost meaningless for Indian linguistics. Well into the 18th century, in India Panini's grammar was considered a virtually complete system that was not amenable to substantial improvements.[85] Linguists outside of India also applied Panini's grammatical method to other unrelated languages, such as Tamil and Tibetan.

But there is additionally a non-Paninian tradition in India. For example, Bhartrhari (6th or 7th century) was the founder of the Sphota school, which focused on the question of how the human mind organizes linguistic units into a coherent whole such as a conversation or discourse. With regard to the doctrine of meaning, the Sphota school promoted *semantic holism*, which means that the meaning of the whole cannot be derived from the meaning of its individual parts.[86] During the 7th century, itinerant monks from China made the Indian linguistic tradition available in Chinese. The monk Xuanzang (600–664) accomplished more as a traveler and translator than anyone else in the exchange between India and China.[87] A great number of texts were also translated from Sanskrit into Chinese in the late 7th and early 8th centuries by Yijing and Fazang. However, outside of these translations, the Indian linguistic tradition had scarcely any influence on Chinese linguistics.

Musicology: Algorithms for All Compositions in a Certain Style

Musical rule systems that generate the melodies of a certain musical genre based on principles had already been developed by Aristoxenus (see chapter 3.3). In the postclassical era we find the first music algorithms not in the Islamic world but in Christian Europe and India, while it is in China that we encounter the oldest history of music.

Europe: Musica Enchiriadis *and a System of Rules for Polyphonic Compositions*

Until well into the 8th century, there appears to be no development in musicology in Europe. However, many musicological insights appear to have been transmitted orally for a long time before they were written down. This can be deduced from 9th-century manuscripts that refer to known practices. Starting around 900, new insights and discoveries were made in blindingly rapid succession.

The end of the 9th century sees the appearance of one of the most remarkable texts in musicology, entitled *Musica enchiriadis* (Music of many parts),[88] attributed to Pseudo-Hucbald, who is most likely also the author of a hilarious ode to the baldness of King Charles the Bald. The *Musica enchiriadis* provides an algorithm that can produce all polyphonic compositions of a certain genre: the *organum*. Under the heading "Symphoniae," the *Musica enchiriadis* describes two different forms of polyphony, known as the *parallel organum* and *modified parallel organum*, the rules of which can be summarized as follows.[89]

Parallel organum:

(1) Take a given Gregorian melody as the *vox principalis*.
(2) Have a second voice double the *vox principalis* with a fifth or a fourth, the *vox organalis*.
(3) The *vox principalis* and the *vox organalis* can be further doubled as desired in a higher or lower octave resulting in a three- or four-part parallel organum.

Modified parallel organum:

(1) Take a given Gregorian melody as the *vox principalis*.
(2) The second voice, the *vox organalis*, maintains the initial note of the *vox principalis*.
(3) Only when the *vox principalis* reaches an interval the size of a fourth or a fifth with the *vox organalis* do both continue in parallel fourths or fifths.
(4) At the end (the cadence) of the organum, the *vox principalis* and the *vox organalis* come together again in the reverse order on the same note.
(5) The *vox principalis* and the *vox organalis* can be further doubled as desired in a higher or lower octave, resulting in a three- or four-part modified parallel organum.

Using a Gregorian melody as input, these two algorithms provide all possible parallel and modified parallel organa for that melody. Just as with the astronomical and mathematical algorithms discussed previously (see chapter 3), all patterns—in this case, patterns of a certain musical style—can be predicted using a small number of principles and an input.[90] But a striking difference with astronomical and mathematical algorithms is that the system of rules is not an arithmetical system but a grammar that defines polyphonic relationships in a similar way that a linguistic grammar like Panini's does. Every piece of music that meets the above rules is an example of the musical idiom in question.

It is important to note that the text of the *Musica enchiriadis* refers to a well-known and established music practice. So, the *Musica enchiriadis* appears to be not a prescription for making organa in general but rather a procedural system for an existing musical practice. Did this system of rules, initially intended to be descriptive, come over time to be interpreted as prescriptive, as we encountered in previous practices? We do not know for sure, but if the system of rules was ever interpreted prescriptively, that situation certainly did not last. It didn't take long for people to start composing organa that did not comply with the *Musica enchiriadis* system of rules. Musical compositions dating from the 10th and 11th centuries show that the two melodic voices were becoming more independent: in addition to parallel movements, alternating *lateral* and *countermovements* also started to appear, known as *melismatic organum* and *free organum*, respectively. At first glance, the polyphonic melodic lines do not appear to be rule based but seem almost free, except that they comply with the rule of consonance (and even this rule is not always adhered to). But on closer inspection, it appears that there are again rules underlying these organa, making them less free than the term "free organum" might suggest. However, the rules are more complicated than those in the *Musica enchiriadis*, and they become further complicated when, at the end of the 11th century, polyphony develops to a point where composers combine two melodically independent lines seemingly with all the possible descending movements. Nevertheless, this new organum style is again described on the basis of rules in *De musica* by Johannes Cotto (Johannes Afflighemensis)[91] and the anonymous *Ad organum faciendum* (How to make an organum).[92] However, these tracts from around 1100 give only a snapshot of the tradition of organum construction, because after the system of rules appears for this more complex type of organum, new, even more complex organa arise, and the rhythm is also set free. In a sense, the tracts resemble a description of a living language without taking into account the fact that a language will change from generation to generation.

So, European music was in no way stifled by rule-based descriptions. The rules could barely keep up with the wealth of musical forms emerging from the Notre Dame school in the 12th century with Leoninus (ca. 1150–ca. 1200) and Perotinus (ca. 1160–ca. 1230). The theoretical works that were still being developed applied to a limited substyle or period.

Discoveries also took place in the field of music theory. We already encountered Nicole Oresme (1323–1382) in mathematics and mechanics, but he was familiar with almost all the domains of study.[93] Oresme's most important musicological contribution is perhaps his discovery or rediscovery of overtones. This phenomenon, which Pseudo-Aristotle called the upper octave, is a sound component whose frequency is higher than the fundamental note of that sound perceived by the ear. Oresme found that overtones play an important role in timbre: two different instruments, such as a lute and an organ, can play the exact same note yet sound different. Some overtones are harmonic; that is, their frequencies are whole multiples of the root. But with less "ideal" instruments, such as a chime, overtones are not harmonic. Oresme's discovery was the first true extension of the theory of harmony since Pythagoras, but his discovery would only be taken up again in the 17th century.

The Islamic World: Transformation of Greek Music Theory

From the beginning of Islamic civilization, Arab scholars compared their musical practice with the musical theories they came into contact with, and it was the theoretically and experimentally elaborated Pythagorean music theory of Ptolemy that made the biggest impression (see chapter 3.3).[94] One of the challenges in early Islamic musicology was providing the empirical intervals in the Arabic 24-tone system with a scientific basis. Al-Kindi (ca. 801–873) may have been the first to apply mathematical Greek music theory to the Arabic idiom. In addition to a model for intervals and scales, al-Kindi also provided an overview of the different rhythmic cycles that are so characteristic of Arabic music.

The most important Islamic musicologist is al-Farabi (ca. 872–ca. 950), although he was also active in many other fields, such as logic, poetics, and philosophy. In his three-part *Kitab al-musiqa*, *The Book of Music*, he examines music theory, instrument theory, and the melodic and rhythmic typology of Arabic music.[95] After an introduction with some speculation about the origin of the music and the nature of musical talent, in the first part al-Farabi gives an overview of Pythagorean harmony in terms of four-note sequences, or tetrachords. Since in practice many of these tetrachords were not used, this work indicates a renewed interest in the study of music as a purely mathematical field. His treatment of rhythm is also largely theoretical: he elaborates a mathematical framework within which all possible rhythmic cycles are defined. In the second part, al-Farabi tackles a number of concrete problems with his theoretical approach, in particular the technical peculiarities of musical instruments, such as the correct placement of the frets on a lute in a way that allowed one to obtain acoustically consonant intervals. The third part bears the promising title "Musical Composition," but al-Farabi is often less explicit here. He describes how the human voice can express literary and poetic forms and how it can stimulate feelings and the soul. Vocal music is presented by al-Farabi as superior to instrumental music. He becomes most specific when he summarizes the different possible melodies in a typological scheme. However, this scheme again is composed mainly of abstract sequences of notes rather than concrete pieces of music or melodies created from them. Just like his treatment of the possible rhythmic cycles, al-Farabi's theory of melody is again mainly theoretical: it is more a description of possible music than of concrete music. His system is not a rule-based grammar but a sort of schematic classification of rhythms and melodies. In a sense, with this classification al-Farabi fits within the method that we referred to above as an *example-based description*, although in this case the examples were largely constructed by al-Farabi himself.

In addition to impressive research into music theory, encyclopedic collections of Arabic music were assembled, of which the monumental *Kitab al-aghani* (Book of songs) by al-Isfahani (897–967) is the most important.[96] In a work comprising more than 20 volumes, or 10,000 pages, which he claimed to have worked on for 50 years, al-Isfahani summarizes the multifaceted arts of Arabic song and poetry from the 8th and 9th centuries. The songs are accompanied by a description of the corresponding rhythmic cycles and sometimes the melodic mode.

Arabic musicology seems to have been virtually unknown in Christian Europe and vice versa. Although exchanges occurred in the instrumental field (the lute and the rebec are of Arabic origin), not a single musicological treatise seems to have been translated into Latin in the 12th-century Renaissance. However, al-Farabi's *Classification of the Sciences* (*Kitab ihsa' al-'ulum*), which contains a short excursus on music, was translated by Gerard of Cremona. Nevertheless, the example-based descriptions that we find in both Christian and Islamic musicology seem to point to a common *process from rules to examples.*

India: Survival of Natya Shastra *and the* Tala

In the Indian Sanskrit tradition, Bharata Muni's *Natya shastra* represents the dominant musicological tradition up to the 13th century (see chapter 3.3). Some new writings also appear around the year 1100, such as Narada's tract *Sangita makarandha*, but the rules remain primarily based on Bharata Muni's blueprint. The main Indian musicological text in the late postclassical period is Sharngadeva's *Sangita-ratnakara*, from the 13th century. This tract is considered the "definitive" text for both Hindustani (northern Indian) and Carnatic (southern Indian) music.[97] The *Sangita-ratnakara* takes as its basic elements the *sruti* (the relative note), the *swara* (the musical sound of a single note), the *raga* (the mode or melodic formula), and the *tala* (the rhythmic cycle). These elements are further developed on the basis of a precisely formulated system of rules. There are seven families of *tala*, each of which is subdivided into specific rhythmic relationships. In a way, the work seems to be merely descriptive, but it has been used for centuries—up to this day—as a basic textbook for improvisation and composition. With respect to form, the *Sangita-ratnakara* is rule based and declarative: the boundary conditions are defined by the rules, but there is no procedure for generating new compositions.

China: Historical Musicology and the Refutation of Musical Cyclicity

Major musical developments took place in China during the Tang dynasty (618–907), most notably the emergence of Chinese opera and the foundation of the first conservatory under Emperor Xuanzong. The history of music would become part of the court chronicles in the historiographic style of Sima Qian (see chapter 3.3), and official reports of music production would be created. But there is little to be found in terms of a search for underlying musical principles and patterns, even though historical research is still in full swing.[98] At the time of the Song dynasty (960–1279), important works on music history also appear, such as the first musical encyclopedia containing all classical texts concerning music (Chen Yang's *Yueshu* from 1104).

The most important musicological study in the Song period is the work of Cai Yuanding (1135–1198). In his *Lülü xinshu* (A new treatise on music theory), he describes how the tones in the traditional circle of fifths contradict the widespread cosmological interpretation of the 12 standard tones.[99] By this interpretation, the 12 tones were equidistant and cyclic and formed complete octaves. However, Cai shows that this interpretation is incorrect when scales are transposed, that is, when they are moved to a higher or lower note. In this way, on purely musicological grounds, Cai refutes the ubiquitous cosmological notion of *cyclicity* in the Chinese worldview (which also plays an important role in historiography). He proposes using six additional notes, but they did not survive in Chinese music. However, Cai's work does show that the mathematical theory of harmony was practiced in full glory during the Song dynasty.

Poetics: Patterns and Interpretations

Like classical poetics, postclassical poetics consisted partly in searching for patterns and underlying principles for poetic compositions and partly in seeking rules for interpretation, especially of sacred books such as the Bible, the Quran, the Confucian classics, and the Vedas.

Christian and Profane Poetics in Europe: Allegory and Algorithm

We have already encountered the Church Father Augustine (354–430) in his capacity as a historian (see above), but he may have had an even greater influence as a poeticist. Poetics, like historiography, was subject to reassessment in Christian terms. In *De doctrina christiana* (397/426), Augustine argues that any text can be interpreted either literally or figuratively, with the figurative interpretation being preferable to the literal in the case of Holy Scripture.[100] Bible exegetes had a tough job: they had to bring the Old Testament based on the Hebrew Bible into line with the Christian New Testament, a task for which allegorical interpretations were usually the only solution. Figurative analyses of texts were far from new: the oldest known example is found in Theagenes of Rhegium, in the 6th century BCE, who explained Homeric stories by providing them with a nonliteral interpretation. The Pergamonese anomalists (see chapter 3.3) also favored allegorical interpretations, but with the Neoplatonists the allegorical method really gained momentum: the world was to be seen as a text, God's book, and it was full of figurative signs that were not meant to be interpreted literally. Hadn't Jesus himself used parables to express the deeper and universal meaning of a story? Slowly but surely a system was established that allowed the Bible to be read on multiple levels. The immensely influential Italian theologian and philosopher Thomas Aquinas (1225–1274) established four distinct levels of biblical exegesis: (1) the literal, (2) the allegorical, (3) the moralistic, and (4) the anagogical (spiritually uplifting).[101] In addition to the literal meaning of a Bible passage, the various figurative meanings could all be true at the same time. The only criterion that limited possible interpretations was the "principle of charity,"[102] according to which all interpretations should be consistent and coherent with Christian doctrine. I referred above to this principle as the principle of biblical coherence.

In addition to this principle, can we perceive any system of rules in the nonliteral interpretations, or is it a case of *anything goes*? Just as with the anomalists of Pergamon, the interpretations of the Christian exegetes got out of hand, deriving encouragement from passages in the Bible itself. For example, in Galatians 4:21–31, Paul interprets the Old Testament story in which Abraham's wife Hagar is expelled in favor of Sarah: *Hagar* is the Arabic term for Sinai and therefore represents the Old Covenant with Moses, while Sarah is the symbolic mother of the Christians and thus represents the New Covenant with Christ. In short, Paul did not shy away from far-fetched etymologies and analogies as long as the desired result is achieved: the foreshadowing of the New Testament in the Old.

The allegorical interpretation method was used not only for religious but also for secular texts. Although Augustine and Thomas Aquinas thought that nonliteral interpretations applied only to God-inspired texts, such as the Bible, the classical writers were subjected to reinterpretation through Christian patterns by others. Origen, for example, found that the Fourth Eclogue by the poet Virgil could be understood as a messianic prophecy of the coming of Jesus.[103] In the 6th century CE, Fulgentius even gave an allegorical interpretation of the entire *Aeneid*: from the first sentence to the last, every word is provided with a Christian reference.[104] The result sometimes makes for fascinating reading, but it is hard to identify a critical method or underlying system. Dante Alighieri (1265–1321) also argues in his *Convivio* that Thomas Aquinas's four-part biblical exegesis can be perfectly applied to profane narrative art. Dante uses Ovid to show how all four forms of interpretation (literal, allegorical, moralistic, and anagogical) can be valid and that they are not necessarily mutually exclusive.

We find a rather different approach to poetics in Guilhem Molinier's *Leys d'amor* in the 14th century.[105] This work contains an empirical study of the poetry of the Provençal troubadours, with an attempt at an algorithmic system of rules. How can this tract be explained amid an ocean of anti-empirical, allegorical interpretation practices? The explanation is as fascinating as it is tragic: at the beginning of the 14th century, the art of the Provençal troubadours was dying out, owing to factors including the terrible massacres in the independent kingdom of Occitania, where the French king persecuted the heretical Cathars, or Albigensians. Entire cities were massacred by mobs shouting "Kill them all! God recognizes his own."[106] After four Albigensian Crusades, little was left of Occitania's wonderful poetic culture. To salvage what remained, a huge work was undertaken between 1332 and 1356 under the leadership of Guilhem Molinier, in which as many troubadour verses as possible were collected and in which their poetic system was described in explicit detail. This was the *Leys d'amors*, the *Laws of Love*. It was a final attempt to save a dying art for posterity. In addition to providing a grammar, which formed the basis for all later grammars of Occitan,[107] the work deals extensively with the rules for prosody, the structure of lines, couplets, and poetic genres. Above all, some procedures are given for the poetic system, such as for generating rhymes. Suppose you wanted to find a word that rhymes with *-ori*. You would start alphabetically with the schematic *a_ori* and subsequently fill in the open position with each letter: *abori*, *acori*, *adori*, *afori*, and so forth. From this list you would then select the words that were actual words of Occitan. And so on with the *b_ori*, *c_ori*, *d_ori*, and so on. Nothing could be more algorithmic, if extremely time consuming and exhausting. But as we read in the wonderful stanzas of the *Leys d'amors*: *mays dura anta que sofracha*, "shame lasts longer than suffering."

Islamic Civilization: Takhyil *and Literary Criticism*

In the Arab world, poetry was seen as an important source of knowledge. Indeed, the Quran was written in verse form.[108] In the 11th century, Ibn Sina (Avicenna) introduced the principle of *takhyil*, which he defined as poetry's power to evoke images in the memory (the mind) of the audience.[109] Imagination and memory are closely interconnected here, since the image evoked did not necessarily have to be a stored image but could alternatively come about as a result of a complex interaction among memory, fantasy, and emotion. Ibn Sina contrasts this ability to evoke images, the *takhyil*, with logic. According to Ibn Sina, logical proof does not stir the soul, whereas *takhyil* does, and that is why the audience is more moved by it. However, Ibn Sina's principle of *takhyil* cannot itself produce poetry; it is better described as an underlying principle that defines the conditions for good poetic recitation.

In the 12th century, Ibn Rushd (Averroes) also sought to find poetry's underlying principles.[110] He regarded poetics as a way of discovering universal canons that apply to all peoples, or at least to most of them. But Ibn Rushd did not take the step of actually identifying these canons. Although he searches for the underlying nature of poetics, he ultimately seems more interested in defending logic and reason than in understanding the inner workings of poetry.

There is also a rich tradition of literary criticism in the Arab world. Tha'lab's 9th-century treatise *Qawa'id al-shi'r* (The rules of poetry) is especially noteworthy. In this work, Tha'lab approaches poetry from a completely linguistic perspective, in particular by analyzing the words instead of more poetic features, such as meter or rhyme.[111] We encounter a broader vision of poetry in Ibn Rashiq's 11th-century reflections on diction, meter, rhyme, and meaning, but we find few if any principles. The same applies to other, quite interesting literary critics, such as the Afro-Arab scholar al-Jahiz (781–868), who makes the notion of *coherence among all parts of a poem* the cornerstone of his literary criticism.[112] Al-Jahiz was of East African origin, probably Ethiopian, and was also the author of one of the most notable works of postclassical times, the *Risalat mufakharat al-sudan 'ala al-bidan* (Treatise on the superiority of blacks over whites). He argues that while blacks had conquered and ruled various lands of white peoples (from Arabia to Yemen), whites had never conquered any black people's land.

Indian Poetics: Mimamsa

Attempts to formalize the interpretation process were strongest in India. The Mimamsa school dates back to classical antiquity and aimed to achieve a rule-based exegesis of the Vedas. Such exegesis had become a current issue, as the Vedic rituals became increasingly marginalized because of "new" Indian worldviews, such as Buddhism. As a counterbalance, Hindu scholars strove to demonstrate the validity of the Vedas on the basis of well-established rules of interpretation so that they could be interpreted by anyone. The most important work in this area consists of Jaimini's *Purva Mimamsa Sutras*, which are, however, strongly prescriptive.[113] Jaimini's attempt seems to have been successful: a long period of stagnation in Indian Buddhism set in. The most important development after Jaimini is an influential commentary by Kumarila Bhatta from around 700 CE that would allow the Vedic tradition to continue in India for centuries.

A Masterpiece from Chinese Poetics: Chen Kui

In China, Chen Kui (1128–1203) created an extremely interesting work, *The Rules of Writing*, or *Wen ze*, which is considered the first systematic treatment of Chinese poetics and rhetoric. Of course, *The Literary Mind and the Carving of Dragons*, by Liu Xie (5th century CE) is older, but it is primarily a work of literary criticism that does not attempt to derive a system of rules for beautiful writing (see chapter 3.3).

Born during the Song dynasty, Chen Kui became the registrar of the imperial library at an early age. He seems to have had a particularly critical and independent spirit. For example, he brought the issue of waste to the attention of the court and once stated that there were many more civil servants than necessary. Chen was sidelined with a promotion in the provinces, but that did not prevent him from writing his masterpiece. To appreciate Chen's tract, an outline of the historical context is in order. After the fall of the Tang dynasty in 907 and the rise of the Song dynasty in 960 (after an interval of five brief dynasties), the attitude of the ruling class toward government officials changed. Instead of a selection system based mainly on privilege, as had been common during the Tang dynasty, officials were now selected on the basis of a competitive exam. The most important part was the essay in which candidates had to prove their originality and skill.[114] The exam was the first step to highly sought-after government careers. With his *Rules of Writing*, Chen produced the first proper manual of Chinese poetics and rhetoric that could also serve as preparation for the exam. The Chinese printing press did the rest.

What makes *The Rules of Writing* so special is that Chen explicitly derives the rules of writing from existing texts. These texts constituted the crème de la crème of Chinese literature. For example, Chen derives the rules for "clear language" from the *Book of Rites* (a Confucian classic), while the rules for "colorful language" are deduced from the *Book of Songs*. With Chen, rules are never simply prescribed; they are first derived from the data, in the form of patterns that are then tested on new works, after which they can be modified if necessary. Chen also studies in detail sentence length in the classical *Tan gong*, in which he believes he has discovered a regularity, concluding that "the language of the *Tan Gong* is simple but not austere."[115] Next, Chen selects the "exquisitely beautiful" sentences from the *Tan gong* and describes them on the basis of a number of characteristics. He tests the regularities he has discovered on sentences from the *Spring and Autumn Annals* and the *Book of Songs*, but when he discovers that his rules cannot be generalized to these new works, he does not immediately reject his own approach (as Dionysius of Halicarnassus had with his rules of natural word order; see chapter 3.3) but instead adapts them. Although this attempt did not lead to a positive result, it did lead in a direction that could be explored further.

Chen combined two goals—an empirical poetics and instructions for ambitious young men—and incorporated them into a single handbook. Chen Kui showed how stylistic analysis can work together with rules for "good" writing but that it is extremely difficult to identify general rules, let alone underlying principles, for "beautiful" writing. Although this result was also reported by Dionysius, Chen provided suggestions for further research into the rules of "beauty." Chen's *Rules of Writing* is without a doubt the most original work in postclassical poetics, yet it is virtually unknown outside of China, apart from a few translated excerpts.[116]

4.5 Medicine: The Inhibitory Effect of Scholarly Medicine

Islamic Medicine and Science of the Soul

Nowhere in the postclassical world did medicine reach a standard as high as in Islamic civilization. But before this could happen, Muslim scholars first had to translate and interpret many Greek, Roman, Persian, and Indian texts. Greek medical knowledge soon came to be regarded as superior. The most important translator of Greek texts was the Nestorian Hunayn Ibn Ishaq (ca. 820–873), who, with more than a hundred translations into Arabic, laid the foundation for Islamic medicine and other disciplines.

So far, Islamic medicine seems to be a repeat of what we saw before: scholars from the Islamic world translated, commented on, and interpreted Greek works, after which they made new discoveries and suggested new principles, some of which reduced earlier principles. But there is an important difference: while Islamic scholars made several discoveries in medicine, we observe no pursuit of new theories, let alone a reduction or unification of principles. The Islamic contribution lay mainly in the way existing medical knowledge was systematized. We see this above all in Ibn Sina's *Al-qanun fi al-tibb* (The canon of medicine) from 1025, which provided a complete overview of all medical knowledge of his time in a text comprising more than a million words. The work is often seen as a concise summary of Galen's 20 works (see chapter 3.5), which Ibn Sina refashions in an Aristotelian way.[117] Indian medical insights were also added to the *Qanun*, and Ibn Sina himself made contributions as well. The *Qanun* was widely distributed and became the standard work of Islamic medicine. When it was translated into Latin by Gerard of Cremona in the 12th century, the work also attained the highest status as a medical text in Europe. In his *Divine Comedy*, Dante honors Ibn Sina with a place between Hippocrates and Galen in limbo.[118] The *Qanun* becomes the culmination of Galen's ideal of the physician as a philosopher, resulting in a field of medicine that is scholarly rather than empirical.

Attempts to systematize medical knowledge occurred in the Islamic world long before Ibn Sina. More than a century earlier, al-Tabari (838–923) compiled the first compendium in which Persian, Indian, and Greek surgery were summarized. In addition to being a medical scholar, al-Tabari was also a historiographer (see above), who with his isnad-based historiography can be considered one of the most accurate historians of postclassical times. We find a sharp critique of Greek medicine with al-Razi or Rhazes (865–925), who on the basis of cases and clinical observations denounced Galen.[119] Yet al-Razi eventually declared himself a student of Galen. Another major contribution to Islamic medicine is the development of ophthalmology, a field in which the translator Hunayn Ibn Ishaq, mentioned earlier, played a major role.[120] Ibn Ishaq provided an overview of the anatomy of the eye, eye diseases, their symptoms, and treatments. He posited that the lens was in the center of the eye, a fact that would become accepted as standard knowledge in the 16th century.

Despite these systematizations and new insights, the contributions of Islamic medicine are of an order different from those of the other Islamic disciplines. Al-Khwarizmi's algebra, for example, led to a flourishing branch of mathematics, and Sibawayh's example-based grammar kicked off a new direction in linguistics, not to mention the isnad method in historiography and the Tusi couple in astronomy. Virtually nothing of this kind can be found in Islamic medicine: no new principles were conceived, and no existing ones were reduced.

It is not easy to explain this difference between medicine and the other disciplines. It may have to do with the fact that Islamic scholars used Greek knowledge as a starting point. While this Greek basis was quite fruitful for areas such as mathematics, astronomy, musicology, and poetics, this was less the case for Greek medicine, which had a strongly learned, philosophical approach. Although there were also empirical schools, such as in Hellenistic medicine, the dominance of Galen's ideal of arriving at a synthesis of philosophy and medicine was so great that the field remained stifled by it for centuries: to treat the sick the physician had to restore their balance, with an analysis of the humors and an understanding of the relationship between the microcosm and macrocosm considered more important than any surgical intervention. This preoccupation kept Islamic medicine from gaining a strong empirical basis in the field of diagnosis and treatment. There was certainly no question of anatomical dissection of human corpses because that was considered a desecration of the body.

One of the few physicians who did develop a new empirical practice and a corresponding theory was Ibn al-Nafis (1200–1288). He was the first to provide a description of pulmonary circulation. Educated in the famous Nuri hospital in Damascus, Ibn al-Nafis's interests were in logic, grammar, theology, and medicine. But today he is best known for his commentary on Ibn Sina's *Qanun*, in which he observed that, contrary to what Galen claimed, blood could not flow directly from the right ventricle to the left ventricle by passing through "invisible pores." Ibn al-Nafis determined that the material making up that part of the heart is impermeable. How he came upon that knowledge remains unclear, but considering the fact that autopsies were highly unusual, he may have observed this during a heart surgery. Whatever the case, Ibn al-Nafis found empirically that the blood passes through the arteries to the lungs where it is mixed with air, after which it flows back into the left atrium through the pulmonary veins.[121] Ibn al-Nafis's description of pulmonary circulation could have inspired a new direction in medicine—including the insight that blood circulation involves more than just the lungs—if only his description had not disappeared into obscurity. It would not be until the 16th century that the Spaniard Michael Servetus rediscovered the pulmonary circulatory system, unaware of Ibn al-Nafis's pioneering work.

In addition to physical medicine, mental health also enjoyed great interest in the Islamic world. The Islamic physicians wanted to integrate the study of the soul, or *nafs*, meaning "self," with the study of the body. Galen had already paved the way for this with his theory of the humors, but this medicine of the soul was further elaborated among Islamic scholars. For example, al-Razi and Ibn Sina described insomnia, neurosis, depression, and mania, along with possible treatments. The integration of mental and physical health is mainly found with Abu Zayd al-Balkhi (850–934). Al-Balkhi criticized doctors who overemphasized physical illnesses and claimed that body and soul were intertwined. He held that if a body becomes ill, the *nafs* loses much of its cognitive capacity, and if the *nafs* becomes ill, the body does not enjoy life and physical health will eventually decline.[122]

Europe and the Rise of Empirical Medicine

Like almost all European science and scholarship, learned medicine in the Latin West showed a sharp decline after the fall of the Western Roman Empire. Many of the works by Galen and Hippocrates were lost, and Greek medical knowledge was transmitted only through the summaries of Isidore of Seville (560–636). Yet medical care went on, especially in the thousands of monastic hospitals that were established starting in the 5th century. It is only after the translation of Islamic sources that the Latin West comes into contact with classical medicine again. This takes place first at the Schola Medica Salernitana in Salerno, where Greek, Latin, Arabic, and Hebrew sources had been studied since the 11th century, with Trota of Salerno at its core. Trota is one of the few medieval female physicians whose names have been preserved for posterity, but her works were for centuries attributed to her male colleagues.[123] The medical curriculum was primarily theoretical, with the doctrine of the humors being the only accepted medical theory.

In addition to this learned medicine, there was also a more empirical medical practice in Europe, but this took place mainly outside university walls. Many medical practitioners were not trained physicians but priests, barber-surgeons, midwives, and herbalists. They had knowledge not taught at the universities. The emergence of gardening was perhaps the main cause for the development of empirical medicine in Europe. Monasteries were active in growing plants and herbs, and they became centers of herbal medicine. This medicine developed further with the rediscovery of ancient herbal books such as Dioscorides's *De materia medica* (see chapter 3.5). These herbal books, or *herbaria*, were systematically expanded with new herbs from folk medicine, while old herbs that were found ineffective were removed.

The German abbess Hildegard of Bingen (1098–1179) is the most prominent example of a trained physician who also used insights from folk medicine. She was undoubtedly the greatest female scholar and scientist since Hypatia of Alexandria (see chapter 3.4). Hildegard was active in philosophy, music, poetry, botany, linguistics, and, above all, medicine.[124] In her first medical work, *Physica*, she examines the healing properties of plants, herbs, minerals, fish, and reptiles. Her most important medical work is *Causae et curae*, in which she describes the human body in relation to the rest of the natural world.[125] According to Hildegard, knowledge of the natural could contribute to the treatment of diseases. She elaborates on the treatment of accidents typical of farming life, such as cuts, fractures, dislocations, and burns. She seems to have made extensive use of practical knowledge that was not available in Latin. In addition to this empirical approach, her work also contains some rather imaginative remedies, such as the use of the unicorn's liver for leprosy. She couches all this knowledge in the classical framework of the four humors, where diseases are understood to be an imbalance among the four bodily fluids: blood, phlegm, yellow bile, and black bile. But ultimately Hildegard reduces the origin of all illnesses to the relationship between humans and God. This gives her medicine a strong theological character. Nonetheless, in her extensive Bible commentaries, on more than one occasion she strikes a feminist tone, for example, "Woman may be made from man, but no man can be made without a woman."[126]

After the 12th century, the rise of empirical medicine occurs mainly in Italy, where the difference between surgical practices and learned medicine was less pronounced than in northern Europe. In his Kingdom of Sicily, Frederick II stipulated that only those who had studied surgery as part of their academic training could practice medicine professionally.[127] Surgery became a compulsory subject at Italian universities, and slowly but surely the taboo on the dissection of corpses was lifted. While in the rest of Europe surgery was banned at universities, in Italy new manuals appeared, such as the work of Roger Frugardi at the end of the 12th century. In 1350, Theodoric Borgognoni began his colossal standard work, the *Cyrugia*, which not only brings together all surgical knowledge but, as with Hildegard of Bingen, also supplements that knowledge with herbal medicine.[128]

All in all, European medicine also provided new insights, such as bringing together folk medicine and the theory of the humors, the further development of herbal medicine, and the emergence of surgery in Italy. But no new theory can be found in the Latin West to compare with that of Ibn al-Nafis on pulmonary circulation. All we encounter is cases of disease patterns being reduced to principles within the tradition of Galen, but there the learned physicians did not accomplish much more than informally linking disease phenomena to the theory of humors. No search was undertaken to establish formal relations between medical patterns and principles. The only such attempt I have found, albeit a failed one, was made by the Spaniard Arnaldus de Villa Nova (1235–1311). Arnaldus took on the challenge of giving Galen's medical theory a mathematical, testable basis.[129] One of the most colorful scholars of the European Middle Ages, Arnaldus calculated with the greatest accuracy that the world would perish in 1378. According to him, medicine should make equally precise predictions and establish medical treatments. However, the University of Paris declared his work heretical and ordered that all his books be burned. So much for providing medicine with a mathematical basis. Despite his eventual downfall, Arnaldus's patients included several kings and no fewer than three popes, including Boniface VIII.

India: Unani versus Ayurveda

In India, Ayurvedic medicine (see chapter 3.5) was dominant until the 11th century, after which, as a result of the Turkish-Afghan invasions, Islam introduced new medical practices, especially in and around the Gujarat region. These medical practices became known as *unani*, which is a corruption of the Arabic word for "Greek." Unani medicine harks back to the ideas of Galen and Hippocrates, especially as interpreted by Ibn Sina in his *Qanun*.[130] Unani and its Ayurveda counterpart have much in common: both incorporate classical herbal medicine, and both have a theory of the humors. But where Unani assumes the four humors of blood, phlegm, yellow bile, and black bile (as with Hippocrates), Ayurveda relies on the three humors of air, phlegm, and bile. And in both traditions we find a desire to bring the humors into balance. The main difference was and is in the patients: Muslims follow Unani, whereas Hindus follow Ayurveda. Both medical practices are still used in India, where they are supported by university degree programs, in combination with Western medicine. So, humorism is still very much alive in India and parts of the Islamic world.

China: Continuity and Sun Simiao's Oath

We see a great deal of continuity in Chinese medicine: the universal principles of the life force *qi*, of the five phases of *wu xing*, and of the *yin* and *yang* dyad are alive and well. Harmony among these principles is requisite for good health (see chapter 3.5). But we also see surprising new works, such as that of the physician Sun Simiao (581–682), who was active during the Tang dynasty. Sun was the first to develop a Chinese version of the Hippocratic Oath, entitled On the Absolute Sincerity of Great Physicians. The text of the oath was included in Sun's *Essential Formulas for Emergencies, Worth a Thousand Pieces of Gold* (*Beiji qian jin yao fang*), written in 652.[131] In it Sun describes the general principles of medical practice. Topics covered include ethics, the various treatment methods, diagnostics, preparation of prescriptions, medical applications, and epidemiology. The book comprises 30 chapters, in which over 5,300 remedies are described, divided into 232 categories. Sun's eclectic work brought together earlier Chinese medical knowledge, ranging from breathing exercises, acupuncture, and herbal medicine to moxibustion (also known as moxa therapy, where a plug of dried mugwort is lit on top of a needle inserted into the body, allowing the heat to be conducted deeper into the body through the needle). Sun's work became an indispensable resource for many generations of doctors and earned him the title King of Medicine. His oath, like its Hippocratic counterpart in the West, is still compulsory for Chinese doctors.

After the invention of the printing press during the Tang dynasty, we see the widespread distribution of medical works. In the Song dynasty that followed, a peak occurred in the release of medical texts, which were published under the auspices of the Agency for the Editing of Medical Handbooks. In this way, the Song government achieved an encyclopedic overview of all of Chinese medicine.[132]

4.6 Jurisprudence: A Massive Reduction in Rules and Sources

Byzantium and Europe: Flourishing and Rebirth of Roman Jurisprudence

While all other European disciplines experienced stagnation for centuries or even disappeared from the scene, jurisprudence flourished. But this success took place almost entirely in the Eastern Roman, or Byzantine, Empire. Because of the conquests by Emperor Justinian I (482–565 CE) during the first centuries of its existence, this empire became the largest power in Europe, encompassing southern Spain, Italy, the Balkans, Greece, Asia Minor, the Middle East, and North Africa. Here, the Greco-Roman tradition was continued, or preserved, without any actual innovation. That is why I have largely skipped the Byzantine disciplines (aside from a few brief mentions above).[133] We do find principles, patterns, and reductions in the Byzantine sciences and humanities, but in many cases they add nothing to the existing Greek insights. We see this in the continuation of the Aristotelian theory of spheres, in the medical ideas of Hippocrates and Galen, and in the way Thucydides's principle of eyewitness testimony was used by the historian Agathias, who also imitated his style.[134] The notion of innovation even had a negative connotation in the Byzantine world, and the same applied for the medieval Latin West.

But Byzantine jurisprudence is different. Here we find both a continuation and a revival of classical legal studies. This activity is largely due to Justinian's initiative to overhaul Roman law. Today, this initiative is known as the "codification of Roman law," resulting in the *Corpus iuris civilis*. However, the rather dry term "codification" conceals the fact that a new principle-based method was developed to compile this codex, allowing the many law books of the Roman Empire to be evaluated. Each law had to be examined to determine whether it was still consistent with the legal system as a whole. If not, one had to ascertain whether the existing procedural principles (such as the *lex specialis* and the *lex posterior*) were still sufficient to allow the many mutually inconsistent laws to be applied in a consistent manner. A task of this magnitude, in which the entire millennium-old legal system was examined, had never before been undertaken. However, private jurists had compiled two earlier codices that arranged laws chronologically: the *Codex Gregorianus* with laws from between 130 and 290 CE, and the *Codex Hermogenianus* with laws and constitutions from the first tetrarchy, from 285 to 313. These codices were extremely useful for lawyers and judges because they allowed them to easily apply the *lex posterior* principle, which states that a chronologically newer law prevails over an older law (see chapter 3.7).[135]

Theodosius II (401–450) was the first emperor to set up a commission to assemble all laws in force since Constantine the Great in 312 CE to arrive at a fully formalized legal system. As we have seen, Roman legal scholars had formulated multiple procedural rules and principles for resolving conflicting laws. While these principles resolved most inconsistencies, as the number of laws grew, one had to check whether previous legal provisions, such as for donations or sales, dealt with the same subject. In theory, all laws had to be examined, and the most recent regulation would be considered valid. But it was much more often the case that laws were similar but did not concern the same subject. In such cases, the *lex posterior* or *lex specialis* could not be invoked, and one had to reason using analogies. However, the interpretations could vary so much that Theodosius came up with an additional procedural principle in 426: if the law did not provide a definitive answer, the second source of Roman law, namely, the writings of private legal scholars, or *ius*, had to be consulted—those of Papinian, Paulus, Ulpian, Modestinus, and Gaius. If their opinions did not agree (which was the usual case), the majority decided. If the votes were tied, then Papinian's opinion prevailed. If Papinian had not commented on the matter, the judge would decide.[136]

Although this procedure was precisely formulated, it did not offer a real solution. Increasingly, one had to appeal to the sound but unpredictable opinion of the judge. And for this reason, a mere three years later Emperor Theodosius II assembled a new commission to embark on a comprehensive approach to legislation. In the resulting code, the *lex posterior* principle would become superfluous. Each legal rule had to be brought into line with the rule on the same subject elsewhere in the codex. There would no longer be a need for a principle of preference such as the *lex posterior*. This second, perfect codex would be called the *Codex Theodosianus*. But like any Herculean project, this also proved impossible to implement. The committee established by Theodosius soon abandoned the plan and subsequently retreated to its initial task: collecting and organizing the generally applicable laws or constitutions.

A Unifying Legal System: Tribonian

With the *Codex Theodosianus*, an inconsistent but orderly system of laws was established, along with the *ius*—the writings of the classical legal scholars. These writings were consulted when the codex did not provide a definite answer, after which reasoning was the sole remaining recourse if the *ius* also failed to provide an answer. This was the general state of affairs in jurisprudence at the end of classical antiquity. Soon after the appointment of Justinian in 527, a commission was established to take up the legal system once more.[137] This time, the scope of this legal system was not limited to the constitutions starting from Theodosius but encompassed all constitutions, including those of two private codices—the codices *Gregorianus* and *Hermogenianus*—as well as all previous legislation up through the Twelve Tables from the 5th century BCE. The reason for taking all this law into account was that recent laws could build upon older ones, something that Theodosius had overlooked.

The committee established by Justinian was headed by the prefect John the Cappadocian and included among its members the exceptionally diligent and intelligent Tribonian (500–547).[138] The committee completed its work within 14 months, and the laws were enacted in April 529. But just three years later, a new commission was created, because this first codex had proven insufficient. John the Cappadocian had not taken the second source of law, the *ius*, into account. This meant that jurisdiction was quite incomplete, because in addition to opinions and rules, private legal specialists had drafted some frequently used legal principles, such as that proposed by Paulus Prudentissimus (ca. 200 CE) that the burden of proof was on the accuser and rather than on the accused (see chapter 3.7).

There was thus a great need to arrive at a new legal system as quickly as possible, one that also included the *ius*. The new committee was chaired by Tribonian, who in three years plowed through the private jurisprudence in its entirety. Tribonian selected only those parts that had not yet been discussed in the codex and that did not contradict each other, or that at least did so as little as possible. What Tribonian accomplished is hard to fathom: he reduced 3 million rules to "a mere" 150,000 rules and 2,000 books to 50. These 150,000 rules as a whole were subsequently elevated to the status of law by the emperor. As a result, every rule had equal legal force; no given rule had priority over the other, so the *lex posterior* did not apply. Although Tribonian had not succeeded in making 3 million rules consistent, he was successful with a select 150,000 rules, or at least that is what the emperor would declare. Tribonian even reduced the *ius*, or a selection thereof, to a single law, which was simply incorporated after the other laws, a solution as neat as it was ingenious.

On December 16, 533, Justinian issued the most comprehensive law ever published: the *Digests*, incorporating the 150,000 rules assembled from the ancient legal scholars. It was also stated at the time of enactment that this immense legal code was free of internal contradictions and inconsistencies. We now know that it is almost impossible to select 150,000 internally congruent rules out of the 3 million that had developed over time. The claim that these *Digests* were consistent must therefore be seen as a personal exhortation from the emperor to interpret any contradictions as merely *apparent* contradictions. This marked the birth of an age-old tradition of legal interpretation. Finally, only the emperor himself had the authority to enact new laws, after which they were listed under the "Novellae" category.

So, in postclassical legal scholarship as well, we initially find a drive to reduce the number of principles, such as the attempt to eliminate the *lex posterior*. But it was soon realized that law could not do without such procedural principles. Although this insight did not lead to a reduction in the number of principles, it did result in quite a massive reduction in the number of rules and sources of law: Tribonian forged the opinions of legal scholars into a greatly reduced whole to be appended to the earlier laws as one big law. This did not obviate the need for the principle of *lex posterior* itself, but it did reduce the principle's effect.

Rediscovering the Codex

No code of law has had such a great impact on European culture as the *Corpus iuris civilis*. Yet it took many centuries for this effect to take shape. Although some copies of Justinian's codification were present in Italy since the 6th century, Byzantine rule was too short and the text too long for it to be accepted in the West. Some legal codes, such as the Visigoth code in Spain, retained some characteristics of the *Codex Justinianus*,[139] but the volumes in libraries containing the actual Justinian code remained untouched for centuries. After the expulsion of the Byzantines, the last remains of jurisprudence in western Europe disappeared. The Germans and Franks had no legal education, and there are no descriptions of principles either. But we can find an implicit principle in the Germanic *Lex salica* (Salic law), which corresponds to the ancient principle of retaliation, the *ius talionis* (see chapter 2.4).[140] So we do find a sort of meta-pattern, a tendency, of a recurring retaliation principle in all "incipient" legal systems: in the Sumerian legal text of Ur-Nammu, in the Babylonian legal system of Hammurabi, in the Jewish laws, and in the Salic law of the Germanic Francs.

Around 1070, a copy of Justinian's *Digests* was found in Italy. Less than a generation later, Irnerius of Bologna (ca. 1050–1130) would teach the *Digests* and the other parts of Justinian's code at the glossators' school he founded in Bologna.[141] Irnerius also contributed to a standard edition of the text that would be used in all European schools of law. Together with the codex, which was also found, and the novellae and the institutes, the so-called *Littera Boloniensis* was created: the Justinian text canon that would come to be known as *Corpus iuris civilis*. Law schools produced a new class of lawyers, who achieved formidable status. The best were consulted by the emperor of the Holy Roman Empire, and analyzing, interpreting, and commenting on Roman law became a profession in its own right. The first task for Irnerius and his followers was to make the Justinian code accessible. In this context, the work of Accursius (ca. 1182–1263) in particular deserves mention. In his extensive *Glossa ordinaria*, Accursius attempted to collect and present in an orderly fashion all earlier commentary (the "glosses") on the *Corpus iuris civilis* of his predecessors. This was so important for later legal practice that if a given Roman legal source was not discussed in Accursius's *Glossa*, it was deemed to have lost its legal force.

These so-called glossators were followed by a long tradition of commentators. Cinus de Pistoia (1270–1336) was one of the founders of this school. In his *Lectura in codicem* (1314) he mixed Roman law with contemporary statutes and customary law.[142] One of Cinus's students was the eminent poet and later humanist Francesco Petrarca (Petrarch), who can be seen as a link between the legal scholars and humanists (see chapter 5.1). Cinus himself was also known as a poet of merit: his 200 love poems for his muse Selvaggia were praised extensively by Dante.

Logical Solutions for Conflicting Laws: Bartolus de Saxoferrato

The greatest legal scholar from the Latin West was undoubtedly Bartolus de Saxoferrato (1314–1357). Not only did he develop new legal concepts that were incorporated into almost all forms of private law; he also had a refreshing view of the ever-increasing problem of conflicting laws.[143] Bartolus took to heart Emperor Justinian's urging to interpret contradictions in the codex as merely *apparent* contradictions. To this end he applied logic to the legal texts, especially the *Digests*: potentially conflicting texts and fragments were compared in a step-by-step fashion, formulating an often ad hoc principle for each contradictory fragment. But distinctions were also drawn up that allowed the "apparent" contradiction to be explained. Later legal scholars considered the harmonization of legal texts to be jurisprudence's greatest challenge. Although these activities did not lead to fewer principles—just the opposite—they did lead to the most precise inference possible from a legal case to the corresponding laws. One needed to be able to link every case to a legal text without contradictions. That this was by no means always successful does not detract from the fact that there was a passionate search in European legal scholarship for the most precise possible relations between cases and laws. This challenge has remained a part of jurisprudence.

Cinus de Pistoia and Bartolus de Saxoferrato marked the beginning of a new philological-comparative research method. Although this method initially applied only to legal texts, a generation later the method was adopted by humanists such as Petrarch and was extended to all classical texts (see chapter 5.1). The empirical and comparative analysis of the classics even became the basis of Italian humanism and is now attributed to 14th- and 15th-century philologists. But we already find this approach in the 13th-century study of Roman-Byzantine legal texts. What the Italian humanists added was a study of Classical Latin, which in their view was the only correct variety. In contrast, the Medieval Latin of legal scholars was scorned and rejected. This renewed knowledge of Classical Latin enabled humanists to study all ancient texts and to revive some of the disciplines described in them. Unlike the glossators and commentators, the humanists were also the first to read Greek; as a result, they were able to study the only part of the *Corpus iuris civilis* written in Greek, the "Novellae," in its original language. Moreover, these humanists developed techniques for reconstructing an original text from extant copies (see chapter 5.1). But the emergence of humanism cannot be understood without the comparative legal method previously developed by Bartolus de Saxoferrato. Although Bartolus has now largely been forgotten, for centuries he enjoyed such a great reputation that his Italian name, "Dr. Bartolo," lived on in countless plays and operas, including Mozart's *Le nozze di Figaro* and Rossini's *Il barbiere di Siviglia*. He disappears into oblivion in the 19th century.

Islam: Analogical Deduction in al-Shafi'i's Fiqh

Unlike Roman law, where there was a centuries-long accumulation of legal sources, the Islamic sources came about in a much shorter time. Although Islamic law is constituted not by a single legal system but by several families of legal systems, all of which have their own schools, we can still speak of a shared methodology. The development of this methodology is largely due to the work of the prominent legal scholar al-Shafi'i (ca. 767–820).[144] After the death of the Prophet Muhammad in 632, a colossal problem arose: it was no longer possible to consult Muhammad personally for a decision, as had been customary during his lifetime. Now one needed to resort to other sources, such as the "examples" (the *Sunnah*) of the Prophet through his words and deeds as recorded in the Hadith (see above). In addition, a large body of divergent opinions was being produced by legal scholars, grouped into schools. The tension was mainly between what was considered to be the divine element of justice and what was deemed human. Al-Shafi'i, who was born in Gaza and who studied in Medina at the Maliki school, took on the task of uniting these elements in a strict hierarchical way. After initial resistance and long debate, his formulation became the classic doctrine of Islamic jurisprudence. This doctrine soon became the methodology for all legal schools, known as *fiqh*.[145]

Al-Shafi'i hierarchically organized four sources of law (*usul al-fiqh*, "roots of law"): (1) the Quran, (2) the example of the Prophet as recorded in the Hadith, (3) the consensus of the community (*ijma'*), and (4) analogical reasoning or deduction (*qiyas*). According to al-Shafi'i, the hierarchy is self-evident: the Quran comes first as the word of God, but the Hadith immediately comes second, since the Quran repeatedly states that God and the inspiring example of the Prophet must be obeyed and followed. At the same time, al-Shafi'i allowed for two other sources of Islamic law: consensus and analogical reasoning. He states that human law and human reasoning had an inevitable role, but that they should always be subject to the overarching divine law. This last aspect is the compromise position for which al-Shafi'i became famous: his jurisprudence is theocentric but also assigns a role to the legislative human community. Achieving consensus (*ijma'*) is the primary goal here—after all, the Prophet himself is reported to have said that members of the community couldn't possibly all be wrong at the same time. Analogical reasoning (*qiyas*) is the second aim: this meant deriving rules for cases found in neither the Hadith nor the Quran by analogy to comparable cases described therein.

Ultimately, analogical reasoning is perhaps the most interesting element of Islamic legal scholarship.[146] The Quran and the Hadith were the most important sources, but they are not legal codes. Apart from general basic ethical values, the Quran mentions only a few "serious crimes," such as adultery, theft, gambling, and charging interest. And although the Hadith is more consistent than Roman law, it is far from adequate for everyday legal cases. What's more, traditional written legal opinions were rarely in agreement with each other. So the main activity of fiqh was to link a new legal case to the hierarchically ordered sources of law (from Quran to Hadith to opinions) using analogical reasoning. This led to an exceptionally thriving legal practice in which cases had to be linked to laws based on analogical deduction and induction and sometimes on the basis of categorical syllogisms (see chapter 3.4)—similar to what would arise in the much later European jurisprudence of Bartolus de Saxoferrato. For example, according to Islamic legal scholars, by analogical reasoning the Quranic prohibition on wine could be generalized to all alcoholic beverages, since, like wine, they all bring about an intoxication that inhibits rational behavior. This form of reasoning came to dominate all Islamic schools of law.

We have already encountered the use of analogical reasoning in Babylonian law (chapter 2.4), where "similar" cases were often used to reach a decision. Notions of analogical reasoning can also be found in Roman jurisprudence, but in fiqh this form of reasoning was elevated to an explicit method. Here Islamic jurisprudence bears a surprising similarity to Roman law: the pursuit of an unambiguous relationship between case and law.

China: The Tang Code and Truth Finding

The Tang code, established at the time of the Tang dynasty (618–907), is the oldest Chinese legal system surviving in its entirety. As noted in chapter 3.7, previous legal systems were highly underspecified because the law was considered relevant only for people living outside the bounds of civilized behavior. Civilized people observed the correct rites as described in the *Liji*, the *Book of Rites*, a Confucian classic. Law did not enjoy great prestige: what was "legal" was not necessarily moral or fair. Nevertheless, at the time of the bureaucratic Tang dynasty, further specification of the laws was deemed necessary, as was the practice of jurisprudence. Under the leadership of Confucian minister Fang Xuanling (579–648), the Tang code was established, organized into 12 parts.[147] These parts had titles like "Security and Prohibition," "Office and Hierarchy," "Domestic Affairs and Fidelity," "Stables and Storage," "Accusation and Promotion," "Theft and Robbery," "Fights and Disputes," "Deceit and Falsehood," and "Arrest and Escape." The topics mainly concerned individuals who did not adhere to the code of conduct. For those who did, no law was needed.

But how should civil matters be dealt with, such as a disagreement over a transaction or breach of contract? Who was right if both parties laid claim to something? Unlike in Roman, Islamic, and Hindu law, in Chinese law the burden of proof did not lie with the accuser. The magistrate needed to precisely assess witness behavior through the five interrogations technique: witness statements, expressions, glances, breathing, and reactions to the judge's pronouncements. The magistrate had to carefully observe to determine whether witnesses were telling the truth.[148] If magistrates were unable to settle the case through observation, they could resort to legal torture. Here, too, it was not the case that the accused was presumed innocent until fault was proven by the accuser. According to the Tang code, a defendant could be flogged up to 200 times during a maximum of three interrogations conducted at least 20 days apart. If the defendant was able to withstand the prolonged torture without confessing, the magistrate subjected the accuser to the same torture. Accusers who admitted to making a false charge would be subjected to the punishment that would have been imposed on the accused if the latter had confessed. This approach naturally gave the advantage to the person who could sustain the torture for the longest time. As barbaric as this may seem, torture was not unique to China: jurists introduced the notion of torture in Europe as well. This applied to cases such as those involving "half proof": when suspicion was justified, "complete proof" could be extracted through torture. But as detailed as the Tang code was, in contrast to Roman and Islamic law, in China we see no pursuit of an inference from case to code.

4.7 Conclusion: Reducing Principles in the Postclassical Disciplines

After the explosion of principles in classical antiquity, in the postclassical era we observe a drive to reduce those same principles, especially in the Islamic world, but also in Europe, Asia, and Africa, albeit to a lesser extent. Although there is a tendency, there is no region where this principle-reducing goal can be found in all disciplines.

Reduction of Principles as a Global Phenomenon?

The desire for fewer principles came in the Islamic world first. In the field of history, for example, individual principles for selecting the most reliable source were made obsolete by the unifying isnad method. In astronomy we see the elimination of the equant and the eccentric due to the Tusi couple. And in mathematics, an age-old preoccupation with eliminating the parallel postulate. Linguistics is founded on just two principles: analogical substitution and lexical dependence. Poetics is based on the principle of *takhyil*, and legal scholarship is based on the method of analogical deduction. Although the drive for a reduction in principles was not always successful, it did lead to unexpected benefits, such as Omar Khayyam's steps toward a non-Euclidean geometry (whereas what he was looking for was a way to get by with fewer axioms in Euclidean geometry). The pursuit of reduction is absent in Islamic medicine: the four humors of Hippocrates and Galen seem to remain inviolable.

In Europe, efforts to reduce the number of principles are less dominant but are still prominent in a number of disciplines. In Modist linguistics we see a search for a single universal grammar for all languages. And in jurisprudence an attempt was made to eliminate the principle of *lex posterior* (while later European legal scholarship shows a proliferation of principles). In musicology we observe a search for a single algorithm for all musical compositions in a certain style (the organum). And in poetics a single algorithm is proposed for finding rhymes. Astronomy and mathematics flourish only after the translation of Islamic works, while in medicine the search for a single principle is virtually absent (with the possible exception of Arnaldus de Villa Nova's failed attempt).

What about China? In the mathematics of Zhu Shijie was a desire to find a general model (of four unknowns) for solving polynomial systems. And Shen Kuo attempted to rid astronomy of the Chinese algorithmic tradition by developing a single geometric willow leaf model for the planetary movements. Elsewhere, the situation is quite different: history, musicology, and poetics may have reached great heights, but a search for reduction was absent. The same applies to medicine, where the principles of *yin* and *yang*, *qi*, and the Five Phases remained dominant.

India evinces the pursuit of reduction in certain disciplines. Just as in Islamic astronomy, Indian astronomers endeavored to eliminate the equant. With the introduction of the decimal positional system, Indian mathematicians succeeded in bringing about a tremendous simplification. And in the Sphota school there was a tendency toward a single linguistic principle in the form of semantic holism.

In North Africa we find an overall reinterpretation of both history and poetics in terms of the single principle of biblical coherence. The wildest allegorical interpretations are permitted, provided they lead to the desired result (agreement with the Bible). This also occurred in Ethiopia, in the field of history, while little is known about other disciplines. We know from pre-Columbian America that there was a search for principles (in astronomy), but there are no indications of efforts for a reduction in principles. It is difficult to say anything about other disciplines in this region, since most surviving sources date from after 1500 CE. We find a similar situation in Oceania, where the sources are so scarce that we can say something only about astronomy, and even for that field we remain uncertain.

All in all, the quest for a reduction in principles can be found in most postclassical regions and in most disciplines. So there is a trend. Only in pre-Columbian America and Oceania is it not possible to establish a quest for a reduction in principles, but the possibility cannot be excluded.

The Nature of Inferences and Predictions Compared

Another commonality among most postclassical fields of knowledge is the ongoing search for relations between patterns and principles. These relations can be logical inferences, algorithmic procedures, or restrictions that indicate the boundary conditions within which patterns can occur. We saw in chapter 3 that these three types of relations between principles and patterns do not divide along the lines of the natural sciences and the humanities—the study of the natural and the study of the human. Algorithmic procedures are in evidence in astronomy (China), mathematics (multiple regions), linguistics (India), poetics (Europe), and musicology (multiple regions), and even in the historical isnad method (Islamic civilization). But restrictions and conditions are mainly found in legal scholarship. In one instance we find an attempt at a mathematical inference in medicine (Arnaldus de Villa Nova), an attempt that dead-ends.

Some algorithmic procedures do seem to run along the lines of the humanities versus the natural sciences: grammars. While procedures in astronomy are mathematical almost everywhere, the same cannot be said of (European) linguistics and musicology. But note that predictions could be made in all these disciplines. Islamic astronomers used the principles of the Tusi couple and the epicycle to predict the planetary motions in a way similar to how musicologists and linguists used a system of rules to predict a new organum or sentence. But these predictions are of a different kind: the grammar of a language cannot be used to predict what a particular person will say at a given time; one can predict only which combinations of words will yield acceptable (grammatical) sentences. In contrast, in astronomy one can calculate what the configuration of the planets, sun, and moon on the firmament will be at any time, not merely which configurations are possible. Interestingly enough, astronomical models such as Ptolemy's and al-Tusi's can do everything a grammatical system can do, but not vice versa. That is, we can use the astronomical principles and models to determine whether certain hypothetical configurations of celestial bodies are *possible* configurations (just as a grammar can determine whether a certain sentence is a *possible* sentence). But conversely, the grammars of linguistics, musicology, or poetics cannot predict what someone is going to produce at a given time, only what someone could possibly produce. The astronomical models are thus more powerful in this respect than the linguistic and musicological models.

Examples versus Rules

Another possible distinction between the study of the human and the study of nature is that the study of language and music sometimes (but by no means always) used an example-based approach, while such an approach was absent in astronomy and mathematics. So, according to Sibawayh, language could best be described using examples. Instead of a system of mere rules, he employs a large number of examples of linguistic expressions together with an analogical substitution operation. Such an example-based approach can still involve an inference;

namely, from a linguistic pattern to a collection of examples. However, we find no such example-based approach in the natural sciences, such as astronomy, unless we consider the non-derivable constants of celestial bodies, such as their orbital times and their specific epicycles, as "examples." From this perspective, Sibawayh's pursuit of only two linguistic principles—analogical substitution and lexical dependence—bears some resemblance to Islamic and Indian astronomers' efforts to reduce the three Ptolemaic principles for planetary movements to a mere two, but the number of non-derivable constants in the linguistics of Sibawayh, with its 900-page list of peculiarities of the Arabic language, is many times greater than the number of constants in postclassical astronomy with its seven planets.

CHAPTER FIVE

The Discovery of Patterns in Deductions

The Modern Era

1500–1800/2000: All Regions

In the modern era a new form of awareness appears in the search for systematic knowledge: deductions from patterns to principles turn out to also display patterns. One of these deduction patterns consists in the systematic repetition in the relation between empirical observations (the phenomena) and theory (the principles). After an empirically observed pattern is linked to theoretical principles via deduction, these principles are used to predict new events or phenomena, after which a new deduction is established, which either succeeds or fails. In the latter case, either the principles are modified, or the patterns to be accounted for are restricted so that the deduction is still valid without having to modify the principles (see below). A deduction can then result in a new test, followed by new modifications, ad infinitum. This is an iterative interaction between theory and the empirical observations, or a pattern in successive deductions, which is also known as the *empirical cycle*.[1]

Ever since the 19th century, the invention of the empirical cycle has been attributed to the natural sciences, and it is taken for granted that the empirical cycle is a Western invention. In this chapter I will show that the former assumption is incorrect, while the second is highly doubtful. My findings do not detract from the importance of what is called the Scientific Revolution, the complex transformation during which the all-encompassing religious worldview gave way to modern scientific ideas,[2] but the insight that the empirical cycle did not originate in the natural sciences and, moreover, did not develop exclusively in Europe, leads to a different history of knowledge than has long been customary.

Deduction patterns, like the empirical cycle, seem to be ubiquitous in modern science and humanities, which makes awareness of them important as well for today's knowledge practices. But they come into full swing by the 17th and 18th centuries. It is for this reason that I end my more comprehensive treatment of the history of knowledge at around 1800, when the notion of the deduction pattern crystallized. I give only a sketch of the period from 1800 to 2000.

5.1 Awareness of the Empirical Cycle in the Humanities: Philology, Historiography, Linguistics, Art Theory, and Musicology

Philology and Historiography: A Pattern in Source Reconstruction

The Beginnings of Humanist Philology: Petrarch

The first seeds of humanism can be found as early as with the medieval legal scholars (see chapter 4.6), but above all with Francesco Petrarca (1304–1374), known in English as Petrarch. His ambition was to revive ancient Rome in a Christian community.[3] Petrarch traveled around western Europe looking for ancient manuscripts hidden in monasteries and cathedrals. He lived for a long time in Avignon, which served as a point of cultural contact between north and south after it became the papal residence from 1309 to 1377. The city came to be one of Europe's leading intellectual centers, with monastic and cathedral libraries within easy reach. Of direct interest to philology were the papal commissions that stimulated the practice of commenting on Roman classics, such as the works of Livy and Seneca. Petrarch arrived in Avignon at just the right moment: he found a community with an interest in texts that had hardly been read for centuries.[4]

Petrarch's philological fame rests largely on his reconstruction of the historical works of Livy. He assembled the various fragments from European libraries and managed to forge them into a coherent whole using comparisons of differences and similarities.[5] Some parts of Livy's text were copied by Petrarch's own hand from libraries he had visited. This is one of the most important characteristics of humanist philology: the humanists were manuscript hunters convinced that they were making empirical discoveries in the world around them, which they viewed as a world of texts. But at this point their discoveries were little more than loose or even inconsistent observations from which a coherent whole could be forged only with great inventiveness.

The Studia Humanitatis*: From Salutati to Bracciolini*

Coluccio Salutati (1331–1406) managed to pass Petrarch's torch to the wave of 15th-century humanists. In addition he introduced a new educational curriculum in 1369, which he called the *studia humanitatis*.[6] This curriculum focused on *humanitas* (humanity), a term that Salutati had taken from his great exemplar, Cicero. In Cicero's view, what distinguished humans from animals was language, with the consequence that the focus of education and upbringing should be its study. Salutati's *studia humanitatis* included grammar, rhetoric, poetics, history, and moral philosophy.[7] In this curriculum, the linguistic disciplines in the *artes liberales* were freed from their propaedeutic straitjacket, which for centuries had relegated them to the task of preparing students to study theology. The *studia humanitatis* found its way into several 15th-century Italian universities, and in student jargon their supporters were called *umanisti*, from which the word "humanist" and the later 19th century term "humanism" both derive. It is remarkable that the five disciplines comprising Salutati's *studia humanitatis* are the same as those of the earlier Arabic curriculum, the *studia adabiya*, as we saw in chapter 4.1, yet there is no indication that Salutati was familiar with it. Further research into this issue is needed, because although the subjects differ in content, the selection of disciplines is identical and therefore quite astonishing.[8]

There is one humanist who stands out above all others in terms of his urge for discovery: Poggio Bracciolini (1380–1459), the man responsible for the humanists' reputation as ruthless and even unscrupulous manuscript hunters.[9] As secretary to the pope, Poggio managed to assemble a remarkably diverse number of classical texts, ranging from polemics to pornography. Poggio's 1415 expedition to the Cluny monastery in Burgundy provided him with previously unseen speeches by Cicero, thanks to a manuscript that had remained untouched for more than six centuries. Poggio's second poaching expedition took place in the summer of 1416 and brought him to the monastery of St. Gallen. The result was a number of unprecedented influential discoveries, followed by a new expedition to St. Gallen in early 1417, this time with official papal sanction. In the summer of 1417 he undertook even more extensive journeys in France, England, and Germany, discovering now-famous texts that had hitherto been completely unknown. At the end of Poggio's life, the lion's share of the Latin literature known today had been discovered. Discoveries continued to be made after him, but the greatest gems had been found. That did not, however, bring the "century of discoveries" to an end. Poggio and his contemporaries only rediscovered manuscripts that had been lying for centuries in monastery and cathedral libraries. The real research had yet to begin: finding a theoretically and empirically motivated method to reconstruct the original source text from the many inconsistent surviving copies.

Philology Becomes an Influential Discipline: Valla's Historical Textual Criticism

With the fourth generation of Italian humanists (after Petrarch, Salutati, and Bracciolini), the study of the classics branched out into new areas. For example, Flavio Biondo was active in archaeology, numismatics (the study of coins), and epigraphy (the study of inscriptions). The large number of texts allowed one to compare many different varieties of Latin. The introduction of the printing press in Europe led to unprecedented access to classical works. Libraries brought books into the public domain, and an international forum emerged for debating text reconstruction and textual criticism.

Classical Latin grammars were also written, with strict rules regarding proper use, form, and style. The most influential of these was the *Elegantiae* by the Italian humanist Lorenzo Valla (1406–1457). This work is both descriptive and prescriptive: Valla extracted his prescriptions from the classical texts he studied, with which he subsequently sought a revival of Classical Latin as the only correct variant. In a few decades the *Elegantiae* went through no fewer than 59 editions, hastening the extinction of Medieval Latin, which was still spoken. Classical Latin was much more difficult than Medieval Latin, with the consequence that many authors resorted to using local vernacular languages such as Tuscan or Occitan. At the same time as Valla's humanist Latin, a new literary language emerged that we now know as Neo-Latin, which was used by an elite group of humanists.

It was Valla's outstanding knowledge of Classical Latin that allowed him to make one of his most important discoveries: in his text *De falso credita donatione* from 1440 he showed that the document known as *Donatio Constantini* (*Donation of Constantine*) was a forgery from the 8th century.[10] This document stated that the Roman emperor Constantine the Great (280–337) had bestowed authority over the Western Roman Empire upon Pope Silvester I out of gratitude for Constantine's miraculous recovery from leprosy. The *Donatio Constantini* document was the church's main justification for its worldly power. In 1433, Nicholas of Cusa (Cusanus) had already concluded in his *De Concordantia Catholica* that the document was apocryphal, but it was Valla who applied a strict critical methodology to the text and identified it as a forgery. Valla was familiar with Cusanus's work, and there are some striking parallels between the two authors,[11] which suggests that northern humanism had an early influence on its southern counterpart (Cusanus, like Erasmus, was educated at the Latin school in the Dutch city of Deventer). Yet it was Valla who first tackled a text on the basis of philological-linguistic criteria and employed these criteria to expose it as a forgery.

Although Valla did not explicitly describe his method anywhere, we can easily distill it from his text. He employs three criteria of consistency; namely, chronological, logical, and linguistic consistency.

Chronological consistency: Valla notes that the date of the document (as stated in the *Donatio Constantini* itself) is inconsistent with its content because it refers to both the fourth consulate of Constantine (315) and the consulate of Gallicanus (317). This chronological or historical inconsistency is an indication that the *Donatio* was either corrupted or is a forgery.

Logical consistency: Valla used an indirect, counterfactual mode of reasoning. He states that if Constantine had given the Western Roman Empire to Silvester, it would certainly have been mentioned in the Acts of Silvester. Since that is not the case, as Valla ascertains, it is extremely unlikely that the donation took place.

Linguistic consistency: Valla's most powerful evidence is linguistic. He notes that the document contains terms that could not have been known in Constantine's time, such as those related to the feudal system, which did not come into existence until after the fall of the Western Roman Empire. Valla addresses the forger directly, pointing out the many linguistic inconsistencies, for example, "Instead of *milites* you write *militia*, which we have inherited from the Jews, whose books were never known to either Constantine or his secretaries."[12]

With these three criteria, Valla developed a type of textual criticism never seen before, giving the humanists an extremely powerful weapon. Valla's refutation was accepted almost immediately and was recorded shortly afterward in a tract from 1453 by Pope Pius II, the humanist Ennea Piccolomini. Yet nothing changed in the church's secular power: the papal state even gloriously outlived the Holy Roman Empire. However, after Pius's death, Valla's work was largely ignored, and when Martin Luther used Valla's rebuttal as an argument for his Reformation, the church placed *De falso credita* on the list of forbidden works. But a few decades later, the church historian and cardinal Caesar Baronius admitted in his *Annales Ecclesiastici* (1588–1607) that the *Donatio* was a counterfeit, settling the matter for good. Valla's refutation was too well thought out to be refuted.

So we see that during the first century of humanism (roughly from 1350 to 1450) the attitude toward texts changed dramatically. Whereas with Petrarch we find uncritical reverence for everything that hinted at antiquity, with Valla this had turned into skepticism.[13] For him no text was sacred. Any source could be corrupted or forged, and it was up to the humanist to separate the wheat from the chaff.

Text Reconstruction as an Empirical Cycle: Poliziano and Erasmus

However brilliant, Valla's textual criticism adds little to the problem of reconstructing a source from surviving copies; it is hard to identify a theoretical foundation for it until roughly 1480. Although reconstruction skills were widely used among humanists, the practice of text reconstruction was more a matter of subjective guesswork than of well-founded emendations. And if a certain emendation had already been substantiated, then the focus was on (the number of) mutually consistent copies, without investigating the genealogical relations among them. Precise references to manuscripts were completely lacking. This all changed radically with the work of the Italian scholar Angelo Poliziano (1454–1494). In his *Miscellanea*, written in 1489, he describes a theoretically and empirically motivated method that allows for a precise comparison and evaluation of sources.[14] Predictions made according to this method could then be tested against newly discovered manuscripts so that additional improvements could be made. Poliziano realized that a set of fully consistent sources could still pose a problem. Suppose we have four sources—A, B, C, and D—all of which agree on a certain point, and B, C, and D are completely dependent on A for their information. Should B, C, and D be included as additional proof of the authenticity of A? Poliziano said that they should not: if derived sources are mutually consistent, they should be identified and eliminated.[15] Sources should be arranged genealogically so that their dependence on an older source could be made clear. A single deviant manuscript could refute dozens of consistent manuscripts merely on the basis of its position in the genealogical order.

The almost obvious preference for an older source had existed well before Poliziano: older manuscripts were considered more reliable than newer ones because they entailed fewer stages of transmission between the source and the author. But Poliziano's method consisted not only in identifying the oldest possible source but also in determining the complete genealogy of the sources. Once this genealogy has been established, one can begin eliminating derived sources. For this reason, Poliziano's method is known as *eliminatio codicum descriptorum* (elimination of copied books). This method will be further elaborated in the 19th century with Karl Lachmann (see below) to become the cornerstone of modern philology.

Poliziano illustrated his genealogical method on Cicero's *Epistolae ad familiares*, of which he possessed a 9th-century manuscript from Vercelli and a 14th-century manuscript originally made for Coluccio Salutati. Poliziano also consulted an unknown number of more recent manuscripts of the text. He then showed that the 14th-century manuscript, in which a piece of text had been inserted due to a binding error, was the source of all more recent manuscripts, because they all contained the same erroneous insertion but not the binding error. He also found that the 14th-century manuscript itself was a copy of the 9th-century manuscript, and that all these later manuscripts therefore had no value for the reconstruction of the original text and all work should be based solely on the 9th-century text.

Poliziano was the first to provide a theoretically and empirically motivated method for text reconstruction. According to his genealogical method, sources should be *weighted* instead of counted. And with this method, Poliziano established a formal relationship between transmitted manuscripts and the source text, indeed a deduction of the latter from the former. Yet Poliziano's method was not immediately embraced. Why should one particular manuscript have more weight than hundreds of others? It was not until the first half of the 16th century that a shift occurred in philological practice, and by 1550 Poliziano's method was applied almost everywhere in Europe. Poliziano himself used his method with exemplary precision and enthusiasm. His search for the oldest surviving manuscript led to extremely precise reconstructions of Terence, Virgil, Seneca, Propertius, and Flaccus.

As obvious as Poliziano's procedure may sound today, no historically founded reconstruction had ever before been proposed in European philology. His method is perhaps closest to the Arab isnad method (see chapter 4.1), which also reconstructs the genealogical chain of transmission back to the source itself (usually a statement of the Prophet). But where the isnad attempts to reconstruct a chain of oral sources, Poliziano's method concerns the reconstruction of the chain of *written* sources. This is a remarkable similarity, nevertheless, and it is not inconceivable that Poliziano was influenced by the many (translated) Arabic works in circulation in Christian Europe, even though we have no evidence for this whatsoever. But it could well be that the European empirical-theoretical tradition can be partly attributed to Islamic civilization.

Whatever the case, Poliziano's theory of manuscript kinship was more than just a theoretical foundation of an empirical practice. His genealogical chain could be used to make predictions testable. A new discovery of, say, an even older manuscript could support or refute earlier hypotheses regarding emendations and could even require a theory to be rejected or modified. This is precisely what happened when the Dutch humanist Desiderius Erasmus (1466–1536) discovered that a more recent but *untranslated* manuscript was more reliable than an older but translated manuscript.[16] The case at hand was a Greek manuscript of the New Testament that was not as old as a Latin translation but that, because it was in the original language, ultimately contained fewer corruptions than the older Latin version.[17] This empirical finding led to a modification of Poliziano's approach, in which *untranslated* manuscripts take precedence over *older* manuscripts. In this way Poliziano's theory was not so much refuted as it was transformed into a better theory by modifying its underlying principles.

So there is a pattern in these successive deductions from transmitted manuscripts to the original source, and this pattern indicates a recurring interaction between theory and empirical facts, where a theory not only provides a foundation for empiricism but can also be tested against new manuscripts (which in philology constitute empirical observations), which in turn have an effect on the theory of text reconstruction. This further leads to an adaptation or refinement of the underlying principles, which are again tested against empirical observations. And so on. This deduction pattern, today referred to as the *empirical cycle*, is one of the most fascinating aspects of early modern philology.[18] We will encounter this deduction pattern, this cyclic interaction, in practically all other fields in the humanities, from art theory to musicology.

The Empirical Cycle in Chronology: From Scaliger to Spinoza

The philological achievements of the 15th century are raised to great heights in the later 16th century by the French humanist and Leiden professor Joseph Justus

Scaliger (1540–1609). Scaliger was without a doubt the greatest philologist of his time.[19] His immense erudition quickly became evident when he deciphered comprehensible content in Manilius's *Astronomica* (1st century CE). This text was so corrupted that large parts of it were completely incomprehensible. Scaliger turned Manilius into a readable author whom others had not been able to make sense of. He was also the first person capable of handling the work and its author as an organic whole, taking into account the author's intellectual background in addition to the text itself. Scaliger's fame spread rapidly, and he was asked to succeed Justus Lipsius as professor at the University of Leiden. A few years after its foundation (1575), this university became the most prominent in Europe, and as a Huguenot, Scaliger seemed to be the ideal candidate for the vacant position, for which not only excellence but also adherence to the Calvinist faith was a precondition. After initial hesitation and several negotiations, the persecution of the Huguenots by the French Catholic government became perilous for him, and he accepted a position as professor without teaching duties in Leiden. He had an excellent group of researchers under his wing, including the prodigy Hugo Grotius (1583–1645) (see below).

Scaliger used the philological empirical cycle not only to reconstruct texts but also for what he saw as the ultimate goal: establishing the definitive chronology of the world from Creation to his own time. Chronology was a controversial subject, but it was also the field that brought all of the disciplines together: from astronomy to philology and from history to theology. Scaliger was in a better position than others before him: he had mastered more languages than anyone else, and his knowledge of Syriac, Aramaic, Ethiopic, Arabic, Hebrew, and Greek sources was unparalleled. In *De Emendatione Temporum* (1583) Scaliger produced a new timeline of classical antiquity in which he placed Greco-Roman history in the context of Babylonian, Egyptian, Persian, and Jewish history through calendar comparisons.[20] To this end, Scaliger developed a new unit of time, the Julian period, which enabled him to include both lunar and solar calendars using a single overarching frame of reference in his calculations. Both the Julian period and the Julian day derived from it are still used as a reference point in time in astronomy. Just as Bede had done centuries before him (see chapter 4.2), Scaliger showed how important astronomy was for dating—and conversely, how important philological preoccupations could be for astronomy.

In the remaining 24 years of his life, Scaliger put the philological method into practice in an exemplary manner, especially in his *Thesaurus Temporum*, written in 1606. In this work he collected, restored, and arranged almost every surviving historical fragment. This included reconstructing some very important chronological texts, such as Manetho's history of the earliest Egyptian dynasties (see chapter 2.6). Based on extensive information about the duration of the various dynasties, Scaliger succeeded in dating the beginning of the first Egyptian dynasty to 5285 BCE. To his dismay, this date was nearly 1,300 years before the generally accepted Creation date, which biblical chronology approximated to 4000 BCE.[21] According to the empirical cycle, this would mean that Scaliger needed either to modify his theory or to conclude that parts of the Bible could no longer be taken literally. Scaliger opted for the former solution. In order to "save the phenomena"—at least for the time being—he introduced a new notion of time, the *tempus prolepticon*, a sort of time before time, and placed all events that occurred before Creation, such as the early Egyptian kings, in this "proleptic" time. So, Scaliger did not disregard the empirical cycle. Rather, he adapted his theory by introducing an imaginary era in what for us may seem a surprising move. But at the beginning of the 17th century, calling the Bible into question would have been unthinkable for a devout Calvinist, and yet this is what he did in an implicit way.

Scaliger's chronological dates from the earliest Egyptian dynasties were hardly accepted in his own time. The meticulous Gerardus Vossius (1577–1649) thought he could solve the problem of proleptic dating by assuming that various Egyptian dynasties were not consecutive but simultaneous (reigning in different places). But apart from an analogy with Babylonian history, he had no empirical evidence for his position. Vossius's proposal looks like a return to the principle of biblical coherence (see chapter 4.1), according to which all historical facts must be reconciled with biblical doctrine. Others, such as the theologian Jacob Revius, argued that everyone was wrong, referring to the usual biblical excerpts, while the Irish archbishop James Ussher once again argued in his *Annalium Pars Posterior* from 1654 that Creation had taken place at 6 p.m. on Sunday, October 23, 4004 BCE.[22]

But times were changing, and less than a generation later the powerful philological deduction pattern—the empirical cycle—was fully applied to chronology. In 1655, the French legal scholar and theologian Isaac La Peyrère (1596–1676) claimed that several Creations of humans had taken place at different times. So, there had been people living before Adam and Eve, the so-called *pre-Adamites*.[23] However, he seems to have pulled his claims out of a hat, stating, for example, that the Egyptian kings reigned for millions of years. Isaac Vossius (1618–1689),

Gerardus's youngest son, did provide philological argumentation. Instead of saying that humans had lived before Adam and Eve, he showed in *De Vera Aetate Mundi* from 1659 that the earth must have been at least 1,440 years older than previously assumed. Isaac was able to substantiate his argument with additional evidence from geographical studies and Chinese texts. His work became widely known in learned Europe.[24]

With the Dutch philosopher Baruch Spinoza (1632–1677), biblical criticism is elevated to a secular political philosophy. In his anonymous *Tractatus Theologico-Politicus* from 1670, he argues with unprecedented passion that the books of the Bible are texts that developed historically, created by people and transmitted at a specific time. The biblical method that Spinoza used for his purposes was based on the historically founded textual criticism of his illustrious philological predecessors.[25] In Spinoza's hands the destructive power of philology exploded: no text was absolute. He extended the results of philologists and historians to their limits and then claimed the right to free use of reason, without interference from theologians, with democracy emerging as the preferred form of government. Spinoza succeeded in deploying the historical-philological paradigm for a new, secular worldview, giving an important impetus to the 18th-century Enlightenment.[26]

In this context, Scaliger's discovery that world history was at variance with biblical chronology had far-reaching consequences. It came at the beginning of a chain of radical changes that led to a worldview in which the Bible could no longer be taken seriously as a historical source.[27] Whereas Scaliger failed to take this decisive step, Isaac Vossius and Spinoza did just that: they took the early Egyptian kings together with other historical actors out of the pre-Creation proleptic era and simply adjusted the age of the earth. The empirical cycle was restored, and the path to a new view of the world's age opened up.[28]

Yet Scaliger's dating of the Egyptian dynasties was contested well into the 18th century. Even Isaac Newton (see below), that icon of the Scientific Revolution who aspired to the most far-reaching integration of empiricism and theory in his scientific work, held firmly to biblical accounts in his chronological work, stating, for instance, as Gerardus Vossius had also said, that the pharaohs had existed simultaneously.[29] But these views gradually died out.

The Status of Philology—Female Philologists

Because of its impressive discoveries, philology achieved such a high status that no scientist could avoid the field. For example, many icons of the Scientific

Revolution, including Copernicus, Vesalius, Kepler, and Galileo, had training in philology (see below). The empirical cycle was thus transplanted almost directly from the humanities to the natural sciences, as we shall see.

Yet a philologist's main activity was not so much the development of new methods for reconstructing texts or of improved chronologies as it was in editing the classical works and making them available. In the Dutch Republic, meticulous editions of Latin authors were produced by philologists including Gronovius, Heinsius Sr. and Jr., and Vossius and his children. Gerardus Vossius's entire family was active in philology. When he was appointed a professor in Amsterdam, Vossius Sr. negotiated a large building with 20 rooms, enabling his home to develop into an international center where foreign students and visitors found a welcoming place to live.[30] The most talented of Vossius's children, apart from the aforementioned Isaac, were Dionysius (1612–1633) and Cornelia (1613–1638), both of whom died far too young. Cornelia was one of the first female philologists in the Dutch Republic, but she was preceded by brilliant Italian colleagues such as Isotta Nogarola (1418–1466), Alessandra Scala (1475–1506), and Cassandra Fedele (1470–1558). Yet when these 15th- and 16th-century philologists aspired to equality with their male colleagues, they were scorned. Their only options were seclusion or marriage.[31]

This changed over the course of the 17th and 18th centuries. As difficult as it was to become a celebrated philologist as a woman, some attained rock star status nonetheless, as if they were representing the classical author in person. Such was the happy lot of Anne Le Fèvre Dacier (1647–1720), better known as Madame Dacier, who achieved cult status in France because of her passionate translations and editions of Homer, Aristophanes, Plautus, and above all of Sappho's erotic poems. As a child, Anne was not instructed in the classics, but she quickly mastered Latin and Greek by secretly listening in on her brothers' lessons. And when she showed her father how much she had learned, she was immediately given the same training as her brothers. With her editions, she participated in the philological controversies of her time, often showing herself to be superior to her male colleagues. She became a living attraction who brought in visitors from far beyond France.[32]

Madame Dacier was not the only woman to achieve fame. The Dutch scholar Anna Maria van Schurman (1607–1678) also studied philology, and in 1636 she became the first woman to enroll at a Dutch university (Utrecht). She was required to sit behind a curtain so as not to distract her male classmates. Anna

Maria was dubbed the "Star of Utrecht," and rightly so: less than four years later, in 1640, she had mastered 14 languages and produced editions in Latin, Greek, Hebrew, French, Arabic, Persian, Ethiopic, German, and Dutch.

Women had no opportunity to pursue an academic career until the 20th century, with Italy being the only exception. Since the Middle Ages, women had been able to enroll at the University of Salerno, where the 11th-century Trota of Salerno had translated the Greek medical texts (see chapter 4.5). But credit for the first female professors goes to the University of Bologna. It was here the philologist Clotilde Tambroni (1758–1817) was appointed professor of Greek linguistics in 1793 and also of Greek literature starting in 1800.[33] Other female professors in physics and mathematics had gone before her (see below), after which Bologna became known as "Paradiso delle Donne."[34]

Starting in the 19th century, we find the most innovative work in philology not in translations or editions but in empirical philology, especially that of Karl Lachmann (1793–1851). His stemmatic philology, in which surviving texts are placed in a family tree (a *stemma*) on the basis of mutual differences and similarities, in such a way that one can derive the original text,[35] set an example for comparative linguistics, evolutionary biology, and even for genetics much later.[36] This impact of philology, which found its way into 20th-century DNA analysis via the stemmatic method, is almost forgotten. However, the interaction between philology and genetics is one of the most fascinating cases of cross-pollination between the humanities and the sciences.

Philology also served as the basis of the method of historical source criticism. Although the method can be traced back to humanist philology, it was Leopold von Ranke (1795–1886) who gave a systematic explanation of the application of philological criticism to historical material. It was not only the content of the source itself that needed to be subjected to a philological analysis but its external aspects as well, such as its form and the material on which it was written. Adherence to this method was intended to guarantee the historian's objectivity, to achieve Ranke's goal of reconstructing *wie es eigenentlich gewesen* (how it actually was).[37] Although this sort of historical objectivity is not considered feasible today, philologically informed source criticism does constitute the basis for testing the reliability of a historical document. Today, the method is included in history courses around the world with unprecedented impact. It is used in court cases, to establish the historical truth (such as identifying bogus sources in Holocaust research), and in the forensic sciences.[38]

China and the Empirical School: Gu Yanwu and His Followers

Analytical and skeptical studies of classical texts were also underway in China, especially during the Ming dynasty (1368–1644). Chen Di (1541–1617), for example, showed that Old Chinese had its own phonology with rules of pronunciation distinct from those of contemporary Chinese. In doing so, he refuted the existing practice of systematically changing the characters in ancient poems in order to preserve the rhyme.[39] While Chen was both a linguist and a philologist, Gu Yanwu (1613–1682) was mainly a philologist. He perfected Chen's work, using it as a basis for the study of the Chinese classics. Gu contended that such a study had to be preceded by philological, linguistic, and historical research. This is why he is considered the founder of the Empirical School of Textual Criticism, the Kaoju Xue, also known as the Empirical school.[40] Gu argued that a philologist and historian should use both internal and external evidence to determine the authenticity of a text. A judgment should be based on the highest possible probability by comparing as many sources as possible. Knowledge had to be derived from facts and independent observations not only of manuscripts; texts could also be compared with epigraphic remains, such as stone and bronze inscriptions. Although we have no direct evidence of an empirical cycle with Gu, it is clear that he was promoting a strong empirical approach. It is difficult to determine to what extent this approach also triggered a backlash to the theory, but such an effect certainly seems likely.

The Empirical school has some impressive philological discoveries to its name. For example, based on an analysis of Chinese characters and the pronunciation associated with them, Yan Ruoqu (1636–1704) demonstrated that 26 presumed chapters from the *Book of Documents* were actually counterfeits from the 4th century BCE.[41] And Cui Dongbi (1740–1816) analyzed to what extent Confucius actually compiled the works attributed to him, such as *Spring and Autumn Annals* and the *Book of Songs*. Doubt ensued, and today the attribution to Confucius is considered apocryphal.

One explanation for the rather sudden emergence of this empirical philological school in late imperial China can be found in the economic prosperity of the Yangtze Delta at the end of the 16th century. Merchants and intellectuals sought out ancient works of art, early manuscripts, and rare editions, often paying huge sums of money for a single manuscript. This encouraged the production of imitations and fakes, which in turn furthered the study of manuscript authenticity. There was also a renewed interest in (reprinting) classical works, which found their way to Vietnam, Korea, and Japan in large quantities.

It could be that Chinese empirical philology was boosted by the arrival of the Jesuits, who began introducing Western scholarship and science to China in the 16th century.[42] The Jesuits were obsessed with the dream of creating a Sino-Christian civilization similar to Roman-Christian civilization. Although they failed to spread the Catholic faith, the Jesuits did contribute to a scientific and scholarly exchange between China and Europe. The Jesuit Matteo Ricci (1552–1610) was even convinced that Confucian teaching included the monotheistic concept of a supreme being. Ricci believed Christian doctrine was already laid out in the Chinese classics. Although only a very small percentage of Chinese intellectuals converted to Christianity, Jesuit influence on the practice of science and scholarship in China was substantial, as will be apparent from their influence on Chinese astronomy (see below).

Whatever the precise contribution of the Jesuits may have been, the Chinese philological tradition is surprisingly similar to its European counterpart: the rediscovery of classical works and their reconstruction can be found in both regions, as well as counterfeits and their exposure. Although we have not found a formal theory of manuscript transmission in China like Poliziano's genealogical theory, material sources such as inscriptions are used in both regions. In addition to this striking similarity, there is a significant difference in the way philology was received: textual criticism does not seem to have brought about radical social changes in late imperial China in the way it had in early modern Europe, where philological historiography promoted a new secular worldview (see above). This is striking because the philological method could be just as "destructive" for Confucius as for the Bible. However, we must remember that the Chinese Empirical school operated under a single solitary (super)state, whereas European philologists were working in a variety of countries. When the situation became too precarious in France, Scaliger was able to move to the Netherlands. Conversely, Grotius was able to flee to France when he was persecuted as a Remonstrant in the Dutch Republic.[43] But where could someone like Gu Yanwu go? When intellectual repression in China took on grotesque forms in the 18th and 19th centuries under the Manchus, Chinese philologists and historians avoided topics that had any moral or political overtones. The Empirical school turned to annotations and commentaries that soon led to academic hair splitting.

Chinese Chronologies: Li Zhi's Relativism

As in previous eras, dynastic chronology was the principal historiographic activity during the Ming dynasty. Court historiography was carried out in a very dry way by the Bureau of Historiography (Shi Guan). Yet during the Ming era, one historian clearly came to the fore: Li Zhi (1527–1602).[44] In addition to his many biographies focusing on the dynastic histories, Li sharply criticized earlier historical works because they were subject to the norms and values of their compilers. For this reason, past judgments needed to be revisited and revised. He showed how historical figures who had been considered depraved in early works can quite simply and utterly convincingly be portrayed as heroes in his own biographies. Considering his method, we can characterize Li as a historical relativist. Li even dared to question Confucius's position, leading to a prison sentence and ultimately to his suicide in 1602. Li would not be rehabilitated until the People's Republic.

African Chronologies: Djenné, Timbuktu, Kilwa, and Pate

Over the past 60 years, a tremendous wealth of manuscripts has emerged from the Niger Valley, of which the manuscripts from Timbuktu are the best known.[45] These works, which number in the hundreds of thousands, were written in a variety of languages, including Tuareg (Tamasheq), Songhai, and Fulani.[46] They deal with an extremely diverse array of matters, including theology, linguistics, logic, astronomy, and musicology. But because of their vast chronologies, it is the historical works that appeal most to the imagination. Two chronicles stand out in particular: the *Tarikh al-fattash* from Djenné with a history of the Songhai Empire, and its continuation, the *Tarikh al-Sudan* from Timbuktu. Djenné and Timbuktu were among Africa's major intellectual centers. Djenné was known for its architectural opulence (including its famous adobe mosque), while Timbuktu had the largest mosque schools and libraries south of the Sahara. In 1550, the Andalusian traveler and merchant Leo Africanus wrote in his *Descrittione dell'Africa* about Timbuktu's fabulous wealth. The city maintained contacts with book markets in Morocco and Spain, and Ibn Khaldun's works (see chapter 4.1) and many other writings were available.

The *Tarikh al-fattash* chronicle was written by three generations of the Kati family in Djenné, whose library was recently discovered. In 1519, Mahmud Kati initiated the chronicle that was completed by his grandson around 1591. The work provides a summary of the Songhai Empire up to the Moroccan conquest in 1591. As with Polybius (see chapter 3.3), the contemporary part of the chronicle is based on personal experience, while the earlier historical periods described are based on (centuries of) oral tradition that, as was usual in this region, was maintained by the family itself. The *Tarikh al-Sudan* of Abderrahman al-Sadi from Timbuktu includes the later history of the Songhai Empire until 1655.[47]

This form of chronicle based on a combination of personal experience and oral tradition spread from Djenné and Timbuktu to the south and west. And there, the long-existing orally transmitted king lists, biographies, clan genealogies, and local chronicles were written down over the course of the 18th century in Arabic or one of the local languages. The *Kitab al-Ghunja*, a chronicle of the Gonja kingdom on the northern Gold Coast (modern-day Ghana), is one of the most compelling examples of this tradition. The manuscript can now be admired at the Institut Fondamental d'Afrique Noire library in Dakar, Senegal.

East Africa has a similar tradition, with chronicles written in either Arabic or Swahili (but in Arabic script). Most chronicles cover the history of individual coastal cities, such as the *Chronicle of Pate* and the *Kitab al-Sulwa fi Akhbar Kulwa* (Chronicle of the city of Kilwa). Pate is located in present-day Kenya and was an important trading post in eastern Africa, in competition with the Portuguese until the end of the 18th century. Kilwa is on an island off the Tanzanian coast and was the largest port city on the Indian Ocean from the 9th to the 19th century. The city is now famous for its medieval ruins such as the Husuni Kubwa palace, with its public market, housing complexes, mosques, city walls, and cemeteries. The *Chronicle of Pate* survives only through oral tradition (the original text was lost during the English conquest). The Kilwa chronicle was written between 1520 and 1530 and was translated into Portuguese as early as 1552, possibly due to the city's strategic importance. The famous 17th-century poem *Paradise Lost* by John Milton refers to Kilwa (i.e., Quiloa).[48] In terms of methodology, the chronicles of both Pate and Kilwa are strikingly similar to those from western Africa: they combine the oral tradition of ancient king lists and genealogies with the authors' personal experiences.[49] Oral culture—alongside written culture—was and still is rich and prestigious in Africa.

Ethiopia also has an extensive tradition of chronicle writing, as we saw in chapter 4.1. From the ascension to the throne of the Solomonid dynasty in 1270, the kings' chronicles have been a continuous line, initially written in Ge'ez and subsequently in Amharic. According to tradition, the genealogy of this

Solomonid dynasty can be traced back to Menelik I, the son of Solomon and the Queen of Sheba, as described in the *Kebra nagast.*

These chronicles represent just the very tip of the iceberg. Many manuscripts from Senegal, Ghana, Nigeria, Cameroon, and other regions have yet to be inventoried. In the region around Timbuktu alone, the number of manuscripts is estimated at 700,000, usually kept by families. Several thousand manuscripts have been cataloged by the Ahmed Baba Institute.[50] The vast majority are still waiting to be made accessible, and the most urgent matter is rescuing and conserving existing manuscripts. It is slowly becoming clear that African written culture has been underestimated for centuries, a situation due in no small part to European colonial prejudices.[51] Although only a fraction of the chronological works are currently accessible, it can be concluded from the African chronicles available thus far that they hardly ever used the formal isnad method of tracing how sources are transmitted (see chapter 4.1), regardless of the extent to which the regions concerned were under Islamic influence. Instead, we find a combination of personal experience (for contemporary historiography) supplemented with an even more intensive use of oral tradition (for earlier historiography), which can be traced to the enormous prestige that oral tradition had and still has in all parts of Africa. Age-old knowledge of the past was preserved by "specialists of the word" (the *griots*) as well as by broad segments of the population.[52]

The value that was and is given to oral history in Africa immediately drew the attention of the first Western anthropologists to visit the continent.[53] This led to a scholarly discussion in Europe and the United States about the value of "oral history" and "life histories" at a time when the Rankean paradigm of textual sources was dominant (see above). It is not implausible that the discovery of oral history outside of Europe, especially in Africa, was what sparked interest in oral history, for example among the Chicago school of sociologists in the 1930s.[54] In any case, research into life histories and oral history was incorporated into 20th-century historical scholarship, an intriguing example of African influence on European knowledge practices.

India, the Mughal Empire: The Present as a Variation on the Past

Chronicles were also written in the Islamic Mughal Empire in the form of universal histories (from Khvandamir around 1528) and court chronicles (from

Abu'l-Fazl around 1600). Under the third Mughal, Akbar (1556–1605), a historical method emerged that resembles the Chinese court chronicle with its cyclic patterns.[55] Abu'l-Fazl (1551–1602), a minister under Akbar, brought together many sources in his *Akbarnama* (Book of Akbar). He believed he could trace Akbar's parentage back to Adam through his family tree on the basis of 52 biographies.[56] The book was illuminated with the famous Mughal miniatures by as many as 49 artists from Akbar's studio. The hagiographic character of the Mughal chronicles is especially evident in the *Padshahnama* (Chronicle of the king of the world) about the fifth Mughal emperor, Shah Jahan (died 1666), best known as the builder of the Taj Mahal. The chronicle is nearly 3,000 pages long and gives an account of almost every public appearance by the emperor, all richly illustrated by the best imperial artists.[57]

The Mughal method is mainly prescriptive: the description of the present always includes earlier texts, usually in the form of summaries, with the aim of presenting *the present as a variation on the past*. The past serves as a primal model used to interpret the present. Unfortunately, this procedure and the resulting cyclic variations and patterns hardly got a chance to crystallize: after the death of Emperor Aurangzeb in 1707, the Mughal Empire became bogged down in chaos, taking with it the Indian practice of writing court chronicles.

Chronicle Writing in Pre-Columbian America

The tragic commonality of the intellectual activities of pre-Columbian peoples such as the Incas, Aztecs, Mixtecs, and Mayans is that they disappeared within a few generations of the arrival of the Spanish conquistadors. Not only did the Europeans wage a terrible war; they also brought along viruses to which the pre-Columbian peoples had no resistance. Successive outbreaks of smallpox, typhus, flu, diphtheria, and measles wiped out 95% of the native population in a short time.

The surviving pre-Columbian chronicles are pictorial rather than textual, and they constitute one of the most spectacular forms of historiography. An example of this is the 15th-century *Codex Zouche-Nuttall*, written in Mixtec glyphs, which tells of the achievements of the 11th- and 12th-century Mixtec rulers and their alliances with other peoples. The codex can be unfolded like an accordion and has a total length of more than 36 feet (11 meters).

As for the Aztecs' pictorial histories, they are mainly post-Columbian.[58] The *Codex Mendoza*, for example, combines text and images to tell a story dating from 1541 in which the conquests of the Aztec rulers are described, as well as everyday life. The *Aubin Codex* is even more impressive: in the form of a beautifully colored pictorial narrative, it tells of the departure of the Aztecs from Aztlán as a result of Spanish rule. This 81-page historiography contains one of the most dramatic moments in Aztec history: a testimony of the massacre and destruction of the temple in Tenochtitlan in 1520.

Now that the Mayan script has also been almost completely deciphered, it has become clear Mayan historiography is extensive, complete with dynastic chronicles, life histories, and descriptions of political controversies and battles.[59] One of the most important works of Mayan historiography is the historical-mythological book *Popol Vuh* of the Quiché people of Guatemala.[60] Pre- and post-Columbian chronicles are sometimes compared to medieval European chronicles, where "myth" and "history" do not divide along clear-cut lines. The notion of truth in these chronicles is what is true for the writer: regardless of whether a narrated event actually took place. However, we know nothing about any critical treatment of these Meso-American sources with respect to their truthfulness in their own time.

Polynesian Genealogies

The first written sources in Polynesia date from the time after the arrival of Europeans in the 17th century, with a notable exception: the wooden tablets from Easter Island written in rongorongo script. Rongorongo has yet to be deciphered, but from the few fragments that have been decoded, we can deduce that the 17th- to 18th-century tablets contain a wealth of genealogical knowledge (as well as astronomical information, see chapter 4.2).[61] But at this point, there is still little to be said about the nature of the historiography produced on Easter Island. The same goes for the rest of Polynesia, where we do not find local history until the relevant languages were equipped with writing systems after the arrival of the 19th-century European colonizers. In any case, this development of writing systems occurred too late for Easter Island, considering that most of the native population was decimated and deported in the 19th century, including the last people who could read rongorongo. For the Tonga Islands we do have some early records of the oral genealogies. These were compiled after the arrival of the Dutch explorers Willem Schouten and Abel Tasman in the 17th century.

Linguistics: The Empirical Cycle in Comparing Languages

Humanist Grammars for the Vernacular

In the 15th century, European linguistics partly overlapped with philology. Lorenzo Valla's Latin grammar, the *Elegantiae*, was actually a philological exercise, and the Latin grammars of Thomas Linacre (1460–1524) and Julius Caesar Scaliger (1484–1558), the father of Joseph Justus Scaliger (see above), were situated within the humanist program to revive antiquity. Of course, humanism with its *studia humanitatis* also strove to liberate the linguistic *artes liberales* from the university curriculum. But it remains striking how frequently medieval achievements got swept under the rug. Linacre and Scaliger, for example, peppered their grammars with Modist terms while claiming to be based on Varro and Priscian. References to Modist linguists such as Roger Bacon (see chapter 4.4) are lacking.

Grammars of vernacular languages also underwent a profound change due to humanist influence. The main aim was often to show the extent to which these grammars corresponded to Latin, the rules of which were considered "universal." One of the first examples of a humanist vernacular grammar was Leon Battista Alberti's *Grammatica della lingua toscana* from 1437–1441.[62] The background for Alberti's grammar was typically humanist. A controversy arose among four humanists, Bracciolini, Valla, Guarino, and Filelfo, as to whether Classical Latin was spoken universally in antiquity.[63] That is, was Classical Latin the language of the male elite in antiquity, or was it also used by women, children, and slaves? Since the humanists themselves struggled to master the rules of Latin, it seemed unlikely to them that this language could have been spoken by everyone. The vast majority must have spoken a simpler vernacular, they thought, the rules of which were arbitrary, while the higher language of Classical Latin had a complex system of rules that required scholarship to master.

Since no text could settle this issue, Alberti tackled the problem from a different angle. He showed that the rules for inflecting verbs in *spoken* Tuscan were just as consistent as those in Latin, including (almost) the same categories for temporal concepts. So while a vernacular might appear arbitrary, upon closer inspection it was just as "precise" as Classical Latin. There was thus no reason to believe that knowledge of Latin would have been limited to a small group of learned speakers. It is clear that here Alberti was giving voice to the empirical facts rather than to "theory."

A New Syntactic Theory Based on Four Operators: Sanctius

In retrospect, Franciscus Sanctius Brocensis (1523–1600), working in Salamanca, has proven to be one of the most influential linguists. With him we see a profound interest in syntax for the first time in centuries, especially from the perspective of logic. In his *Minerva seu de Causis Linguae Latinae*, written in 1587, he states that "the sentence—or syntax—is the object of grammar." In this work, Sanctius forges the grammars of Linacre and Scaliger into a new whole, while one seems to detect the influence of Sibawayh's descriptive grammar—the Arab legacy was still palpable in 16th-century Spain. But Sanctius's grammar is more than the sum of the parts of earlier grammars. In his *Minerva*, he presents a new syntactic theory based on four operators (including Sibawayh's *substitution*) that can be used to construct all sentences:[64]

(1) *Substitution*: makes it possible to substitute words with other words or phrases.
(2) *Deletion*: allows words or phrases to be omitted, as in the shortened phrase *The dog wants out*, which can be derived from *The dog wants to go out*.
(3) *Addition*: can govern inflections (of adjectives, for example) and conjugations (of verbs) when they take the form of prefixes and suffixes.
(4) *Permutation*: can rearrange words or phrases, as commonly happens with compound words, such as the Latin *mecum* meaning *cum me* (with me).

These four operators are bound by rules: not everything can be substituted, deleted, added, or permuted. The rules differ from language to language, but the four operators (the principles, in our terminology) used by the rules are the same for all languages. Sanctius provided rules for a number of syntactic phenomena in Latin.

Sanctius's Theory Catches On: The Port-Royal Grammarians

Sanctius's *Minerva* went unnoticed for a long time, until it was rediscovered in France in 1650 by Claude Lancelot (ca. 1615–1695), a Port-Royal linguist. Port-Royal was a 13th-century monastery near Paris that attracted some Jansenist intellectuals in the middle of the 17th century. The most important among them were Claude Lancelot and Antoine Arnauld (1612–1694). Sanctius's influence in the Port-Royal group is particularly recognizable in Lancelot's pedagogical grammar, the *Nouvelle méthode pour facilement et en peu de temps comprendre la langue latine*, which was even used by Louis XIV to learn Latin at the age of six. The rules of grammar were presented in rhyming form, and everything was geared to making learning as easy and pleasant as possible. When Lancelot came into contact with Sanctius's *Minerva* after the second edition, he decided to completely rewrite his grammar, making it four times longer. To his own surprise, he had largely overlooked Latin syntax—as was the case with most Roman grammarians but not with their medieval counterparts. In the preface to the longer, third edition of his *Nouvelle méthode*, Lancelot writes, "Sanctius deals mainly with the structure and connection of speech, which the Greeks called 'syntax,' which he explains in the clearest way, and reduces it to its first principles."[65] It was because of Lancelot that the Port-Royal linguists became aware of the idea that certain "initial principles" underpinned both word forms and sentences in a language.

The Empirical Cycle in Comparative Linguistics: From Sassetti to De Laet

The first comparative studies of languages begin with the Florentine merchant Filippo Sassetti. During his stay in India in 1585, he established several similarities between Sanskrit and Italian, such as *deva/dio* for "God," *sapta/set* for "seven," and *nava/nove* for "nine."[66] Methodical principles for language comparisons were introduced a generation later by the Dutch polyglot Johannes De Laet (1581–1649). De Laet was one of the founders of the Dutch West India Company, but he spent more time on his ethnological and linguistic interests than on running the company.[67] Before him, Hugo Grotius had argued that the Native American languages must have Hebrew influences considering that all humans descended from Adam and Eve. In his *Notae ad Dissertationem Hugonis Grotii* (1643), De Laet refutes Grotius's view, stating that Native American languages bore no affinity with Hebrew, Greek, Latin, or any modern European language.[68] In De Laet's time it was not uncommon to compare languages, but this was done only in an associative way. Like Grotius, De Laet was schooled under Joseph Scaliger (see above), who, in his *Opuscula Varia* (1610), published posthumously, divided the European languages into 11 families: four major and seven minor ones. The four major language families correspond to the Romance, Greek, Germanic, and Slavic groups recognized today. However, Scaliger's language comparisons were based on just a few words in these languages, especially the word for "God." For this reason, he spoke of *Deus* languages, *Theos* languages, *Godt* languages, and *Boge* languages, respectively. But now, De Laet showed that this type of comparison could not hold up for all words, and he argued for stricter criteria on word comparisons that could be tested unambiguously, where the testing could again have an impact on the underlying hypothesis. To this end, he developed a precise interpretation of the preexisting notion of *permutatio litterarum*, which allows one to equate words from different languages. De Laet formulated two criteria:

(1) A quantitative criterion that the relatedness of languages can be established only if a sufficiently large number of words was included in the comparison.
(2) A qualitative criterion that every claim concerning language kinship had to be supported not only on the basis of phonology and the lexicon, but also on a syntactic basis.

Using these two criteria, he was able to refute Hugo Grotius's position that all languages had descended from Hebrew. Across Europe, the origins of Native American peoples now became a topic of discussion, and De Laet's "evidence" was cited avidly. For example, La Peyrère referred to De Laet's work when defending his renowned thesis that human civilization preexisted the Christian date of Creation (see above). And Isaac Vossius made ample use of De Laet's insights to support his argument that the earth had to be older than the age deduced from the Hebrew Bible. So, alongside Scaliger's philological-historical work, La Peyrère's speculative work, and Vossius's geographic work, De Laet's linguistic work is also part of the complex chain of 17th-century transformations that culminate in a new secular worldview where theologians no longer have the final say.

Art Theory: An Even Older Empirical Cycle?

Interaction between Theory and Empiricism in the Study of Art: Alberti

In the study of art, we encounter the interaction between empiricism and theory in the first half of the 15th century with the humanist Leon Battista Alberti (1404–1472).[69] Alberti's *De Pictura* from 1435 was the first art-theoretical work in Europe since antiquity, and it immediately became one of the most influential works in the history of knowledge.[70] Virtually every idea in this book was adopted and developed in the centuries that followed. As Alberti himself emphasizes, his work is not a history of art like that of Pliny, but a theoretical work. In the best humanist tradition, he endeavored to develop a theory for the ancient rules of art, which he thought he could find in illusionism (see chapter 3.3). But in the elaboration of these rules, he crafts a theory that far exceeds classical art. He develops a method for the illusionist representation of three-dimensional objects on a two-dimensional plane.

In this way, Alberti provides a theoretical foundation for an empirical practice that had existed in Florentine art for at least 10 years: *linear* (or *mathematical*) *perspective*, which he ascribes to the sculptor and architect Filippo Brunelleschi, to whom he dedicates the Italian translation of his work. Alberti is partly prescriptive in his treatise, such as when he states that painting should comply with laws of linear perspective rather than with the rules of thumb developed in the studios (paragraph 19 of *De Pictura*). But at the same time, he is descriptive when addressing an existing tradition: linear perspective had already been used in Donatello's relief *St. George and the Dragon* in 1417 (Bargello, Florence), as well as in Masaccio's fresco *The Holy Trinity* in around 1425 (see figure 10).[71] But besides a description of perspective, Alberti also provides a geometric foundation for the technique. This could be seen as the underlying hidden rule that an artist like Masaccio had already applied and was now precisely defined.

Alberti's analysis of perspective is one of the clearest explanations in the literature: an image of reality should be constructed in such a way that it resembles "the view from a window." This view corresponds to the pictorial image, while the window corresponds to the pictorial plane (the painting). Alberti explains his method using imaginary lines that connect the artist's eye to the subjects in the image and, when intersected by the pictorial plane (the window, the painting), yield the depicted composition. Here Alberti applies the 11th-century Arab scholar Ibn al-Haytham's knowledge of optics (see chapters 4.2 and 4.3) that had been translated into Latin and widely disseminated through the work of 13th-century Franciscans including Roger Bacon.[72] Alberti achieved the *illusion* of a "view from a window" by bringing all lines that perpendicularly intersect the pictorial plane into a single point on the horizon (the so-called *vanishing point*). This method allows artists to determine the relative size of the objects depicted in the image.

Theoretical versus Empirical Perspective: Da Vinci

Alberti's theory of perspective was a resounding success and was immediately adopted by the artists of his time. In 1450, Piero della Francesca even wrote a

Figure 10. Masaccio, *The Holy Trinity*, ca. 1425, Santa Maria Novella, Florence. Photo by John Spike; https://commons.wikimedia.org/wiki/File:Masaccio,_trinit%C3%A0.jpg.

follow-up treatise, *De Prospectiva Pingendi*, in which he made Alberti's method more accessible by projecting a number of key points in the image onto the plane one after the other. All Italian artists keeping abreast of current developments quickly adopted linear perspective, sometimes using quite complex constructions. Alberti's law of perspective initially seemed like a success story that warranted no objections. However, anyone who looks closely at paintings based on pure linear perspective cannot avoid the perception that there is something artificial about images produced using precise application of these laws. But even worse, Alberti's method turned out to be "wrong" for wide-angle perspectives, as would be demonstrated by Piero della Francesca.[73] This situation somewhat resembles the Pythagorean theory of consonant intervals (see chapter 3.3): such intervals are mathematically defined as whole-number ratios, but if we judge with our ears rather than with mathematics, the consonant intervals sound better if the pure ratios are slightly adjusted upward or downward (depending on the instrument and the interval), as proposed by Aristoxenus.

This also appears to be the case with linear perspective. What Alberti could not have foreseen was discovered by Leonardo da Vinci (1452–1519): an image of three-dimensional objects on a two-dimensional plane is not necessarily correct if it scrupulously follows the mathematical laws. Da Vinci still believed that laws underpin the science of visual imaging but just that they were "different" laws. This was partially true: the insufficiency of Alberti's approach for wide-angle views was solved mathematically. However, what da Vinci undertook was a search for the principles underlying the many perceived patterns of perspective. Da Vinci was extremely systematic in this endeavor, as we can see in his *Trattato della pittura* (published posthumously).[74] He investigated all possible perceptual changes that occurred when the relative positions of the objects, the pictorial plane, and the observer were varied. He worked with differences in color and shape, developed machines that could draw in perspective, and tried to take into account how the eye works. As a result, da Vinci no longer embraced Alberti's linear perspective but became aware of the illusions and complications of the visual process. He did not abandon the laws of the mathematical perspective but tried instead to compensate for their shortcomings with small, experimentally derived modifications to make the result look more credible. He also added some achievements of *non*linear perspective, such as the gradual shift of light and shadow as objects become more distant. Although da Vinci did not develop a simple alternative to perspective theory, we do see an empirical-theoretical integration in which new assessments always lead to modifications to the underlying theory, which are then expanded, resulting in an increasingly complex system of rules that are constantly tested ad infinitum. Although da Vinci's system of principles for perspective did not have the elegant simplicity of Alberti's, using the empirical cycle he succeeds in repeatedly testing principles of perspective against observed patterns and improving them.

In the 16th century in northern Europe, the new understanding of linear perspective became well known. Painters in the Southern Netherlands (Flanders) like Jan van Eyck and Rogier van der Weyden were already using an approximation of linear perspective in the 15th century, but they lacked a mathematical foundation and a theory of art deducing principles from patterns. Thanks to Albrecht Dürer's visit to Italy, the theory of perspective was taken north in 1506. However, in his *Four Books on Measurement* (*Underweysung der Messung*), written in 1525, Dürer did not adhere to Alberti's explanation in all aspects, and we could even speak of a construction error that by chance led to a perspective that was empirically more effective.[75]

Musicology: Experiment and Theory in the Study of Music

A Renewed Battle for Consonant Intervals

It is not only in philology, linguistics, and art theory but also in the 15th-century study of music that we encounter patterns in deriving the principles underlying consonant intervals that point to the empirical cycle. As a consequence of the study of Greek music theory by the Italian humanists, renewed attention arose for "harmonious," or consonant, intervals. Pythagorean music theorists held that consonant intervals corresponded to simple ratios of the first four whole numbers, that is 1:1 for the unison, 1:2 for the octave, 2:3 for the fifth, and 3:4 for the fourth, and there were no others (see chapter 3.3). However, this law sparked controversies for intervals that sounded "more or less" consonant, such as the third and the sixth, but that in Pythagorean tuning were considered to be dissonant. Aristoxenus, on the contrary, stated that it is not theory but empirical facts (human hearing) that should have the final say. The controversy was articulated in Boethius's standard work, but this thread would not be picked up again in the postclassical period. We do see extensive musical systems of rules, such as for the organum (see chapter 4.4), but these were based on the Pythagorean consonant intervals.

However, when classical music theory was taken up again in the 15th century, a renewed interest in the laws of harmony arose. Meanwhile, the third and sixth were introduced into music composition, especially in the work of the composer John Dunstaple (ca. 1390–1453), who created elegant harmonies with them. A blind embrace of Pythagorean music theory had become questionable, to say the least, and in the second half of the 15th century, humanists resumed their quest for the theoretical and empirical underpinnings of harmonies. Whereas in the medieval *artes liberales* curriculum the study of music was categorized with the mathematical disciplines (the *quadrivium*), in 15th- and 16th-century Europe this discipline was taken up by humanists. Almost 1,500 years after Ptolemy, one of the oldest questions in learning and science was revived: Is there a system underlying the consonant intervals? Ptolemy's view (which, as noted in chapter 4.4, can be traced back to Aristoxenus) became the focus of attention: because music is a human experience, judgments about consonant intervals should be made using human hearing, assisted by reason. This contrasted with the Pythagorean view, according to which only reason should make the final judgment, since the senses are too easily deceived.[76]

Bartolomeo Ramis de Pareia (ca. 1440–1491) was one of the first humanist music scholars to experiment with the monochord, the single-stringed instrument. In his *Musica Practica* (1482), Ramis reports the problems he finds with Boethius (see chapter 3.3)—he doesn't seem to have read Ptolemy—and then discusses the "imperfect consonances" that correspond to intervals beyond the Pythagorean.[77] While Pythagoras based the consonant intervals on simple ratios of only the first four whole numbers, Ramis goes all the way to the number eight. This evokes fierce criticism from the dyed-in-the-wool Pythagorean Franchino Gaffurio (1451–1522), who considered these intervals irrational. Yet Gaffurio could not deny the practical problem with the Pythagorean system, considering existing music practice. In fact, related to the problem of the consonant intervals, there was the problem of tuning instruments, which could also be based on either hearing or mathematics. However, practice was so unruly that Gaffurio in his *Practica Musicae* (1496) acknowledged that tuning sounded acceptable only if the fifth were allowed to deviate slightly from the pure mathematical ratio.[78] In contrast, a few generations later, Gioseffo Zarlino (1517–1590), the famous chapel master of San Marco in Venice, seems to help the Ptolemaic vision gain the upper hand. In *Le istitutioni harmoniche* (1558) he proposes setting a limit for consonance, the *senario*. This "scenario" corresponds to the first six integers, 1 through 6, accounting for all natural intervals.[79] The thirds and the sixths could now also be represented using the acceptable ratios of 4:5 and 3:5, respectively. He justified his use of the number 6 on the basis of extensive theoretical and practical arguments, but the most important argument was that 6 is the first "perfect" number, the sum of whose divisors is equal to the number itself (1 + 2 + 3). However, reality was headstrong, dooming Zarlino's proposal. His *senario* proved impracticable and was quickly dismissed. In 1585, for example, Giovanni Battista Benedetti demonstrated that it was impossible to sing polyphonically using it.

Vincenzo Galilei: Discovery of New Music Laws

Vincenzo Galilei (ca. 1520–1591), the father of Galileo Galilei, appeared on the scene amid all this musicological confusion. Vincenzo had a great reputation as a lutenist and composer, but he was also active as a music theorist. He believed that so-called equal temperament, which divides an octave into 12 equal parts, was the only solution for instruments like the lute. In addition, in his *Discorso intorno all'opere di Messer G. Zarlino* (1589), he argues that all intervals are natural, going so far as to propose an infinite number of consonant intervals. A solution to the problem of consonant intervals seemed more distant than ever. Vincenzo based his claim on a series of experiments he conducted with strings of different lengths, materials, thicknesses, and tensions, with which he hoped to prove Zarlino's theory of consonant intervals wrong. While Vincenzo may not have succeeded in the latter, the results of his experiments, especially the way he linked them to mathematical regularities, were of tremendous importance.[80] In his essay "Discorso intorno a diversi pareri che hebbono le tre sette piu famose degli antichi musici," published in 1589, Vincenzo discusses his experiments using strings on which he hung different weights. His main discovery was that a string's pitch increases in proportion to the square root of the weight attached to the string. So, for a note twice as high, a weight needed to be attached that is four times as heavy. This regularity may be the first nonlinear mathematical description of a phenomenon in the history of knowledge.

But Vincenzo went one step further: he argued that the pitch of an organ pipe increased in proportion to the cubic root of the amount of air that passed through it. This law is not so easy to reproduce, and Vincenzo may have conducted few or no experiments to support it but instead simply made a guess (even his son Galileo mentions only his father's string experiments and how they related to his quest for regularity; see below). In any case, Vincenzo posited not only one musical law but three: an interval is (1) proportional to the length of the string (which had been known since Pythagoras or earlier), (2) proportional to the square root of the weight attached to the string, and (3) proportional to the cubic root of the volume of air, as in an organ pipe. Vincenzo concluded, among other things, that the pure fifth is produced by the ratio 3:2 when it concerns the lengths of the strings, by the ratio 9:4 when weights are hung on strings of equal length, and by the ratio 27:8 for the volumes of organ pipes.

Now, one would expect Vincenzo to have shouted his laws from the rooftops, but that was not the case. He was convinced that they had been known to Pythagoras. Vincenzo's goal, like that of other humanists, was to revive antiquity, and this antiquity, according to the humanists, was of comprehensive, unsurpassed wisdom. The surviving sources of Macrobius and Boethius also mention that Pythagoras had experimented with strings of different lengths and with different weights. But according to these sources Pythagoras had found the same proportions in both cases: 2:1 for an octave, 3:2 for a fifth, and 4:3 for a fourth.[81] However, Vincenzo showed that these proportions applied only to lengths and not to weights, for which the proportions are 4:1, 9:4, and 16:9 for the octave, fifth, and fourth, respectively. In short, it was Vincenzo who was right, not Pythagoras. But here Vincenzo does something that would seem surprising to us today: instead of refuting Pythagoras's results, he argues that Pythagoras must have been right, but that something had gone wrong along the way as his works were passed down to us.[82] Vincenzo was convinced that Pythagoras in his wisdom knew all the laws of music, many of which had been corrupted or lost over the centuries. Vincenzo's idolization of antiquity was not unique to the humanists; we will see that even Isaac Newton attributed his universal law of gravitation to Pythagoras (see below).

Thanks to Vincenzo, for the first time in centuries we have a new quantitative musical law, one that has been confirmed many times: the pitch of a string increases proportionally to the square root of the weight attached. And although Vincenzo used this string law to an unfeasible end—to find a universal principle underlying the consonant intervals—and although he attributed his string law to someone else, it was nevertheless Vincenzo who reshaped music theory in a new way that was both empirical-instrumental and mathematical-theoretical. From a modern perspective it could be argued that Vincenzo's experiments pertain to the natural sciences rather than to musicology. After all, he worked with strings, weights, and vibrations. But the way Vincenzo framed his question was also historical. Together with other Florentine musicians and poets, he strove to reconstruct and revive the glorious days of ancient Greek music. This is what led him to consult ancient sources and connect his experiments to lutes and other instruments. Vincenzo's research was therefore both humanistic and experimental, characteristics that were by no means mutually exclusive, as we saw earlier in Alberti's and Leonardo's art-theoretical research into perspective and the philological research of Poliziano and Erasmus.

So the old cliché that humanists were interested only in ancient writings is incorrect. Many like Vincenzo experimented enthusiastically, applying the empirical cycle wherever possible. As his famous son Galileo would later boisterously report, the cellar of the Galilei home in Pisa was one big laboratory full of lutes and strings of the most varied lengths, materials, thicknesses, and weights.[83]

Still No Law of Consonant Intervals: From Vincenzo to Mersenne and Huygens

Back to the problem of the consonant intervals, with which Vincenzo's experiments had begun and for which he argued the only solution was to divide the octave into 12 equal parts. Now, a colossal theoretical problem arose: there was no whole number that would allow an octave to be divided into equal segments (there are indeed 12 notes, but their mutual ratios correspond to the 12th root of 2, which cannot be written as a ratio of two integers). Anyone looking for a theory of music based on integer ratios was aware of this problem. This shows how strongly music theory was hitched to the Pythagorean numerological yoke.

In the 17th century actual music practice comes to the rescue. Intervals previously perceived as completely dissonant were now becoming increasingly common in music composition. Even the (minor) seventh was no longer shunned by a composer like Claudio Monteverdi. This raised the question of whether the theory of harmony was universal, or whether it, like the theory of organum construction, was dependent upon time and location.

While there was no conclusive distinction between consonant and dissonant intervals, there could still be a gradual continuum. But here too, complications remained. For Marin Mersenne (1588–1648), the question of whether the fourth was more consonant than the third was one of the biggest problems in musicology because reason and sensory experience seemed so mutually contradictory.[84] Virtually all 17th-century scientists studied the theoretical or empirical justification for the degree of consonance, from Galileo, Kepler, Beeckman, Descartes, Wallis, Holder, and Huygens to Euler. I'll mention some of their ideas here, the main one being a new and shared understanding that consonance could no longer be linked to abstract integer ratios; it depended on the physical category of the vibration frequency (with the proportions remaining the same).[85] For example, according to Galileo's "theory of coincidence," consonant intervals arose when vibrations often coincide, pleasantly caressing our ears.[86] But this theory produces incorrect predictions more often than correct ones. Isaac Beeckman argued that dissonance was caused by pulses generated in the sound vibrations. John Wallis gave a physical analysis of harmonics in 1677 but did not get much further than Mersenne in justifying the continuity between consonant intervals. William Holder (who was also active as a phonetician and philologist) and Christiaan Huygens (see below) developed microintervals and even completely new scales, which did not lead to a conclusive law of consonant intervals either. Ultimately, it was music practice, and the consequent steady acceptance of "new" consonant intervals, that largely negated all the theoretical work.

Although unsuccessful in finding an absolute law of consonant intervals, the quest for a theory of consonants in the 15th through the 17th century was far from fruitless. For instance, it became clear that there was no hard distinction between consonant and dissonant intervals, thus refuting the ancient Pythagorean cosmic harmony.[87] Second, this search led to new laws, such as Vincenzo Galilei's string law. As a result, his son Galileo became familiar with experimentation and the empirical cycle from an early age.[88] Still, as we will see, musicology cannot take all the credit for the transfer of the empirical cycle from the humanities to the natural sciences. After all, the interaction between empirical observations and theory was also being used in philology, linguistics, art theory, and chronology. Whereas Galileo learned about the empirical cycle from musicology, we will see that Johannes Kepler learned about this interaction from philology and that Andreas Vesalius became familiar with it via art theory (see below).

China: Music History Remains a State Affair

While a fierce debate was raging in Europe about the nature of consonant intervals and little was being done in the area of music history, just the opposite was taking place in China: the history of music was being documented extensively while empirical musicology came to a near standstill. The most impressive work from the Ming period is Zhu Zaiyu's *Yuelü quanshu* (Collected works of music theory).[89] It provides a historical overview of the various music-theoretical achievements from China, including the famous and probably oldest theory of equal temperament by Cai Yuanding from the Song dynasty (see chapter 4.4). Zhu Zaiyu does not fail to criticize fellow music theorists, but he does not provide any new insights either. Another important work of music historiography is *Shenqi mipu* (Manual of the mysterious and marvelous), by Zhu Quan, the 17th son of the Hongwu Emperor.[90] It contains an anthology of 64 annotated Song-era compositions, and it gives detailed instructions on how the pieces should be performed, as well as commentary on their historical and theoretical aspects. The transition to the Qing dynasty in 1644 did not change much in Chinese musicology, and there is no indication of an empirical cycle like that in the Chinese philology of Gu Yanwu (see above).

India: Systems of Rules for Gamaka *and* Raga

The Indian musicological tradition remained rule based and declarative (see chapter 4.4): the preconditions for possible pieces of music were defined using principles, but there was no procedure for generating new compositions. This can be seen in the many 16th- and 17th-century tracts such as the *Sangita-parijata*, which gives rules for ornaments, called *gamaka*. However, this work is nothing compared to Somanatha's treatise *Raga-vibodha* from 1609, in which the endless variants of the *gamaka* are recorded using a new, almost pictorial notation system. We find manuals with extensive procedures for the composition of Carnatic music as well. For example, the 17th-century *Sangita sudha* explains how motifs are developed into melodic units, how they can be extended or shortened and then be combined into ascending or descending patterns. The work describes the complex structure of the raga, including the first and second expositions of motifs, the ways these motifs can recur and expand into increasingly long and wide chains (the *brikka* or *phirukka*). And finally, the book discusses how a virtuoso piece can be made to descend again step by step, so that a few phrases can bring it back to end on the root.[91] These works are incredibly fascinating, as they seem to define almost the entire space of a musical idiom. The texts have a pedagogical purpose, but it is not clear whether the systems of rules are descriptive or prescriptive.

Africa: Musicology, Polyphonic Singing Cultures

Although an enormous wealth of music has been produced in Africa, little musicology has survived. It is now known that the recently saved but extremely fragile Songhai manuscripts from Timbuktu (see above) also deal with the music of their time. Unfortunately, most Songhai texts are not yet accessible. The only African music histories currently available are from external sources, except for a few brief musical references in Ibn Khaldun's work. In 1596, while visiting Mozambique on his way to India, the Dutch merchant and historian Jan Huygen van Linschoten provided one of the oldest descriptions of an African musical instrument, a mouth harp, complete with an image. In the 17th century, Italian missionaries active in the African kingdoms of Congo and Matamba described the local music practice. Girolamo Merolla's *Breve e succinta relatione del viaggio nel Regno di Congo nell'Africa Meridionale* (Brief and concise account of the journey in the Congo Empire in Southern Africa) from 1692 is one of the most important sources of African music history. From these and other works, including Peter Kolb's description of South African music from 1719, it becomes clear that polyphonic singing culture is not unique to Europe. There is an ancient, rich polyphonic music in Africa, from the indigenous peoples of the Congo to the Khoikhoi.[92]

Ottoman Empire: Heir to Arab Musicology

Ottoman musicology is often dismissed as pedagogical rather than theoretical, but that is incorrect. Although we do not find music-theoretical explorations of harmony, there are studies on the underlying schemes of the Turkish melodic idiom. These schemes define the class of possible melodies and—as in Indian music—can be seen as a declarative system of rules for Turkish music. For example, the 14th-century Ibn Kurr and the 15th-century al-Ladhiqi define the various tonal steps needed to create a melodic contour. A later anonymous 17th-century work called the *Shajara* gives both melodic and rhythmic cycles for Turkish music. This work is similar to what al-Farabi did for the Arab musical cycles centuries earlier (see chapter 4.4).

The most important musicological work in the Ottoman tradition is a tract from circa 1700 by Dimitrie Cantemir (1673–1723), prince of Moldova. During his exile in Istanbul (1687–1710), Cantemir studied both Turkish language and music and wrote the work *Kitâbu 'ilmi 'l-mûsikí alâ vechi'l-hurûfât* (The book of the science of music through letters).[93] In this work he discusses the different classes of melodies based on simplified tone schemes without any indications of intervals. One could interpret these shortened tone schemes as a notational limitation, considering that musical notation was not widely known among the Turks. But with Cantemir it is more likely that he was aiming to depict the underlying schematic system of the Turkish music idiom. He was quite adept at notation and even developed a specific one for reproducing Turkish instrumental music, the *ebced* notation. It is thanks to this notation that hundreds of Ottoman musical pieces from the 17th century were preserved for posterity. The extent to which Cantemir tested his system of tone schemes and whether he modified it according to the outcome of these tests is unknown.

The Empirical Cycle in the Humanities

What was the state of the empirical cycle in the early modern humanities? In all the disciplines discussed here—philology, history, linguistics, art theory, and musicology—the empirical cycle came into use in the 15th or 16th century, but not in all regions. Outside of Europe, we have some evidence for the empirical cycle only in philology (in China), and possibly in historiography and musicology as well. So in the humanities, the empirical cycle seems to have been utilized mainly in Europe, where efforts to revive antiquity were stronger than they were elsewhere. In their enthusiasm, however, the humanists went well beyond the classics they admired: they developed new text reconstruction methods, improved historical dating, designed methods for establishing linguistic kinship, analyzed linear and empirical perspective, and discovered new string laws—all based on the cyclic interaction between theoretical speculation and empirical observation.

5.2 The Empirical Cycle Migrates from the Humanities to Astronomy

While the humanities at the beginning of the early modern period were flourishing as never before, the situation in 15th-century astronomy looked quite different. In Europe, these two domains were almost two different worlds. The 15th century was the century of humanism, and the humanists largely ignored the study of nature. It was only in the course of the 16th century that astronomers adopted the empirical cycle from the humanists, largely owing to the fact that they themselves had received training in philology.

In most books on the history of science, the 15th century is swept under the rug, treated almost as if it were a "middle age" unto itself, a century in which little of interest took place. However, this century was a period of flourishing with one discovery after another being made in the humanities if not in the natural sciences (see above). Without the pioneering contribution of the humanities, the rise of science in the 16th and 17th centuries would be almost inexplicable. Although the humanist contribution is increasingly recognized by historians of science, this recognition is often limited to their translations of classical texts, while their other innovations go unmentioned.[94]

Astronomical Tables Central: Peuerbach and Regiomontanus

Like other early modern scholars, Regiomontanus (1436–1476), whose original name was Johannes Müller, received a humanist education, first at the University of Leipzig and subsequently at the University of Vienna.[95] Together with his teacher Georg von Peuerbach (1423–1461), he subsequently focused on improving the commonly used *Alfonsine Tables* (see chapter 4.2). For centuries, these tables formed the basis for calculating planetary positions and solar and lunar eclipses using rules of thumb. But Regiomontanus was also instructed in Ptolemaic astronomy. On the recommendation of Peuerbach, he traveled to Rome with the Byzantine scholar Bessarion. There he gained access to the many manuscripts from Byzantium, including a copy of Ptolemy's *Almagest* (see chapter 3.2). Regiomontanus soon realized that all existing translations of the *Almagest* were unreliable, leading him to take the initiative to produce a critical translation complete with commentary. It would serve as the foundational text for later astronomers, including Copernicus, whose teacher had once been Regiomontanus's apprentice.

Although Regiomontanus was aware of the shortcomings of the Ptolemaic model, especially of the equant, in his short life he never developed his own alternative theory. However, he did show that the epicycles for the inner planets used by Ptolemy could be accounted for just as easily using eccentrics (which Ptolemy had demonstrated only for the outer planets). Regiomontanus's focus was rather on improving the tables and their calculation rules. In a way, Regiomontanus's approach, like that of his medieval predecessors, resembled the older Chinese astronomical models, but the latter were based on an explicit principle of "ultimate origin": the moment in time when all celestial bodies were set in motion and from which one could use an algorithm to calculate any future configuration (as with Liu Hong; see chapter 3.2). No such principle can be found in European "tabular astronomy," although with a bit of effort we might be able to distill at least some *arithmetical* principles. Regiomontanus was not acquainted with the work of his Chinese colleagues, but he was familiar with Islamic astronomy, a subject that he taught.

Regiomontanus was most innovative in astrology, where he developed an explicit system for calculating astrological houses.[96] A house is an area of the firmament within which a planet operates astrologically. Regiomontanus divided the equator into 12 equal parts and drew auxiliary circles through the resulting dividing points. The houses are positioned at the intersections of these auxiliary circles with the ecliptic. Together with the astronomical tables and astrological handbooks, such as Ptolemy's *Tetrabiblos*, Regiomontanus made more refined astrological predictions. Astrology disappeared from the field of science in the course of the modern era, a fact I will return to in the conclusion.

In addition, Regiomontanus also has a major feat to his name in tabular astronomy: on the basis of his tables, he published an astronomical almanac listing the dates and times he had calculated for future lunar and solar eclipses, among other things. Due to the recent invention of the printing press, his almanac was widely disseminated, reaching a readership that extended far beyond professional astronomers. For example, it was Regiomontanus's almanac that allowed Columbus to impress the indigenous people of Jamaica by "predicting" the lunar eclipse of February 29, 1504.[97]

Astronomical Models Central: Copernicus

Owing to Regiomontanus's improved translation of the *Almagest*, the shortcomings of the Ptolemaic model were brought to the attention of European astronomers in a penetrating way. It seemed to be only a matter of time before astronomers started endeavoring to improve the Ptolemaic system, as had their Muslim colleagues centuries earlier when the *Almagest* was critically translated into Arabic. However, the question arises as to why in early modern Europe the study of the *Almagest* led to a revolution in worldview—a transition from a geocentric worldview to a heliocentric one—while it had not in the postclassical Islamic world.[98] There is a complex of factors here, but as we have seen, the age-old, predominant Aristotelian worldview had come under increasing pressure in Europe, not least because of the "discovery" of the New World, which was not mentioned in any of Aristotle's works. Aristotle came under further pressure from the new insights of chronology regarding the age of the earth (see above). All this put the carefully forged unity between theology and philosophy seriously to the test. The search for an alternative model of the cosmos was considered by many scholars and scientists to be not only acceptable but urgent as well. All the more so because alternative models could be found in antiquity. Why limit oneself to the geocentric views of Aristotle and Ptolemy when heliocentric views of astronomers such as Aristarchus were also available (see chapter 3.2)?

The founder of the "new" heliocentric worldview, Nicolaus Copernicus (1473–1543), had received a humanist education at the Universities of Krakow, Bologna, and Padua.[99] His principle motive was not the refutation of the Aristotelian geocentric model. What didn't sit well with Copernicus and many other astronomers was the way Ptolemy implemented this geocentric worldview, especially regarding the equant. The equant is perhaps the most detested principle in the entire history of knowledge: practically all mathematical astronomers from the 4th century to the 16th pulled out all the stops to eliminate it by looking for a simpler model (see chapters 3.2, 4.2, and 4.3). Copernicus shared this ideal: his goal was to bring Greek astronomy back to its original purity with its uniform circular motions. And Copernicus realized that as long as the equant was used, the planets would move at nonuniform speeds (which, incidentally, was the whole point of the equant, allowing the planets to have a constant *angular* velocity; see chapter 3.2). But because the planets in the sky have different apparent velocities, with his ideal, Copernicus faced the colossal challenge of creating a model with uniform velocities of pure circular motions. Copernicus recognized that a heliocentric model would make this possible.

But a *conceptual* heliocentric model, as had been proposed by Aristarchus in antiquity, is not the same as a fully elaborated mathematical theory that can be used to make quantitative predictions. For the latter, it is extremely difficult to work out a single model that can account for all the planets. Copernicus ran into this problem as well: as early as 1514 he wrote an anonymous manuscript of about 40 pages, the *Commentariolus* (*Little Commentary*), which he distributed among his friends and in which he set out the heliocentric hypothesis. It was not until 20 years later, around 1534, that his opus magnum, *De Revolutionibus Orbium Coelestium* (*On the Revolutions of the Celestial Spheres*), was completed, presenting a completely new theory of planetary motions. But Copernicus was not yet ready to publish this work. A mere year before his death in 1543, his only pupil managed to convince him to do so. In the first version of this manuscript, Copernicus referred to Aristarchus's heliocentric theory. But where Aristarchus (or what was passed down to Copernicus through Archimedes) proposed a conceptual model, Copernicus presented a mathematical model in the tradition of Ptolemy, one that could be used to make precise predictions.

When viewed realistically, no observation argued in favor of Copernicus's heliocentric theory. His choice was aesthetic. In the style of Euclid, he posited a number of heliocentric axioms and worked out the propositions that could account for the planetary motions under the conditions assumed. Not only was Copernicus's model simpler than Ptolemy's in some respects; it also explained the apparent retrograde motions without introducing epicycles. However, Copernicus still needed epicycles to explain the other properties of planetary motion: while the first part of *De Revolutionibus* is beautifully constructed, the remaining five parts are a mathematical monster in which Copernicus eventually has to introduce even more epicycles than Ptolemy had. This is because the planets *do not* move at a constant speed, as would later be demonstrated by Johannes Kepler. So Copernicus's main axiom, which had been Plato's starting point (see chapter 3.2), proved incorrect from the very beginning! Even more problematic was the fact that when tested against the existing empirical data, Copernicus's model did not perform any better than Ptolemy's. Actually, the discrepancy between the empirical evidence and theory only grew larger, and thus we cannot speak of an empirical cycle when discussing Copernicus.

The empirical cycle started to be used substantially in astronomy only when the superior observations of the Danish astronomer Tycho Brahe (1546–1601) became available.[100] Although Tycho is best known today as a great astronomical observer, he also has an intermediate model to his name. This model resembles the system suggested in ancient times by Heraclides and Martianus Capella (see chapters 3.2 and 4.2). According to Tycho, the planets revolved around the sun, but this system—the sun and planets—in turn revolved as a whole around the earth (whereas with Heraclides and Martianus, only the inner planets Venus and Mercury revolved around the sun). Tycho reasoned that if the earth itself orbited the sun, the positions of the stars in the sky would have to change over the course of a year, since they would be observed at different angles. This parallax movement was not observed by Tycho, even with his unprecedentedly accurate observations of positions of planets and stars, so he concluded—fueled in part by religious considerations—that the earth did not move. As we now know, stars are so far away that their parallax motion can be observed only with powerful telescopes.

Although Tycho's hybrid system enjoyed considerable support as a compromise between the geocentric and the heliocentric frameworks—especially among Jesuit astronomers, who even exported it to China,[101] his accurate observations fueled further skepticism about the Aristotelian worldview. For example, Tycho's observation of the Great Comet of 1577 led to the insight that it was (much) farther away from earth than the moon was. This refuted Aristotle's premise that everything in the heavens above the moon was perfect and unchanging (chapter 3.2). Although this refutation did not mean that one also had to reject the geocentric system, the Aristotelian worldview came under increasing pressure.

With a model that was not more accurate than that of Ptolemy and that moreover did not align with the visible planetary movements (not to speak of its false premises), in the 16th century it seemed unlikely that Copernicus would unleash a revolution, but the pursuit of a model that seemed simpler and more aesthetically pleasing than previous models, and that above all could overthrow the Aristotelian worldview, had such appeal that it caught the attention of some of the most brilliant minds of the early modern period.

Kepler Brings the Empirical Cycle from Humanism to Astronomy

Without Johannes Kepler (1571–1630), heliocentrism could have suffered an early death. Initially, Kepler's education was thoroughly humanist.[102] He attended the Latin school and seminary at Maulbronn and received an excellent education in philology at the University of Tübingen. In addition, Kepler showed himself to be a brilliant mathematician, who also drew up horoscopes for his fellow students with the greatest of ease. In Tübingen he was introduced to both Copernican and Ptolemaic astronomy by Michael Maestlin, who himself was largely a supporter of Copernicus.

Although Kepler made significant contributions to both philology and astronomy, he is rarely, if ever, mentioned as a philologist in overview histories of science.[103] However, Kepler could compete with the best philologists and chronologists of his time. He reconstructed Tacitus's *Historiae* using translations, and he communicated with Joseph Scaliger, whose *De Emendatione Temporum* he even managed to improve (see above).[104] One of the highlights of Kepler's philological-chronological work is *On the True Date of Birth of Christ* (1614), whose full title reads, *De Vero Anno Quo Aeternus Dei Filius Humanam Naturam in Utero Benedictae Virginis Mariae Assumpsit*. In this text, Kepler shows on the basis of a combination of, among other things, philological and historical indications, that Christ must have been born not in 1 CE but in 4 BCE. The Christian era introduced earlier by Bede in western Europe was based on chronological sloppiness attributable to Dionysius Exiguus (see chapter 4.1). The influence of Kepler's discovery was tremendous: his dating of Christ's year of birth had implications for other historical dates, and it is considered valid today.[105] Despite this feat, Kepler's philological-chronological work fell into oblivion while his astronomical insights have become part of the canon of the history of knowledge. Twentieth-century historians didn't know what to do with intellectuals who, in today's terms, were both humanities scholars and scientists.

But for Kepler himself, one couldn't exist without the other. Philology, chronology, physics, mathematics, musicology, optics, and astronomy—Kepler worked in all of these disciplines, and he strove for the greatest possible accuracy everywhere. Admittedly, Kepler rebelled against the authority of the classics and humanist hair splitting (like Galileo; see below),[106] but in the way he went about his work, he was a follower of those same humanists. He generalized the empirical cycle of testing and adaptation from philology to other knowledge practices. For Kepler, textual and nontextual phenomena were not epistemologically different. Although we cannot rule out the possibility that Kepler reinvented the empirical cycle in astronomy, the fact that he was trained in philology and that he also used the empirical cycle in that field makes it plausible that he transferred this cycle from philology to astronomy.

In his first book, *Mysterium Cosmographicum*, written in 1596,[107] Kepler endeavored to explain the mutual distances of the five planets based on the five regular polyhedra (the so-called Platonic solids). Since it had been proven since Plato that exactly five of these regular polyhedra existed, Kepler proposed that this insight explained the number of the five planets in Copernicus's heliocentric hypothesis; after all, God thinks like a mathematician. In a sense, Kepler's theory was a continuation of the theory of the spheres but now with the sun in the center and with regular polyhedra as the spheres. Although Kepler's statement did not meet with much approval, his view was interesting enough to be noticed by Tycho Brahe, after which Kepler was invited to Prague, where Tycho served at the court of Rudolf II.[108] Tycho was reluctant to share his hard-won planetary observations, but he allowed Kepler to work on the portion of his observations of the orbit of Mars. Like the other planets, Mars moves across the sky not at a constant but at a variable speed. The challenge was to find the underlying principles that allowed the planets to move at a constant speed while also accounting for the well-known apparent movements in the sky (the patterns), whose speeds were not constant. This challenge had gone unmet since antiquity.

Using Tycho's accurate observations, Kepler quickly achieved success. Kepler was convinced that the sun affected the movement of the planets. For this reason, he searched not only for a mathematical model but also for a physical explanation. And if the sun wasn't exactly at the center of Mars's orbit—as suggested by the notion of the eccentric—Mars was sometimes closer to the sun and at other times farther. The influence of the sun on the motion of Mars could then cause Mars to move faster when it was closer to the sun and slower when it was farther away. With this reasoning, Kepler accepted the nonuniform motion—that is, inconstant—velocity of Mars in its orbit around the sun. This assumption was a step too far for most astronomers, but it was only a first step for Kepler, because when he tested his underlying model against Tycho's data, it turned out that his model was quite accurate—up to 8 arc minutes, about a seventh of a degree. But that was still not accurate enough, because Kepler knew that Tycho's observations were accurate to 2 arc minutes. Kepler took the accuracy of Tycho's data so seriously that he felt compelled to adopt a new principle: Mars's orbit could not be purely circular.

So what shape did the planet's orbit take? In Kepler's quest, we see the empirical cycle at work in its purest form. Kepler tried an egg-shaped orbit, but that did not give him the desired accuracy. Kepler continued to search and described his many trials and failures in great detail—from principles to patterns and back to the principles—until he stumbled upon the ellipse. Until that point he had simply assumed—erroneously—that such an obvious shape, which after all was one of the classical conic sections (see chapter 3.4), must have already been considered by earlier astronomers. After lengthy calculations—Kepler produced approximately 900 pages of them, devoted solely to finding an appropriate curve for the data—it turned out that what best matched the accuracy of Tycho's observations was an elliptical model. The result has been called Kepler's first law, although Kepler himself did not use the word "law." Kepler also discovered what has come to be known as Kepler's second law, or the area law, according to which a line segment joining a planet and the sun traverses equal areas during equal intervals of time. It follows that a planet moves faster when it is closer to the sun. These two laws were published in *Astronomia Nova seu Physica Coelestis* in 1609.

Ten years later, Kepler discovered a third law, which states that the ratio of the cube of the average distance r of a planet to the sun (the semimajor axis of the ellipse) and the square of its orbital period T, that is, r^3/T^2, is constant. Kepler published this law in 1619 in his *Harmonices Mundi*. The laws found by Kepler would still require an explanation based on underlying principles (and could better be called "patterns"). Although Kepler did look for such an explanation (see below), he considered the mathematical regularities he had found to be the expression of God's mathematical ingenuity and his plan for the cosmos.[109]

Kepler's pursuit of the highest accuracy achievable set a scientific standard that we seldom encountered before him. Although the practice of the empirical cycle already existed in 15th-century philology and art theory, Kepler's preoccupation with precision is of a different nature: in Kepler's view, the theory needs to be as accurate as the observations themselves. In philology it was sometimes possible to establish with certainty that one manuscript was older than another or that a corrupted word corresponded to an existing one, but in astronomy—and in chronology as well—accuracy is quantitative and therefore dependent on the accuracy of the measured positions of celestial objects (or of a point in time in chronology). Kepler's pursuit of maximum precision may be comparable to that of Leonardo da Vinci in his earlier quest for the empirical perspective (see above). But while da Vinci tested his perspectival models only against his own subjective judgment—something akin to letting merchants audit their own books—Kepler was able to test his models against data collected by someone else (Tycho Brahe).

Kepler realized that his application of the empirical cycle in astronomy brought about something new. Until then, in Europe, and before that in the Arab world, astronomy was mainly seen as a form of mathematics separate from natural science. It is for this reason that Kepler referred to his activities as celestial physics. He sought not only mathematical models but also underlying physical forces that could explain these models. He attempted this with the help of another deduction pattern: analogical thinking. What works well in one case to link patterns to principles can also work well in another case, and this recurring way of linking patterns and principles becomes a pattern itself. It is this sort of recurring pattern in deductions that constitutes analogical thinking. And the analogy that Kepler envisioned was that of a magnetic force, as described by William Gilbert in *De Magnete* in 1600.[110] Kepler posited that the sun could be thought of as a magnet that exerted a force on the planets in such a way that a planet's velocity was inversely proportional to its distance to the sun. In this way, Kepler linked his laws to an underlying principle analogous to magnetism. This was the first attempt to apply a form of physics to astronomy. Kepler summarized these ideas in his *Epitome Astronomiae Copernicanae* (*Summary of Copernican Astronomy*) from 1621.

Kepler's results were largely ignored by his immediate contemporaries—even those who were fond of the heliocentric worldview. Galileo Galilei (see below) did not have much sympathy for the idea that planets moved in elliptical orbits, but in astronomy Galileo primarily observed rather than performing calculations. With later, mathematically oriented astronomers, the main problem was that Kepler had distanced himself from the idea of constant ("uniform") motion.

For example, the French astronomer Ismaël Boulliau (1605–1694) accepted the notion of an elliptical orbit but replaced Kepler's second law with uniform movement,[111] which did not lead to the same degree of accuracy. Another astronomer, Seth Ward (1617–1689), proposed a solution that is among the most bizarre in the history of astronomy: he accepted Kepler's elliptical orbits but supplemented them with a Ptolemaic equant.[112] And then we have the anti-heliocentric astronomers, who often continued to defend Tycho's hybrid model, sometimes extended with Kepler's elliptical orbits (as was the case with Jean-Baptiste Morin in 1650). But no extensive empirical testing was involved here either. This changed when astronomers had the opportunity to observe the transits of Venus and Mercury across the sun as predicted by Kepler and to compare them with the predictions made by alternative models. In 1631 Pierre Gassendi (1592–1655) demonstrated that the observed transition confirmed Kepler's theory, while the prediction of the alternative Ptolemaic model was significantly inferior.[113] As more of Kepler's predictions came true, appreciation of his work grew.

After 1631, Kepler's *Epitome* quickly became the most widely used astronomical textbook. An increasing number of astronomers switched to Kepler's ellipse-based astronomy, while few adopted his physics-based astronomy. Giovanni Borelli (1608–1679) and Robert Hooke (1635–1703) again began to assume forces of attraction between the sun and the planets. But no one succeeded in creating an underlying formula for this force that would allow Kepler's laws to be derived and explained. Only Isaac Newton was up to that feat (see below).

Galileo's Spectacular Astronomy: A Comparison with Kepler

The most ardent supporter of the heliocentric worldview was Galileo Galilei (1564–1642).[114] Until the age of 45, Galileo was a relatively unremarkable and underpaid mathematician at the University of Padua. That changed when in 1609 he learned about the newly discovered telescope, for which Hans Lippershey from Zeeland had applied for a patent a year earlier.[115] In no time, Galileo built a better telescope himself. And when he aimed it toward the heavens, he made one discovery after another: the moon turned out to be covered in craters, the planet Jupiter had no fewer than four moons, the galaxy was not a nebula but consisted of thousands of stars, the sun exhibited spots, and later he also discovered that Venus had phases. The discovery that Jupiter was orbited by moons showed, for Galileo, that a (presumed) argument against Copernicus, according to which everything revolved around the earth, didn't hold water, putting further pressure on the Aristotelian worldview. In 1610, Galileo published these observations in a concise work entitled *Siderius Nuncius* (*Starry Messenger*), which made him famous throughout Europe.[116] When after initial skepticism his observations were confirmed by the prominent Jesuit astronomer Christopher Clavius (1538–1612), Galileo received a hero's reception in Rome.

But Galileo's rejection of the geocentric system went a step too far for Rome. Moreover, others showed convincingly that the phenomena observed by Galileo, such as Jupiter's moons and the phases of Venus, could be explained equally well using Tycho Brahe's geocentric system. In Tycho's hybrid model (see above), all planets revolved around the sun, but this system as a whole—the sun and planets together—revolved around a stationary earth. Tycho's model was near and dear to the hearts of the Jesuit mathematicians and astronomers because the geocentric model was consistent with church doctrine, but Galileo rejected this model. For him it had to be either Ptolemy or Copernicus, nothing in between. But Galileo overplayed his hand when he stated that the Copernican model was not only a correct calculation model but that it corresponded to reality itself. Without solid empirical evidence, that was unacceptable to the Catholic Church. Objections were raised that were both physical (the absence of the star parallax, already noted by Tycho) and theological (according to the Bible the earth was stationary), after which the pope forbade Galileo from defending and teaching the Copernican system.

But when in 1623 a friend of Galileo's was elected Pope Urban VIII, Galileo was given new hope and began his most controversial work, the *Dialogo sopra i due massimi sistemi del mondo, Tolemaico e Copernicano* (*Dialogue concerning the Two Chief World Systems, Ptolemaic and Copernican*).[117] He assented to the pope to write a strictly impartial treatise on the two major world systems: the Ptolemaic system and its Copernican counterpart. However, the person in the dialogue who defended the Ptolemaic system, Simplicio, was portrayed by Galileo as stupid and simplistic, and the reader could not help but recognizing him as the pope himself. Although the book did not reflect the most recent astronomical state of affairs—for example, Galileo still defended the circular motion of the planets and failed to mention Kepler's work on this point—it was an extremely legible, wide-ranging book. The pope was not amused, and, their friendship notwithstanding, Galileo was obliged to appear before the Inquisition at the age of 68. He was placed under house arrest and was no longer allowed to publish, but that did not prevent him from writing the most important mechanical treatise since antiquity in 1638 (see below).

It should be noted that Galileo does not seem to have used an empirical cycle in his astronomical work, whereas Kepler had. While Galileo's impressive observations put great pressure on geocentric assumptions about the cosmos and supported the Copernican system, Galileo did not test the predictions of the Copernican system against planetary positions. The question is why Kepler followed the empirical cycle in astronomy while Galileo did not (at least not demonstrably), especially since Galileo seems to have followed the empirical cycle extensively in his other work, the study of mechanics (see below). It could be argued that Galileo's aim was only to reject the geocentric worldview, which does not require an empirical cycle (since in principle a single powerful counterexample would be sufficient). In his *Dialogo*, Galileo's main purpose was to show that the earth moved, and he was therefore less interested in calculating the exact, quantitative consequences of the Copernican model. Kepler's project, on the contrary, was to discover the divine plan for the cosmos, which he said was accessible to human reason.[118] Kepler thus had theological motives and strove for the greatest possible accuracy in calculating the planetary positions, making an empirical cycle indispensable. But there is an additional explanation for the fact that Galileo did not follow the empirical cycle in astronomy while Kepler did: Galileo came from a different scholarly tradition.

Whereas Galileo grew up with the musicological experiments and string laws of his father, Vincenzo Galilei, where the patterns could be generated (see below), a large part of Kepler's background was in philology, where the patterns in the available manuscripts (just as in the astronomical phenomena) could not be generated but could only be observed. Galileo was thus familiar with the empirical cycle in the experimental discipline of musicology, while Kepler was familiar with the same cycle in the nonexperimental discipline of philology. In other words, Galileo had little interest in the empirical cycle in astronomy because he could not generate new patterns there. What he could do was link the observed patterns to an underlying theory, namely his theory of the heliocentric system with a moving earth.[119] So we see that two different starting points, *disciplines with manipulable patterns* versus *disciplines with non-manipulable patterns*, can lead to two different approaches: following the empirical cycle with manipulation of patterns (musicology, which with Galileo leads to this approach in mechanics but not in astronomy) versus following the empirical cycle but without manipulating the patterns (philology, which with Kepler leads to this approach in astronomy).

It is therefore interesting to see that taking the humanist background of historical actors into account—musicology and philology, in this case—leads to a broader perspective on Galileo's and Kepler's ways of working and possibly to a better understanding of them, in addition to their other motives. (Recall, as we already explained in the first section of this chapter, that musicology as practiced by the humanists was not merely a mathematical discipline but was also viewed as a historiographical one.) We must keep in mind that Galileo and Kepler were both critical of the humanists' zealous attachment to old books,[120] but they were eager to adopt the humanist practice of the empirical cycle, each in his own way.

Astronomy Flourishes Elsewhere in the World: Samarkand and Istanbul

What did early modern astronomy look like in other parts of the world? With some notable exceptions, the once-thriving astronomy of India and China was falling into decay. Still, many manuscripts from these regions have not yet been translated, let alone studied, and the image of an ailing astronomy may be subject to change as more resources become available. It is now known that some disciplines in these regions were flourishing rather than decaying. That is the case with Indian mathematics with its famous Kerala school (see below), and it is also true of Chinese philology and history (see above). In Chinese astronomy, we see little change during the Ming dynasty (1368–1644) until the Middle Kingdom was introduced to European astronomy by the Jesuits.[121] The astronomical models they brought with them, which were mainly Ptolemaic or Tychonic, provided predictions that were more accurate than those of their Chinese counterparts, as shown by a competition among astronomers under the Kangxi Emperor.[122] Chinese observatories following the European model were promptly built, which is not surprising considering that knowledge of the heavens was paramount in China: he who had mastered heaven had mastered everything.

Large observatories were built in the Islamic world, especially in Samarkand (ca. 1420) and later also in Istanbul (1574), before they became fashionable in Europe. The former was erected by the Mongol ruler and astronomer Ulugh Beg (1394–1447), grandson of the great warlord Timur Lenk.[123] Ulugh Beg had a sextant of marble built with a radius of curvature of no less than 40 meters. This allowed astronomers to achieve an accuracy in the observed positions of the sun, moon, and planets that was unmatched in Europe until Tycho Brahe a hundred years later. By the time the astronomy of Samarkand became known in Europe around 1665, the Scientific Revolution was already in full swing there.

Ulugh Beg's astronomy was primarily empirical. As far as we know, it produced no improved models, let alone quantitative laws. Unfortunately, Ulugh Beg's observatory was short lived; he was more interested in astronomy than in running an empire. Ulugh Beg lost so many regions that he was eventually killed by his own son.

The observatory in Istanbul was commissioned by Sultan Murad III to the polymath Taqi al-Din. This observatory was very similar to that of Tycho Brahe and was even built at the same time. It is sometimes claimed that Tycho knew of Taqi's observatory, but there is no evidence for that.[124] What is certain is that Taqi, like Tycho, also observed the Great Comet of 1577. This coincided with the first day of Ramadan, which must have come at an unfortunate time, as the sultan was planning on marching his troops to Persia. The sultan asked Taqi to use his observatory to ascertain the significance of this comet. Taqi determined that the comet's tail was pointing toward Persia, which he considered a bad omen for the Persians but not for the Ottomans. Taqi also believed that the comet's movement toward Aquarius was a sign of peace. All in all, he predicted that the comet was a good omen for the sultan to attack Persia. But these predictions did not come true: plague broke out in the Ottoman army, and Taqi's observatory came under great pressure and was torn down a few years later, in 1580, signaling the end of an era in Ottoman astronomy.

Neither in Samarkand nor in Istanbul was astronomy given the opportunity to develop. In Europe it was easier to escape the whims of local rulers and despots, as we saw with the humanists (see above). For example, Tycho moved from Denmark to Prague, and Kepler always kept one step ahead of oppressive rulers by traveling between Tübingen, Prague, Linz, and Silesia. But where could the Chinese and Ottoman astronomers go in those vast empires? In many cases, their activities were nipped in the bud.

5.3 Mechanics and Its Integration with Astronomy: The Empirical Cycle Gains Ground with Analogical Thinking

Galileo: Balls on Inclined Planes and the Empirical Cycle

After his turbulent appearance on the scene as an astronomer (see above) and his subsequent house arrest and publication ban, in 1633 Galileo immersed himself in continuing the experiments he had started 40 years earlier and developed them into a new theory of mechanics. This led to one of the most fascinating works in the history of science: *Discorsi e dimostrazioni matematiche intorno a due nuove scienze* (*Discourses and Mathematical Demonstrations Relating to Two New Sciences*).[125] Although Galileo was under house arrest, he managed to smuggle this work into the Dutch Republic, where it was published by Elsevier in 1637.

One of the most important results in the *Discorsi* was his law of free fall. Galileo began with Aristotle's statement that a body falls at a speed proportional to its weight. But—speaking through Salviati, one of the book's fictive interlocutors—Galileo doubted whether Aristotle ever actually tested this claim. He suspected that Aristotle was mistaken, reasoning as follows. Since air has a much lower density than water, and the lower the density the less the resistance of the particular medium, Galileo predicted that in a medium without any resistance (a vacuum), any body, whether a feather or a ball of lead, would fall at the same speed. Galileo then defined the notions of uniform motion and uniform accelerated motion in which the speed increases proportionally to the increase in time. According to Galileo, falling bodies begin falling slowly, accelerating in equal steps with time.

Galileo next describes a famous experiment in which a steel ball rolls down a groove on an inclined plane. On the basis of an ingenious water clock, he demonstrates that these and other balls all roll down at a uniformly accelerated motion; in other words, the speed increases proportionally with time, which means that the distance traveled increases with the square of the time.[126] While Aristotle claimed that heavier bodies fall faster than lighter ones, Galileo states that this only happens when other forces are involved, such as air resistance and friction. By making the inclined planes progressively smoother and conducting one experiment after the other, Galileo succeeded in generating a new pattern: bodies of different weights still rolled down at the same speed and acceleration. This allowed him to substantiate his initial principle but only by carrying out increasingly improved experiments that yielded ever more stable patterns, which he tested against his principle that all bodies fall at the same speed. In a sense, Galileo's procedure was the reverse of Kepler's, which was an adaptation of not the patterns but the principles to the constant patterns in Tycho's observations (see above).

Galileo then provides a mathematical derivation showing that the velocity v of a ball rolling downward can be calculated as the product of the effective acceleration a and the elapsed time t. Galileo expresses this relationship between speed and time in geometric terms, which in modern notation would be expressed as $v = at$.[127] Galileo also gives a derivation for the distance traveled s, which is equal to half of a multiplied by the square of t, that is, $s = \frac{1}{2}at^2$ in today's notation (a is usually represented as g when it concerns the acceleration of a body in free fall on earth, and g is approximately equal to 9.8 meters per second increase in velocity per second, which means that with every second the speed of a falling body on earth increases by 9.8 meters per second).

The insight that distance increases with the square of time had been suggested earlier in the 14th century by Oresme (see chapter 4.3). But experiments involving falling or rolling, showing that bodies of different weights indeed descend at the same speed, cannot be found with Oresme, nor do we find an empirical cycle with him. Galileo left behind descriptions of his extensive experimentation, not with balls dropped from the tower of Pisa—as his student Vincenzo Viviani claimed—but with balls rolling down inclined planes and with swinging pendulums.[128] It was only after many cycles of experiments and mathematical considerations that Galileo believed he had sufficient evidence for his mathematical laws of falling objects that not only express the location of a moving object in time but that could also reveal his deeper generalization: *all bodies fall with the same acceleration*.

Although this law of falling is one of the most important results from the *Discorsi*, the work is full of other insights. For example, 1,700 years after the appearance of Archimedes' law of leverage (see chapter 3.6), Galileo provides a new proof for it. And he deduces that projectiles move in the form of a parabola. He studied the motion of an oscillating pendulum, which he showed to be proportional to the square root of the pendulum's length and independent of its amplitude. A completely different part of the *Discorsi* deals with materials science, with an investigation into the properties of various materials and their surfaces and volumes.

Where did Galileo get the empirical cycle, which is so essential for bringing together experimentation and mathematical deduction? Although this question is addressed briefly above, we must ask ourselves whether Galileo perhaps learned of the empirical cycle through Kepler. This seems unlikely because although the two communicated in writing, Galileo does not seem to have ever read the relevant portion of Kepler's work. He also disagreed with Kepler's ellipse-based astronomy, and in their correspondence, Galileo shows no understanding of Kepler's results. Moreover, the way the two went about their work was quite different, as discussed above: Galileo worked with manipulable patterns and (more or less) "fixed" principles, whereas Kepler worked with non-manipulable patterns

(Tycho's observations) and "flexible" principles. It is therefore more likely that Galileo was familiar with the methodology of his father, Vincenzo Galilei, whose experimental-theoretical work he had experienced up close.[129] While Kepler was acquainted with the empirical cycle from philology, Galileo learned this cycle from his father's work on the musical string laws—humanistic domains of study in both cases. We have already asked whether we can consider the study of music in the 16th century to be a humanistic activity. After all, in the curriculum of the *artes liberales*, music was grouped with the mathematical disciplines (the *quadrivium*). But no research was carried out in this curriculum, let alone experiments. As I showed above, the musical research of Vincenzo and his colleagues was not only experimental but also had a historical purpose: to revive the music of ancient times. Their study of music thus pertained to Renaissance humanism, as has been argued by music historians.[130]

Galileo and Kepler's generation applied the empirical cycle with even more enthusiasm and self-awareness than the humanists before them had ever done. According to the "new scientists," it was not so much a matter of promoting the classics as it was a matter of the experimental study of nature. To this end, an inspired research program was developed, first by Galileo in *Il saggiatore* from 1623; then by René Descartes, who claimed that the whole world should be understood in terms of natural laws; and Francis Bacon, who argued that experimentation was the only method for revealing the secrets of nature (see below). We do not find such a coherent program with the humanists: their research material turned out to be too fragmented. Although they began by studying classical texts, the humanists soon turned to all manner of human production, from paintings, musical compositions, plays, poetry, and architecture to sculpture. The new scientists, by contrast, mainly studied nature, which afforded them an unparalleled focus.

Descartes: Deductions without an Empirical Cycle

Not all "new scientists" applied the empirical cycle in natural science. The Frenchman René Descartes (1596–1650) had a different starting point: in his *Discours de la méthode* (1637) he focused on the notion of "clear and evident ideas."[131] Although Descartes was one of the greatest advocates of laws of nature, which he believed could be used to derive all phenomena, he remained mainly theoretical and was not much of an empiricist. One of his most influential "clear and evident ideas" was that all action in nature is causal, that is, based on cause-effect relationships. Descartes held that the Aristotelian notion of purposefulness (see chapter 3.6) does not occur in nature. To the contrary, the world is constructed "mechanically," consisting of particles that interact as a huge machine through pushing and bumping. From a falling stone to a moving planet, everything could be explained using the forces of pushing, bumping, and pulling.[132] All of the philosophy of nature before Descartes could best be discarded, so he maintained. After all, it was "evident and clear" that the universe was a mechanical creation.

Descartes's ideas had a huge impact on early modern knowledge: they led to a research program that viewed the human body (but not the human soul), as well as the earth, the planetary system, and the universe as a whole, as machines. Everything was subject to causal laws of nature. Noncausal descriptions were rejected.

Descartes attempted to describe in detail the workings of the planetary system in terms of forces of push and pull (in his *Principia Philosophiae* from 1644), and he also used these notions to find laws of collision. His descriptions seem plausible at first glance because he traces all of his patterns (such as planetary motions and collisions between billiard balls) back to the underlying principles of push and pull. Yet few of his results could pass the empirical test: almost none of Descartes's laws of collision and planetary laws have proven correct. They were later empirically refuted by Huygens and Newton. Only his (optical) refractive law and his study of certain concrete phenomena, such as his explanation of the rainbow, have stood the test of time. This shows that the highly regarded *deductive* inferences from patterns to principles in no way guarantee that these patterns and principles will be correct. No matter how "clear and evident" the principles may appear, if the patterns derived are not tested against reality, nothing can be said about their actual correctness. Deductions without an empirical cycle are of little value in physics, which after all is not a deductive science.

But Descartes's philosophical idea that nature is subject to laws that are the same everywhere and at all times has proven enduring, even though his ideas sparked fierce controversies. After all, if everything can be explained by natural laws, without God's intervention, was there still room for a supreme being? The influential 17th-century theologian Gijsbert Voetius from the University of Utrecht argued that Descartes's philosophy would lead to atheism.[133] The discussion got so heated that Descartes, who had moved to the Dutch Republic to enjoy greater liberty, didn't feel free or safe anywhere.

Francis Bacon's Reflection on the Experimental Method

Although the Englishman Francis Bacon (1561–1626) can hardly be called an experimentalist, his ideas concerning the experimental method were of great importance. Bacon was active in the fields of jurisprudence, historiography, and rhetoric, but he is best known for providing the philosophical inspiration for the experimental method. His ideas regarding inductive science are laid out in his *Novum Organum* (1620) and *New Atlantis* (1627).[134] Bacon pushed Plato's deductive approach aside. New knowledge could be obtained only through inductive generalizations about sensory observations. Experimentation was the central method to unravel the secrets of nature.

Novum Organum describes the conditions for making the systematic observations necessary for arriving at new knowledge. Bacon also discusses the process of generalizing over empirical facts, allowing one to derive patterns. New data must be collected so that additional patterns can be identified. Certain factors are particularly useful, such as rebuttals and exceptions. This process is repeated step by step to construct a system of knowledge with an empirical basis. Bacon states that previous knowledge systems were often based not on facts but on deductions and metaphysical conjectures. Even when theories were based on facts, they often got bogged down in unrestricted abstractions from more or less chance observations. Bacon viewed abstracting notions such as first principles as excessive and to be avoided.

Considering his skepticism about abstractions, was Bacon an anti-theorist? Although this appears to be the case, in the end he mainly advocated constructing a theory on a strict, inductive basis. This is an understandable position after all the attempts that had been made to gain new knowledge on the basis of abstract and theoretical considerations. However, there is a limitation to Bacon's approach: it is not always possible to produce underlying principles inductively. Sometimes one simply has to make a guess if no principles are available yet. The second step is to test that guess by deriving patterns from the principles and comparing them with the facts or new experiments, applying the results to the theory, and so on. But Bacon has little to say about this empirical cycle. So although Bacon may be called the father of empiricism, he is not the father of the empirical cycle.

Still, Bacon's plea for a strict empirical approach was extremely refreshing in his time. In fact, Bacon became so influential after his death that in 1660 he was elevated to the status of spiritual father by the founders of the Royal Society under Charles II. Bacon enjoyed greater esteem than Descartes for some time in France as well. Voltaire (1694–1778) even presented him as the father of the scientific method. Bacon did not make any major discoveries himself, although he seems to have experimented sporadically. For example, the story goes that he died of pneumonia, which he contracted during an experiment on how long meat could be stored under ice-like conditions.

Whereas Descartes's approach was too theoretical, Bacon's philosophy was overly empirical. The combination of these two approaches—theoretical and empirical—took place not within a philosophical system but within the practice of studying nature itself, as we saw with Galileo, but especially with Huygens.

Huygens's Integration of Mathematical Deduction and Empirical Observation

When we come to Christiaan Huygens (1629–1695), Cartesian philosophy was at its peak in the Dutch Republic.[135] Huygens' refutation of the laws of collision posited by Descartes one by one led to increasing uncertainty concerning the value of the Cartesian system among both scholars and the general public. Descartes's new natural philosophy was the intellectual conversation of the day in the Republic. Weren't those anti-Aristotelian natural philosophers all wrong? Huygens himself was a supporter of the Galilean and Keplerian approach, in which theory and empirical observations go hand in hand to arrive at reliable knowledge. By the second half of the 17th century, the fact that this empirical cycle had emerged from humanism had already been forgotten, even though Huygens himself was also involved in music theory. But most of his work focused on the study of nature: optics, wave theory, mechanics, and astronomy.

Huygens was particularly important in mechanics owing to his laws of collision, his calculation and derivation of centrifugal force, and his work on pendulums. Huygens's laws of collision consist in the insight that the products of the mass and velocity of colliding bodies before impact are equal to those after impact.[136] Suppose we have two colliding balls, with masses m_1 and m_2 and velocities v_1 and v_2; then, in contemporary notation, it holds that $m_1v_1 + m_2v_2$ before the collision must be equal to that after the collision, in other words: $(m_1v_1 + m_2v_2)_{before\_collision} = (m_1v_1 + m_2v_2)_{after\_collision}$.

In the simplest case, think of a stationary billiard ball being hit by a rolling billiard ball. Both balls will move, but with equal masses, the total speed of both balls after touching can never be greater than the initial speed of the rolling ball.

For balls of different weights, the speed decreases proportionally to the weight of the ball that gets hit. With these laws of collision, Huygens was able to describe the so-called elastic collisions. And he also managed to extend these laws to nonelastic collisions.

Huygens was also the first to devise a formula for centrifugal acceleration. The degree to which an object with mass m changes velocity (the acceleration a) is defined as the force F exerted on that mass: $F = ma$ (although Huygens did not use this equation). In a circular motion, the direction of the object's velocity must be continuously adjusted for it to remain on a circular path. The object is always pulled in the direction of the center of rotation, so to speak. One can feel this force when one spins a stone on the end of a cord, for example. From the second half of the 17th century, one could also measure this force with a spring dynamometer consisting of a coil spring calibrated with weights. This dynamometer was based on Hooke's law, according to which a spring's elongation is proportional to the force exerted on it. But no one before Huygens knew how to calculate this centrifugal force instead of measuring it. It should be remembered that this force is present not only in spinning motions on earth but also in planetary motions spinning around the sun. In other words, if we assume that the laws of physics apply everywhere, we can apply the formula for calculating the centrifugal force both on the earth and to the planetary system.

In his *Horologium Oscillatorium* of 1673, Huygens explains how the centrifugal force of a rotating object depends on its speed and distance from the center of rotation. That is, the centrifugal force is directly proportional to the square of the object's velocity v and inversely proportional to the distance r from the center of rotation. Or, in contemporary notation, $F = mv^2/r$, where v^2/r is equal to acceleration a. In that year Huygens also found a mathematical derivation for this law, which he presented in his *De Vi Centrifuga*, but that work was published only after his death, in the *Opera Posthuma* of 1703.[137]

The significance of Huygens's centrifugal law was enormous: it became clear that maintaining a circular motion required a continuous force and thus an acceleration. This refuted the age-old idea that once an object began moving in a circle, it would continue to do so indefinitely. But as mentioned above, a mathematical derivation does not guarantee that the law is correct: it still needs to be tested empirically, after which patterns or principles can be subject to change. Although Huygens mentions nothing about testing his centrifugal law in a laboratory setting, he did test it using one of his inventions: the pendulum clock. Huygens put his law into practice using a concrete application known today as the conical pendulum. With this sort of pendulum, a weight and the string or rod to which it is fixed moves in the shape of a circular cone so that there is a centrifugal force. Huygens's clock was the most accurate in the world, many times better than any clock developed before him. Whereas other 17th-century clocks could indicate the time only to the accuracy of a quarter of an hour at best, Huygens's timepiece was accurate almost to the second, a method of timekeeping that remained the standard until the 20th century. Huygens's experiments with the pendulum clock showed that a theorem or law could be tested not only in a laboratory situation, as Galileo had done with an inclined plane, but also by means of a concrete application, such as the pendulum clock, which utilizes the relevant law.

The example of Huygens's technological application also makes clear why the study of nature aroused such high hopes in the second half of the 17th century. By mastering the laws of nature, the most accurate devices, architectural structures, and even weapons could be built. To further support this research into the laws of nature, academies were established in Paris and London under royal protection. In Paris, the best scientists were pampered: Huygens was given a spacious apartment with a generous salary and direct access to the library, purely for the purpose of conducting research.

I will not go into detail here on Huygens's many other discoveries, but I will mention that he also derived the pendulum law, according to which the pendulum's period is proportional to the square root of its length l: $T = 2\pi\sqrt{(l/g)}$. He also studied light as a wave phenomenon and discovered polarized light. He discovered Saturn's rings and its moon Titan, and to top it off he also wrote one of the first science fiction stories, entitled *Cosmotheoros*, published posthumously in 1698, in which he freely speculated about life on other planets and which was translated into many languages.[138]

But Huygens's main merit was his methodological rigor with regard to integrating mathematical deductions and the empirical cycle. Galileo had already gone before him in this integration, but Descartes had abandoned it: as innovative as his natural philosophy was, it turned out to be built on quicksand, with mathematical derivations that were deductive but untested. In contrast, Huygens managed to establish a solid foundation for the combination of theory and experiment. While this combination had long existed in philology, art theory, linguistics, and musicology, the empirical cycle now became the prevailing method in all sciences, and even more successfully than before.

Newton: Terrestrial and Celestial Mechanics

Unlike others we have discussed here, the Englishman Isaac Newton (1643–1727) was not trained as an all-around humanist.[139] Of course, Newton had been required to master Latin in secondary school, and he must have studied the linguistic disciplines of the trivium in the *artes liberales* (grammar, rhetoric, and logic). But these disciplines were so old fashioned that during his training, Newton was not initiated into the subtleties of contemporary philology. Although he was introduced to Aristotelian philosophy at Cambridge University, he was more interested in modern astronomers and natural philosophers, such as Copernicus, Galileo, Kepler, and Descartes. Above all, Newton was interested in mathematics, and he was fascinated by the work of Isaac Barrow (1630–1677), who had devised a method for determining tangents at Cambridge.[140] In the 1660s Newton made one mathematical discovery after another, and in 1672 he was elected a member of the Royal Society, founded in 1660.

Newton was also known as an excellent optician and instrument builder. In his *Opticks* he developed a particle model of light that allowed him to explain refraction and reflection. In 1668 Newton built the first working telescope that used mirrors rather than lenses. Above all, he took great pleasure in mathematically deriving laws that had already been discovered by other scholars. Although Newton enjoyed significant status at Cambridge University, universities in early modern England were little more than teaching and networking schools; the prominent research was conducted at the Royal Society. It was three gentlemen in this Society—Robert Hooke, Christopher Wren, and Edmond Halley—who pushed Newton's research to the highest level.

As is known from a letter from Halley, these three had a long discussion on a Wednesday in January 1684 about how planetary motions could be calculated if they were determined by an attraction to the sun.[141] It was almost self-evident to them that this attraction—gravity or gravitational force—had to be inversely proportional to the square of the distance. In Halley's letter we read just how well Galileo's and Descartes's ideas had taken root in the second half of the 17th century. The problem, though, was how to calculate a planet's orbit from an inverse square law. Kepler's now-accepted empirical result that planets move in ellipses had to be explained using such a law. Hooke bragged that he was equal to the task, but the other two had their doubts. This challenge required a new form of mathematics, infinitesimal calculus. Newton had been quietly working on this for some time, without publishing anything on the subject, but the members of the Royal Society suspected that if anyone could solve this formidable problem, with his reputation, it had to be Newton.

Since Halley had reason to be in Cambridge, he went to visit the professor. It must have been a relief for Newton to have the opportunity to communicate with someone of his own caliber (or close to it). When Halley asked him what sort of orbit a planet would have if it were attracted by a force inversely proportional to the square of distance, Newton promptly replied that he had shown years earlier that such a planetary orbit was in the form of an ellipse. But he was unable to find his notes and promised to send Halley the evidence soon. In the following months, Newton worked feverishly on turning his notes (assuming they actually existed) into a legible whole, which culminated in *De Motu Corpurum in Gyrum*.[142] This became the core of his later, renowned work *Philosophiæ Naturalis Principia Mathematica* (*Mathematical Principles of Natural Philosophy*) from 1687, commonly shortened to the *Principia*.

Newton took a rigorous approach: he reformed all of mechanics until his day, basing it on just three axiomatic laws from which (together with precise definitions) all other mechanical phenomena could be derived as if they were mathematical theorems. The result is a virtuosity that still appeals to the imagination. While Newton's three "laws of motion," as they are nowadays called, are well known—physics education in secondary schools is largely based on them—one does not often see them in direct translation:[143]

> First law: Every body persists in its state of being at rest or of moving uniformly straight forward, except insofar as it is compelled to change its state by force impressed.
>
> Second law: The change in motion is proportional to the force impressed and occurs in the direction of the straight line on which that force is impressed.
>
> Third law: To every action there is always opposed an equal reaction: or the mutual actions of two bodies upon each other are always equal, and directed to contrary parts.

The first law was implicitly known to Galileo, the second law is usually expressed as $F = ma$, and the third law as *action* = *reaction*. With these three laws of motion, along with improved definitions of force, momentum, and acceleration, Newton brought order to the vast and often contradictory concepts that preceded him. Above all, Newton was able to derive all previously discovered mechanical laws from his three basic laws, often improving them. In addition, Newton posited his theorem concerning gravitation, which has become known as the law of universal gravitation, according to which every body attracts every other body in the universe with a force directly proportional to the product of their masses m_1 and m_2 and inversely proportional to the square of the distance r between their centers: $F = Gm_1m_2/r^2$, where G stands for the gravitational constant. With this law, Newton posited that gravitation is not unique to the sun (which keeps the planets in orbit) or to the earth (which keeps the moon in orbit). This force applies to all bodies, although the constant G is so small that the gravitation between two normal masses on earth, such as two cannonballs, is almost negligible. But the gravitation between a cannonball and the earth is quite measurable because one of the masses is sufficiently large. Newton's gravitational law predicts that although the orbit of a cannonball is approximately a (segment of a) parabola, if the cannonball could be fired hard enough, it would orbit the earth.

Everything now seemed to fall into place: with his infinitesimal calculus, Newton could deduce the motions of the planets around the sun using the law of gravitation. This showed that planets do indeed move in an ellipse (Kepler's first law). Kepler's other two laws could also be derived with Newton's laws, as could the motions of the moons around Jupiter and the orbits of comets, as well as all mechanical phenomena on earth—from the velocities of falling objects and the trajectories of projectiles to lever structures, not to mention the complex tidal phenomena of ebb and flow, caused by the simultaneous attractions of the moon and the sun.

Insights on Regularity versus Deviation

The motion patterns in the sky and those on earth can thus be accounted for using the same principles. In addition, Newton's laws were found to explain not only regularities but also all sorts of deviations from them, such as irregularities in the orbits of the moon and of the planets. These deviations or irregularities initially seemed to refute Kepler's laws, but on closer inspection they again exhibit patterns. These deviation patterns are predicted by the "disturbing" influences of other planets. So, the underlying Newtonian laws explain not only pattern-based phenomena but also apparently pattern-deviating phenomena (although they were not always explained by Newton himself). Regularity and deviation could thus be handled uniformly with the principle of gravitation.

An interesting comparison arises here with the age-old debate on the opposition between the regular and the irregular, or between the "analogous" and the "anomalous," as we encountered in ancient linguistics, philology, and jurisprudence (see chapter 3). And in Chinese astronomy, we saw that irregular or exceptional phenomena played just as much a role as regular, recurring phenomena. Recall that in linguistics and jurisprudence the contrast between the regular and the exceptional was resolved with meta-rules. In a way, patterns were sought out in the exceptions. What had never been seen before, however, was a general law that could bring regular and abnormal cases under a common denominator. And this is what Newton's law of gravitation could do: he showed that the anomalous disturbances in Saturn's orbit can be predicted if not only the attraction of the sun but also that of Jupiter was taken into account. This led to a complex mathematics that would develop into a new subdiscipline called perturbation theory. Thus, with the universal gravitational law and the three laws of motion, not only were the heavenly and earthly patterns unified but also the regular and the irregular.

Additionally, Newton's laws could be used to make all sorts of new predictions. For example, Newton predicted that the earth was not purely spherical but was slightly flattened at the poles by rotation (and the associated centrifugal force). This difference is not large, but it can be measured. Several decades after Newton's death, it was confirmed that the diameter from pole to pole was about 27 miles (43 kilometers) smaller than the equator.[144] And when Halley used Newton's laws to predict the return of a comet (which would later be called Halley's Comet) with great accuracy, its return was a real sensation.[145] In addition, new predictions of phenomena on earth could also be made: frictional laws could be integrated into the second law, $F = ma$, making more precise applications possible in ballistics. Although it had not yet been determined how great the gravitational constant G was, it was known how this constant could in principle be measured: according to the law of gravitation, two large lead balls should attract each other measurably using a very sensitive torsion balance. It was not until more than 100 years later, in 1798, that such an extremely sensitive torsion balance could be built, allowing Henry Cavendish (1731–1810) to estimate the earth's mass, from which the value of the gravitational constant G also followed, now usually expressed as $6.67408 \times 10^{-11}\ m^3kg^{-1}s^{-2}$.

A Newtonian Worldview, the First Female Physicists

As deterministic as his laws may seem, Newton himself argued that God's intervention in the world was still necessary. Newton noted that the irregularities in the planetary orbits would increase over time because of the mutual forces of attraction between planets and comets. This would sooner or later make the planetary system unstable.[146] Contrary to Descartes's system, in the Newtonian system God was needed to put the planets back into order on a regular basis, so to speak. As might be expected, theologians soon had more sympathy for Newton than for Descartes, although Newton was scorned by philosophers such as Leibniz (see below). Moreover, Newton was not completely wrong, because over the course of the 18th century many a mathematician would set out to work on the so-called three-body problem (book 1 of the *Principia*). It turned out that while Newton's laws were capable of solving a system with two celestial bodies (such as the sun and earth, or the earth and moon), when a system of three or more bodies was presented in the form of mathematical equations, it could not be solved analytically but could only be approximated using the perturbation theory mentioned earlier. And then it also became apparent that over long periods, three-body systems, such as the sun, earth, and moon, led to unstable and even chaotic orbits. Several physicists and mathematicians were tearing their hair out over this problem until the French astronomer Pierre-Simon Laplace (1749–1829) showed that the Newtonian universe was stable. Although disturbances would indeed occur in systems of three bodies or more, the orbits of the planets would not devolve into chaos, because the mass of the sun in the planetary system is many times greater than that of all the planets combined. Laplace's calculations showed that, at most, long-term cyclic variations would occur, such as in the moon's orbit around the earth. In an apocryphal anecdote, Napoleon Bonaparte asked what the role of God was, to which Laplace replied that he "had no need of that hypothesis."[147]

Newton's work had since spread throughout western Europe thanks in part to the comments of the Dutchman Adriaen Verwer, the translations of the Italian Laura Bassi—who was also the world's first female professor (in Bologna in 1745)—and the French marquise Émilie du Châtelet. The latter also performed experiments with what would later be known as the "kinetic" and "potential energy" of balls falling on soft clay, an experiment that had been devised by the Dutchman Willem 's Gravesande in 1722.[148] Du Châtelet took the amount of clay that was displaced as a measure of kinetic energy, and that amount turned out to be proportional to the square of the velocity (roughly the kinetic energy) and the height from which the ball was released (roughly the potential energy). Newton and others had assumed that this "energy" of a moving body was proportional to the velocity (momentum) rather than the square of that velocity. This difference is important, since du Châtelet, using her notion, showed that the total energy of an object—its kinetic energy plus its potential energy—remained constant and conserved.[149] In this way, it was she who actually provided the first version of the later law of the conservation of energy, one of the fundamental laws in physics. Du Châtelet was soon considered the most learned woman of her time, a status she shared with the eminent philologist Madame Dacier (see above). It appears that 18th-century men didn't know quite what to make of all this female scholarship. Du Châtelet's lover Voltaire even appears to have called her "a great man whose only fault is that she is a woman."[150] And Immanuel Kant commented, "A woman who has a head full of Greek, like Madame Dacier, or builds on fundamental controversies about mechanics, like Marquise du Châtelet, might as well have a beard." Incidentally, there were also men who wrote more respectfully about women, such as François Poullain de la Barre and David Hume, but they were exceptions.[151]

Despite the success of Newton's theory, some contemporaries greeted his innovations with skepticism. They struggled with the idea of gravity having an effect remotely, without any contact. For example, Huygens and Leibniz believed that Newton's notion of force was, in fact, a return to scholastic concepts from the Middle Ages that rendered matters cloudier rather than clearer. The Cartesian concept of the transmission of force through direct contact remained dominant throughout the 17th century. Newton was well aware of this and was himself also unsatisfied with his law of gravitation. He would have preferred to describe this force on the basis of causal connections, such as Descartes's forces of push and pull. But Newton rejected causal descriptions that could not make precise mathematical predictions. The next-best option was a mathematical description without causal connections. As Newton stated, he did not want to come up with new hypotheses that could explain gravity: *hypotheses non fingo*.[152] During the 18th century, rather than explaining laws causally, physics increasingly focused on representing these regularities using mathematical equations, which had to be reduced to more general principles and tested iteratively with the empirical cycle. In the latter respect, physics was no different from the other disciplines.

The Newtonian theory of motion acquired such an inviolable status that its laws were regarded as mathematical axioms that could be used to explain all natural phenomena. But physics was and has remained an empirical science, and every new mechanical phenomenon or pattern constituted a new test case for the underlying principles of motion. This became clear when phenomena were discovered in the 19th century that did not seem to comply with the three laws of motion. After several attempts to salvage Newtonian mechanics, it became clear that this form of mechanics was not adequate for velocities approaching the speed of light or for strong gravitational fields, such as those near the sun (which the planet Mercury was subject to). New principles were needed and proposed, particularly by Albert Einstein (1879–1955) at the beginning of the 20th century. And eventually these principles were successfully tested against the newly discovered patterns but only after some of Einstein's laws themselves also underwent modifications (with respect to the cosmological constant). The centuries-old deduction pattern from philology, art theory, linguistics, and musicology—the empirical cycle—remained valid for both Newton's theory and Einstein's.

Newton as a Scholar: From Musicology to Historiography

So far, in our discussion Isaac Newton resembles a mathematician and natural scientist with little concern for other disciplines. But that image of him has long been obsolete.[153] In reality, Newton spent most of his life working in disciplines other than physics, and it is in those domains—theology, alchemy, and history—that he wrote the most by far. Newton saw himself primarily as a philosopher, although he approached this field through mathematics, or as he states in book 3 of the *Principia*, "I have set out the principles of philosophy, principles that are not philosophical but mathematical." In Newton's day, the title of philosopher was applied to anyone who engaged in the study of nature, but this was not the case for the study of history, theology, and philology, though the boundaries were vague.

It should come as no surprise that Newton was well versed in Vincenzo Galilei's string laws (see above). When Newton was working on the "Scholium generale" for the second edition of his *Principia*, he tried to find precursors to his theory of gravitation. From this it becomes clear that Newton believed that the ancients had known all the laws of nature, but their true knowledge was largely lost and could only be distilled from surviving fables by the careful listener.[154] Although this view can also be found with 15th- and 16th-century humanists concerned with art theory, musicology, and philology, it is perhaps surprising to encounter this belief in Newton as well. For example, he argued that Pythagoras must have known about the inverse square law for gravitation. It was thanks to Vincenzo Galilei that some of Pythagoras's laws had been rediscovered. Newton elaborated on Vincenzo's string law: "If two strings of equal thickness are stretched by attached weights, then these strings will be in harmony when the weights are reciprocal to the squares of the lengths of the strings."[155] Newton then argued that Pythagoras had applied this string law to the sky and therefore knew that "the planets' weights toward the sun are inversely as the squares of the distances from the center of the sun." Furthermore, Newton believed that Pythagoras had concealed his true law of gravitation from the common people. It was only other wise men who understood that Pythagoras was using his musical laws to refer to the *musica mundana* (music of the cosmos) rather than to the *musica instrumentalis* (music of instruments). In other words, in his musical law—that is, the law assumed to have been rediscovered by Vincenzo—Pythagoras was referring to nothing other than the inverse-square law of gravity.[156]

Newton devoted himself even more extensively to drawing up a chronology of world history.[157] In *The Chronology of Ancient Kingdoms Amended* (published posthumously in 1728), he believed he could refute Joseph Scaliger's hypothesis that the most ancient Egyptian kings had lived before the biblical date of Creation (see above). Based on historical and philological source material, Scaliger had shown that the beginning of the first Egyptian dynasty should be dated to 5285 BCE. But this date was nearly 1,300 years before the date of Creation according to biblical chronology, approximately 4000 BCE. Newton was convinced that Scaliger and others—such as Isaac Vossius and Spinoza—were mistaken, and he spared no effort to create a chronology that could accommodate all the Egyptian kings within the biblical framework. Although Newton supplemented his argumentation with the dating of observed astronomical phenomena, he ignored the fact that the reconstructed historical sources, such as Manetho's king lists, were consistent and mutually coherent, indicating that people had lived before 4000 BCE. This fact was reinforced in the second half of the 17th century by Chinese texts that also showed that the earliest Chinese emperors had ruled before 4000 BCE (see above). Moreover, philology was highly regarded in Newton's time, more than natural philosophy, especially after philologists uncovered many fakes and constructed legible versions of corrupted ancient works. Newton himself maintained ties with the best English philologist of his time, Richard Bentley (1662–1742). But Newton's chronology missed the mark both philologically and historically despite its seeming precision. He believed that only the Bible presented a reliable account of ancient times and that all other chronicles had been inflated by their writers to make their people seem more important (by giving them a longer history). However, in doing so Newton subordinated historical-philological achievements to his own convictions.

So we see that Newton dealt with the humanistic disciplines differently than had his predecessors, such as Kepler and Galileo. While Kepler and Galileo took the empirical cycle from the humanistic disciplines and applied it to astronomy and mechanics, in Newton's time, use of this cycle was already so widespread that scholars no longer had to be both humanists and natural philosophers. And when Newton entered the field of history and philology, he let the empirical cycle fall to the wayside.

Newton and the Occult Sciences: Alchemy and Its Relation to Mechanics

Newton's historical work was thus motivated by his belief that the ancient sources—from the Bible to the Pythagorean works—contained hidden truths that had been lost over the centuries. For example, he believed that the concept of the Trinity was demonstrably incorrect (since it was added to the Vulgate later, as Erasmus had already shown) and that worshiping Christ as God was a form of idolatry. But Newton took pains to keep these ideas secret, and they were only discovered a few centuries after his death when his surviving notes were sold at auction. Had Newton's heretical ideas become known in his own time, he would have lost not only his professorship but possibly also his life, as happened to his contemporary Thomas Aikenhead, who had denied the Trinity publicly.[158]

Less risky than Newton's theology was his alchemical work, which spanned more than a million pages. This work was omitted by Newton's biographers until well into the 20th century, but alchemy was a widespread activity in the 17th century. Newton wanted to assemble all existing alchemical knowledge and then use experiments to discover the true ancient knowledge of alchemy. Thus, in Newton's surviving work, we find lengthy lists of alchemical authors with over 5,000 page references of more than 900 keywords.[159] In this way, Newton hoped to derive a coherent system, but his endeavor was doomed to failure. Nevertheless, the influence of Newton's alchemical thinking seems to be detectable in his work on physics. The embrace of hidden, "occult" phenomena made it acceptable for Newton to introduce the concept of "force," thereby abandoning the Cartesian notions of push and pull.[160]

Newtonianism: Analogical Thinking as a Deduction Pattern

Starting in the 18th century, the idea that an axiomatic mechanics, combined with deductions and predictions, could explain all motion phenomena—regular and irregular, heavenly and earthly—became the model for the other disciplines. Newton's gravitational mechanism was adopted as a prototype for finding laws in other areas. After all, forces of attraction appeared to occur not only in mechanics but also in other fields. And the Newtonians believed these fields could become just as successful as mechanics if they were only worked out analogically. This thinking in terms of analogy, like the empirical cycle, is a deduction pattern (as we already saw in Kepler's analogy with magnetic forces above): where a given deduction worked in one field, it could also work in another. But this analogical way of thinking turned out to be valid less often than the other deduction pattern, the empirical cycle.

Let's start with a successful case: the theory of electricity. Today electricity is studied under the discipline of physics, but in the 18th century, so-called electricists constituted a separate group of researchers. In 1734, Charles du Fay showed that there were two types of electricity: vitreous electricity generated by rubbing, for example, glass; and resinous electricity generated by rubbing resin, among other materials. These two types of electricity are now called positive and negative charges. There appeared to be an attraction between two charged objects if one was charged positively and the other was charged negatively. If the charges were either both positive or both negative, there was a repulsive force. Despite this difference from gravity, the electrical force between charges was modeled on the Newtonian law of gravitation by Charles-Augustin de Coulomb in 1784. That is, the electric force between two charges is directly proportional to each of the charges Q_1 and Q_2 and inversely proportional to the square of their mutual distance r—or, as expressed in a formula, $F = -kQ_1Q_2/r^2$, where k is the Coulomb constant. The charges Q_1 and Q_2 can be either positive or negative: if their product is positive, the resulting electric force is negative and they repel each other, but if their product is negative, the electric force is positive and they attract each other (hence the minus sign in the formula). Here the analogy with Newtonian mechanics worked splendidly: the analogy is not quite one-to-one—due to the fact that masses assume only positive values, while charges can be either positive or negative—but aside from that, the analogy is virtually perfect.

The Newtonian law of gravitation could not be adapted to new phenomena in all fields. Sometimes it was not so much the notion of attraction that was adopted but the idea of the inverse square law (just as Newton thought he had discerned this inverse square law in Vincenzo's string law above). Thus, it was assumed and subsequently confirmed that the intensity of light decreases proportionally to the square of the distance, analogous to gravitational force. Likewise for sound and heat radiation. The main difference with the law of gravitation, however, was that no attraction by light, sound, or heat could be detected.

In yet other areas, only specific Newtonian concepts were adopted, such as force, weight, and pressure. We see this, for example, in the medical work of Herman Boerhaave (1668–1738), who attempted a mechanical explanation of diseases using hydrostatic equilibrium and fluid pressure, albeit without much success (see below).

The musicologist and composer Jean-Philippe Rameau (1683–1764) was also influenced by Newton's work, but by his *Opticks* rather than by his *Principia*. Just as Newton showed that white light consisted of a spectrum of discrete colors, Rameau in his *Nouvelle système de musique théorique* (1726) demonstrated analogically how a single note consisted of a spectrum of discrete tones: the overtones (something that had previously been described by Mersenne and Oresme, see chapters 5.1 and 4.4). But this comparison remained an analogy, and Rameau was unable to derive any further results or predictions from it. However, Rameau's analogy makes clear that in the 18th century the tide had begun to turn: no longer was music the basis for physics (such as even Newton seemed to suggest with regard to Vincenzo's string law); it was physics that set the example for musicology.

And in the 19th century, the linguist Franz Bopp proposed a law of conjugation that appealed to a force of attraction (*Gewichtsmechanismus*) between the verb stem and the ending of the verb. Heavier vowels attracted the ending more strongly, making the endings shorter. Light vowels, on the contrary, produced longer endings. Although Bopp's conjugation law has not stood the test of time,[161] the analogy between the attraction of physical objects and the attraction of parts of words is still in vogue in certain schools of linguistics, although it is not much more than a metaphor.

The use of analogies was not limited to Newtonianism but was also found in later theories. For example, from the moment that Darwin's theory of evolution took hold, Darwinian arguments and references were adopted in many disciplines: from philosophy to sociology and from literary studies to anthropology.[162] Of course, the use of analogies can already be found in antiquity (such as Ptolemy's analogy between jurisprudence and astronomy, see chapter 3.7, or Cicero's analogy between linguistics and the state, see chapter 3.3), but the notion of analogy as a general applicable empirical pattern, like the empirical cycle, becomes dominant only in the early modern period.

Mechanics in China and the Islamic World

It is one of the great mysteries in our knowledge history that outside Europe mechanics seems to be largely absent in the early modern period. The question as to whether this apparent absence is actually true has yet to be answered satisfactorily. Traditionally, most historians of science have turned their attention to the history of European mechanics, which has left many sources outside Europe understudied. Joseph Needham notes in his magnum opus *Science and Civilization in China* (1962) that "the study of motion (kinetics and kinematics) seems to have been, on the whole, conspicuously absent from Chinese physical thinking."[163] However, this does not apply to ancient Mohist mechanics, as we saw in chapter 3.6. In the *Mo jing*, we find several extraordinarily interesting observations about motion. Although a comparison with early modern mechanics in Europe would be anachronistic, the Mohist observations are so fascinating that we cannot set them aside in this chapter. With respect to some of Galileo's and Newton's mechanical insights, they look like identical twins, in spite of the 1,800 years that separate them. "If there is no opposing force . . . , the motion will never stop," and "The cessation of motion is the result of the opposite force . . ." Today these statements seem like obvious truths, and yet for centuries, even millennia, people in Europe and the Islamic world thought the exact opposite. However, the Mohists did not use their principles to predict new phenomena or patterns, which gives their insights a different character.

After this promising beginning for Chinese mechanics, it is even stranger that nothing more was written in China about the motion of physical objects until a couple of thousand years later. But perhaps we are asking the wrong question when we go searching for a Western category such as "mechanics" in the disciplines of China (or India or Mali). All the more so because the Ming period (1368–1644) has yielded impressive technological artifacts that suggest an in-depth knowledge of mechanics.[164] In Ming-era China the study of motion, which had existed since ancient times, is best explored through their technology.

However, it is extremely difficult to recover any underlying knowledge about patterns, principles, and the deductions made from them on the mere basis of technological artifacts without accompanying texts to record them.[165]

The notable absence of most natural sciences in the Ming period—until the arrival of the Jesuits—has led some to conclude that the Chinese domains of knowledge exhibited a general decline in the late empire.[166] But this verdict is a bit too hasty: if there was any decline at all, it would appear to have been limited to the Chinese knowledge of nature and not to Chinese philology, history, and medicine, which flourished in the Ming period (see above and below). This shows how important it is to consider the full spectrum of knowledge, rather than just the natural sciences, if we are to make claims about knowledge in general.

Yet there was indeed a sense of decline among the Ming and Qing literati.[167] However, the notion of decay has a long history. Since the Song dynasty, knowledge of nature had declined so much that they could hardly be tested in state exams. For example, Shen Kuo (see chapter 4.2) wrote that the essays written by exam candidates were so jumbled and the examiners themselves were so ignorant of the subject that all candidates were passed "with distinction." From the Ming dynasty onward, Chinese literati focused mainly on botany and pharmacology, with the study of nature and mathematics getting short shrift. At the beginning of the Qing period (1644–1911), examinations on the knowledge of nature were completely abolished.

There is not much new under the sun in the field of mechanics in the Islamic world either: if any mechanics was being done at all, it was Aristotelian and not Galilean or Newtonian. New developments appear to be absent. There were still discoveries here and there, such as by Ibn al-Shatir in astronomy, but the decline of Islamic natural science seems undeniable.[168] Many explanations have been suggested: the rise of religious elites, the crusades from the west, and the destruction of Baghdad from the east.[169] But none of these explanations is sufficient; in fact, the decrease in activity does not apply to all disciplines in the Islamic world. Philology and history flourished in the Songhai Empire, astronomy flourished among the Mongols, the humanistic disciplines flourished in the Mughal Empire (see above), and medicine and jurisprudence flourished in the Ottoman Empire (see below).

All in all, we can say that the knowledge disciplines in general did not go into decline outside of early modern Europe.[170] If we can speak of decline, it applied to some of the natural sciences, but not to disciplines such as philology, history, medicine, and jurisprudence. The first of these disciplines—philology—even stood at the cradle of the empirical cycle, both in Europe and in China.

5.4 Mathematics: A Nonempirical Discipline with an Empirical Cycle?

Europe: Arabic Mathematics as a Source of Inspiration

In 15th-century Europe, algebra was not an independent discipline but rather a collection of applications in fields such as accounting, cartography, and perspective.[171] Initially, accounting did not require advanced mathematics, but as trade flows became more complex—with bills, debt securities, and interest calculations—the complexity of the mathematics required increased as well. The Roman numeral system was cumbersome for calculating interest on interest, and the Hindu-Arabic decimal system that had been introduced in medieval Europe was better tailored for merchants (see chapter 4.3). The first textbooks were mainly aimed at merchants' children, who had to acquire the necessary algebraic skills for trade. These textbooks taught them the decimal number system and the algorithms needed to make calculations. Slowly but surely Roman notation was replaced by its Hindu-Arabic counterpart. Here we see a stark contrast with the revival of the Roman and Greek disciplines pursued by the humanists: although they wanted nothing more than to revive ancient mathematics—with Euclid as the great example—for most purposes, Arabic mathematics was superior. The glory of antiquity extended far, but not as far as arithmetic and algebra.

For other applications, it was not algebra that was central but geometry. As we saw above, Alberti wrote his influential study on linear perspective in 1440, and Piero della Francesca developed procedures for perspectival representations in 1474. Many of Piero's ideas were elaborated—and plagiarized—by Luca Pacioli (1445–1517) in his *Summa de Arithmetica, Geometria, Proportioni et Proportionalita* from 1494.[172] In addition to a tract on accounting, this work also included mathematical puzzles and new algebraic notations. For example, Pacioli was the first to use the plus and minus signs in print.

New mathematics could also be found in other disciplines: from astronomy and mechanics to cartography. For example, the mathematical challenge in cartography was to project the globe onto a flat surface in such a way that the angles were not deformed. This would allow compass courses to be faithfully displayed on maps. The Flemish cartographer Gerardus Mercator (1512–1594) was the first to create such a projection that faithfully represented angles, or directions.

New Discoveries: Cubic Solutions and Decimal Notation

In the course of the 16th century, European mathematicians increasingly turned their gaze to abstract topics, such as the age-old problem of cubic equations. The versatile Persian Omar Khayyam (see chapter 4.3) had argued that there was no general method for solving equations with a term to the power of 3, like $x^3 + ax^2 + bx + c = 0$.[173] He believed that solutions existed only for very specific cases. This view was echoed by Luca Pacioli in his 1494 *Summa*. Only a few years later, Scipione del Ferro (1465–1526) found a method to solve general cubic equations, but his method remained unpublished and had no effect on his contemporaries. Independently, Niccolò Tartaglia (1499–1557) also discovered a general solution for cubic equations in 1541. He expressed it in the form of an elegant poem addressed to the humanist Girolamo Cardano (1501–1576). The latter published the method for solving the problem in his own work, the *Ars Magna*, without Tartaglia's permission, which outraged Tartaglia so that he publicly called Cardano a *huomo di poco sugo* (man of little sauce).[174] Tartaglia's solution still bears the name "Cardano's method." Whatever the case, with its solution for cubic equations, Italian mathematics surpassed Arabic mathematics for the first time. Now the way was open to solve even more complex equations with terms to the fourth power, which Cardano's student Lodovico Ferrari (1522–1565) succeeded in.

François Viète (1540–1603) and Simon Stevin (1548–1620) are among the first great early modern mathematicians from northern Europe. Viète devised a general theory of equations using new mathematical notation. This allowed him to represent mathematical problems much more efficiently and also reduced the different types of equations to a smaller number. He introduced the use of letters into algebraic equations, which would become highly influential and is still the practice today.[175] In his *De Thiende* (1585), Simon Stevin also focused on new notations and introduced the decimal representation of numbers for subdivisions of whole numbers.[176] He showed how to extend existing mathematical operations to decimal notations, demonstrating that irrational numbers and negative numbers can be treated in the same way as other numbers. This was in sharp contrast to the Greek view, which rejected the existence of these sorts of numbers (see chapter 3.4). In *L'arithmétique* (1585), Stevin also rejected the time-honored Pythagorean idea that 1 was not a number but a "unit." In his work he wrote, "THAT UNIT IS A NUMBER."

Descartes's Unification of Geometry and Algebra

What René Descartes (see above) set in motion was of a different order than the mathematical innovations of the 16th century. Descartes's goal was to redefine mathematics as a whole, which has become known as *algebraic geometry*. Actually, this new mathematics was the invention of both Pierre de Fermat (1601–1665) and Descartes, but the Cartesian formulation, in part thanks to the elaboration by Dutch mathematicians, became the most common. In *La géométrie*, which was added as an appendix to the *Discours de la méthode* from 1637, Descartes introduced a number of concepts and operations that developed the notion of the coordinate system.[177] In a (two-dimensional) coordinate system, each point on a plane is represented by two numbers—the coordinates—on horizontal and vertical axes, also called the x axis and y axis, respectively. A mathematical equation with two variables can be represented as a collection of points on this plane, where the coordinates of each point must comply with the corresponding equation. The representation of geometric figures through equations meant that while figures could be expressed geometrically, they could now also be calculated algebraically. Descartes's insight that all points on a curve can be expressed as a relationship by means of a single equation[178] also suggests a unification of Euclidean geometry and Arabic algebra. The idea of coordinates was not in itself new—Eratosthenes used it for his atlases as early as the 2nd century BCE (see chapter 3.6), but geometry based on coordinates had never been seen before.

With the Cartesian representation of curves, ancient geometric notions suddenly looked very different. For example, the Euclidean definition of a circle translated into Cartesian terms as the coordinates x and y, where the distance from the points (x, y) to the center M is equal to the radius r of the circle. If we take the origin O of the coordinate system as the center, then according to Pythagoras's theorem the following applies to all points (x, y) on the circle: $x^2 + y^2 = r^2$.

Yet we must take care not to attribute too much to Descartes. As important as his algebraic description of curves is, he himself seems to have had one foot in the Euclidean tradition: nowhere in the *Géométrie* does Descartes introduce a curve with an algebraic equation. He always describes the curve geometrically and then derives the equation. It almost seems as if the algebraic equation is not a defining criterion for Descartes but rather a tool for studying curves.

The notion of the coordinate system together with the new analytical geometry nevertheless turned out to be a gold mine: new propositions could be posited and proved. Above all, algebra and geometry now converged. But this success would have been unthinkable had Descartes's work not been made accessible by his colleagues. While Descartes had wonderful ideas (which he had on occasion adopted from others, such as from Isaac Beeckman; see above), he was not known as someone who worked out his ideas systematically, at least not in science. For example, in his coordinate system he used a single axis, where a point could also be put at a certain distance above that axis without introducing a separate *y* axis. This made calculations in Descartes's system quite difficult. It was Frans van Schooten (1615–1660), a professor at Leiden, who explained what Descartes meant. Van Schooten expanded Descartes's notion of a coordinate system to two axes and, together with his students, wrote the standard works that elaborated the new mathematics.[179] Leiden became the center of European mathematics for some time, and van Schooten's books were popular with Newton, Leibniz, and others. Van Schooten also proposed expanding the Cartesian system to three axes, so that three-dimensional figures could also be represented. Planetary orbits could now be expressed numerically using Cartesian coordinates, so that their position and velocity could be calculated instead of constructed in Euclidean fashion.

Infinitesimal Calculus: An Empirical Cycle in Mathematics?

The greatest mathematical innovation of the 17th century was *infinitesimal calculus*, today simply known as *calculus*. We owe this invention to Gottfried Leibniz (1646–1716) and Isaac Newton. For years the two men contested who was first to make this discovery, but today we know they each largely developed calculus independently.[180]

Calculus is concerned with how a curve changes in response to infinitely small—"infinitesimal"—changes. The study of this type of change was not merely interesting from a mathematical point of view; it arose from the question of how to determine the velocity of an object based on the motion curve. If the curve in a Cartesian coordinate system represents the position of a body over time, then the velocity is the change of that position per unit of time; but if the velocity of a body is not constant, we then refer to the velocity at a specific moment in time, that is, at one point on that motion curve. In that case we have to examine a minute part of that curve and calculate the change in location by the smallest possible unit of time. Such a change can be described by a *tangent line*, or simply *tangent*, of the point on that curve. The fascinating thing is that when the equation of the motion curve is known, the formula for the velocity can be derived from it.

It was already known to the ancient Greeks that an infinitesimal change at a point on a curve can be described using the notion of a tangent,[181] but no one endeavored to find a general method for calculating it. Still, Euclid, Archimedes, and Apollonius of Perga were familiar with infinitesimal notions. Archimedes used it in his study of surfaces and in his approximation of the number π (see chapter 3.4). Indian and Islamic mathematicians also used notions similar to those used in calculus.[182] Modern calculus begins, however, with Newton and Leibniz. Although Leibniz was the first to publish on the subject (in 1684), Newton was the first to discover it. But Newton failed to publish a single word about his mathematical discoveries for years, leading to a bitter, long-standing controversy in which Newton accused Leibniz of plagiarism. Newton's ideas were well known in informal circles, through letters and lecture notes, in particular his *Tractatus de Methodis Serierum et Fluxionum* (*Treatise on the Methods of Sequences and Fluxes*) from 1670–1671.[183] In this treatise, we see how strongly Newton's mathematics was linked to mechanics. He introduces a notion of the *fluent*, which changes at a certain velocity. He calls this velocity *fluxion*. The fluent is the position of a body at a given point, and the fluxion is its velocity at that point. Newton notates the fluent by adding a dot above the letter x, which in contemporary terms we would call the derivative of x.

With Newton, the new mathematics is primarily a form of applied mechanics. Does this mean that this form of mathematics, like mechanics, also exhibits an empirical cycle? This question is more complicated than stated here, because whereas calculus seems to be derived from mechanics, over the course of the 18th century, Newtonian mechanics came to be seen as a form of mathematics. For example, the mathematician Leonhard Euler (see below) dealt with classical mechanics from an axiomatic perspective. Was this mechanics still subject to the empirical cycle? Yes, because any regularity (pattern) derived from Newton's three laws of motion (principles) still had to be tested. This happened, for example, when Newton attempted to deduce that the planetary system would become unstable over time.[184] Pierre-Simon Laplace showed that Newton's calculations were here incorrect (see the previous section). Newton's three principles of motion were confirmed time and time again, however, as new forms of mathematics were continually added. This mathematics could be used to describe mechanical reality, which often turned out to be more complex than expected. Viewed in this light, it is not so much mechanics that exhibits an empirical cycle as it is mathematics.

Of course, mechanics was also subject to the empirical cycle. When at the end of the 19th century Newtonian mechanics proved inadequate for very high velocities and for strong gravitational fields, classical mechanics could be brought into line with these high velocities only using ad hoc correction factors. But these correction factors had little to do with Newtonian mechanics. As discussed above, these developments prompted Albert Einstein to abandon classical mechanics, which also required another type of mathematics: non-Euclidean geometry. This episode shows how closely intertwined much of mathematics and physics was and continues to be.

From Statistics to Number Theory: Pascal to Euler

The above suggests an almost one-to-one relationship between mathematics and its field of application, that is, mechanics. This meta-pattern is also found in other branches of mathematics. For example, Pierre de Fermat and Blaise Pascal (1623–1662) developed the framework for probability theory and combinatorics entirely in the applicational context of gambling and card games. And Johan de Witt (1625–1672) worked out probability theory and statistics in the context of insurance, particularly annuity insurance.[185] Calculations needed to correspond to reality, which proved to be an excellent test field for mathematics.

Real-world problems were also a source of mathematical theory for the greatest mathematician of the 18th century, Leonhard Euler (1707–1783),[186] whether it was the puzzle of the seven bridges of Königsberg or the degree of harmony in consonant intervals. For example, the question that Euler asked was, "Is it possible to take a walk through Königsberg, crossing all seven bridges exactly once, and still end up back at your starting point?" Yet Euler's proof that it was not, kicked off a new branch of mathematics—topology—in which problems explored were detached from that reality.

We owe almost all modern mathematical notation to Euler, and without him modern mathematics would be inconceivable. The theory of complex numbers also comes from Euler. The basis of this theory is the imaginary number i, defined as $i^2 = -1$. While this form of mathematics seems purely theoretical (after all, the square root of a negative number does not "exist," because the square of a number is always positive), it soon led to the most diverse applications, such as in the fields of fluid dynamics and electromagnetism, and later in signal analysis and control theory. The same goes for calculus: Euler derived a function that was equal to its derivative, that is, the direction of the tangent line at any point on the function's curve was equal to the value of the function. This is the function e^x, where e is called Euler's number (2.718281 . . .). To top it all off, Euler also showed that the bizarre number i, together with Euler's number e, resulted in what many consider the most beautiful and remarkable equation in all of mathematics:

$$e^{i\pi} + 1 = 0$$

This equation brings together the five most important numbers in mathematics: e, i, π, 1, and 0.

In addition to the interdependence between mathematics and its fields of application—from accounting and mechanics to insurance—there was also a mathematics that was separate from its applications, such as Euler's work in number theory.[187] What we have seen, however, is that the interdependence of mathematics and its fields of application has led to the use of an empirical cycle in a discipline that was once considered immune to it.

The Synthesis of Agnesi, the First Female Professor of Mathematics

In 18th-century mathematics, almost 1,400 years after Hypatia of Alexandria (see chapter 3.4), we again see a celebrated female mathematician: Maria Gaetana Agnesi (1718–1799). She succeeded at representing early modern mathematics up to Euler as one big synthesis. In her *Instituzioni analitiche* from 1748, Agnesi brought together the results and propositions from calculus, forging it with algebra into a systematic whole. Although her work was intended to be pedagogical ("for the Italian youth"), Agnesi also added new discoveries, such as her research on the curve known as the *witch of Agnesi*, although it had already been described by Newton and Leibniz. The strange name is the result of a mistranslation by the English mathematician John Colson, who confused the Italian word *siera* (curve) with another Italian word, *avversiera* (a woman possessed by the devil).[188] Today, the curve, still known by that name, is widely used in physics to describe the energy distribution of spectral lines. Agnesi was appointed professor of mathematics at the University of Bologna in 1750, making her the world's first female professor of mathematics and the second female professor of any subject, after Laura Bassi, who was appointed professor of physics a few years earlier at the same university (see above).[189]

While Descartes is regarded as a great innovator in bringing geometry and algebra together, Agnesi's merging of calculus and algebra is mostly taken to be a merely pedagogical contribution.[190] In her own time, however, Agnesi's work was praised as a complete overhaul of traditional mathematics. In 1749 the French Academy wrote of the *Instituzioni analitiche*, "There is no other book, in any language, that allows a reader to penetrate the fundamental concepts of calculus as deeply or quickly."[191] After these words of praise, in the 19th century Agnesi's name starts to disappear from historiography just as quickly as female professors would disappear from the University of Bologna. Her fame has since been reduced to the "witch" that bears her name. The fact that the deeply religious Agnesi was herself extremely modest about her work—a quality foreign to Descartes—probably did nothing to help perpetuate a recognition of her contributions. For this reason, further research into Agnesi's influence on the mathematics of her time, as well as the contribution and influence of other female scientists and scholars, would be highly desirable for a general history of knowledge.

Mathematics outside Europe

It was long assumed that mathematics outside of early modern Europe exhibited nothing but decline. On closer inspection, this view appears to be based on prejudice rather than on fact. In recent years it has become clear that the number of mathematical manuscripts in China, Japan, India, the Arab world, and Africa is many times greater than previously thought. Surprisingly, it is especially the manuscripts from before 1500 that have been studied. The standard view was that the regions outside Europe had interesting discoveries to report only up to the early modern period, and, in scientific and scholarly terms, these regions had produced little of value after 1500. This led to a historical shortsightedness that has only recently begun to change.

For example, we now know that in China a new mathematics was also developed during the Ming dynasty, especially within concrete applications. Several kinds of discoveries were made in abacus mathematics, including, for example, perfecting algorithms for the crosswise multiplication of fractions.[192] Yet European and Chinese mathematics constituted two different worlds: when in the 17th century, works of Euclid were translated into Chinese by Xu Guangqi

(1562–1633) and the Italian Jesuit Matteo Ricci, they garnered little interest in China. Euclid remained practically untouched in the Middle Kingdom for a century. Although Joseph Needham has assembled several mathematical discoveries from the Ming and Qing periods,[193] there is still no general overview, perhaps because many of these mathematical innovations took place within other disciplines, such as astronomy.[194] In the Islamic world, too, mathematics and logic had hardly come to a standstill after 1500.[195] What we can conclude is that the Chinese and Arabic mathematics from the period 1500–1900 had little influence on the European mathematics and vice versa.

But there is also a region where early modern mathematics does have a direct link to mathematics elsewhere, which is India. We already saw earlier in the chapter that within the Indian Kerala school important mathematical discoveries were made starting in the 14th century, especially regarding series of trigonometric functions. Many of these discoveries were matched in Europe only two centuries later. The activities of the Kerala school continued well into the 16th century. For example, infinitesimal approximations of trigonometric functions using a method known today as the Taylor series were discovered a century earlier in India than in Europe.[196] Many historians seem willing only to attribute non-Western science and humanities an equal role in the premodern period. This has led to what is perhaps the greatest paradox in knowledge history: while it is relatively easy to consult sources from China, India, the Arab world, and Africa from the classical and postclassical periods, it is much more difficult for the period 1500–1900.

5.5 Medicine: The Long-Awaited Empirical Turn

In many ways, medicine has turned out to be something of an outlier in this book. In contrast to other disciplines, in classical antiquity there was hardly any convergence between principles and patterns (chapter 3.5); and in the postclassical period, again unlike in other disciplines, we find no reduction of principles in medicine (chapter 4.5). As a possible explanation, I have suggested that medicine was largely "learned" and "philosophical" and not particularly "empirical." Theoretical principles (such as the four humors) were so predominant, and the taboo on dissection of the human body so great, that new empirical findings were rare. In the 16th and 17th centuries all this changed in Europe, China, India, the Ottoman Empire, Ethiopia, and West Africa.

From Art Theory and Philology to Medicine: Andernach, Vesalius, and Paracelsus

The humanist study of the classics extended to all texts, scientific and scholarly. Similar to the astronomer Regiomontanus, who had applied the critical philological method to Ptolemy's *Almagest* (see above), Johann Winter von Andernach (1505–1574) was the first to apply the new philological method to the medical texts of Galen and Hippocrates.[197] Thanks to his excellent knowledge of Greek, combined with his training in medicine at the Sorbonne, von Andernach breathed new life into Galen's texts. His students included the Spaniard Michael Servetus (1511–1553), whom I already mentioned as one of the discoverers of the pulmonary circulatory system (see chapter 4.5). However, Servetus died an untimely death when under John Calvin's reign he was sentenced to be burned at the stake in Geneva for his dissenting views on the Trinity.

Von Andernach's most famous student was the Flemish anatomist Andries van Wesel, better known as Andreas Vesalius (1514–1564).[198] Vesalius was unimpressed with von Andernach's anatomy lessons based on Galen, and he identified more than 200 errors in Galen's works. Yet it is to von Andernach's merit that he persuaded Vesalius to support his critique with empirical findings—just as von Andernach himself had done in his philological reconstructions of Galen and Hippocrates. Does this point to a deeper analogy between 16th-century philological analysis and 16th-century anatomical dissection? That is, is it plausible to say that Vesalius dissected a body into anatomical parts in the same way that philologists dissected a text into its constituent parts, where each part is related to the whole? This seems plausible because of the link between the philological method of analysis and its art-theoretical counterpart that was prevalent from the 15th century, mainly owing to Leon Battista Alberti (see the beginning of the chapter). In *De Pictura* Alberti presents not only his famous description of linear perspective but also an analysis of a depicted narrative, a *historia*. For example, according to Alberti, the composition (*compositio*) of a depicted narrative consists in a hierarchical relationship of its constituent parts: how such a story is divided into individuals and how those individuals are divided into body parts in certain postures, in turn divided into smaller constituents such as the hand, fingers, phalanxes, and so forth.[199] In short, an analysis of the *compositio* of an image is not only philological but also anatomical. The concept of *compositio* was known to anyone trained in philology and rhetoric, that is, to any 15th- and 16th-century humanist. As Alberti had already shown, this hierarchical method of analysis was useful not only for philology and rhetoric, but also for art theory and anatomy. The method was even used in the analysis of music and literature.[200] The step from art theory to anatomy must have been practically self-evident for Vesalius, considering that he was artistically trained. Together with the artist Jan van Calkar, he supplemented his work with exceptionally beautiful images of anatomical compositions (see figure 11). So there appears to be a transfer of knowledge from art theory to medicine.

In his magnum opus *De Humani Corporis Fabrica* (*The Construction of the Human Body*) from 1543—the same year as Copernicus's *De Revolutionibus*—we see how Vesalius applies the hierarchical method of analysis: each part of the body is related to a larger whole.[201] In seven books, Vesalius successively describes the connections between the bones and joints, between the ligaments and muscles, the veins and blood vessels, the nervous system, the digestive and reproductive organs, the heart and respiration, and finally the brain. In this anatomical work, the structure of the human body is comparable to the above-mentioned art-theoretical notion of the composition of an image, as we can see in the anatomical representations in the text, which are attributed to Jan van Calkar.

With his *De Humani Corporis Fabrica*, Vesalius surpassed Galen's works in one fell swoop. Nevertheless, to show respect for his illustrious predecessor, Vesalius states in book 2 that the anatomical structure described by Galen was not incorrect; it simply did not apply to humans. And indeed, Galen had used animals instead of humans for his dissections. Nevertheless, after the publication of his work, Vesalius was so fiercely attacked by supporters of Galen that he decided to leave academia and serve as the personal physician of Emperor Charles V and, after his abdication, as that of his son Philip II. Nevertheless, Vesalius's premise that any anatomical claim needed to be tested by dissection came to be widely accepted among the next generation of physicians.

Despite its strict empirical and analytical approach, there is no empirical cycle in *De Humani Corporis Fabrica*: cyclic interactions between empirical findings and theory are lacking, considering that Vesalius's approach is primarily empirical and proposed no theory about how the body functions. However, his approach is based on the notion of the part-whole relationship, which could be seen as a theory concerning the cohesion between the parts of the human body, but this notion is never questioned or tested. For this reason, we should see Vesalius's work primarily as an analysis of the human body in terms of its parts and their interrelationship without deeper principles about how it operates. This

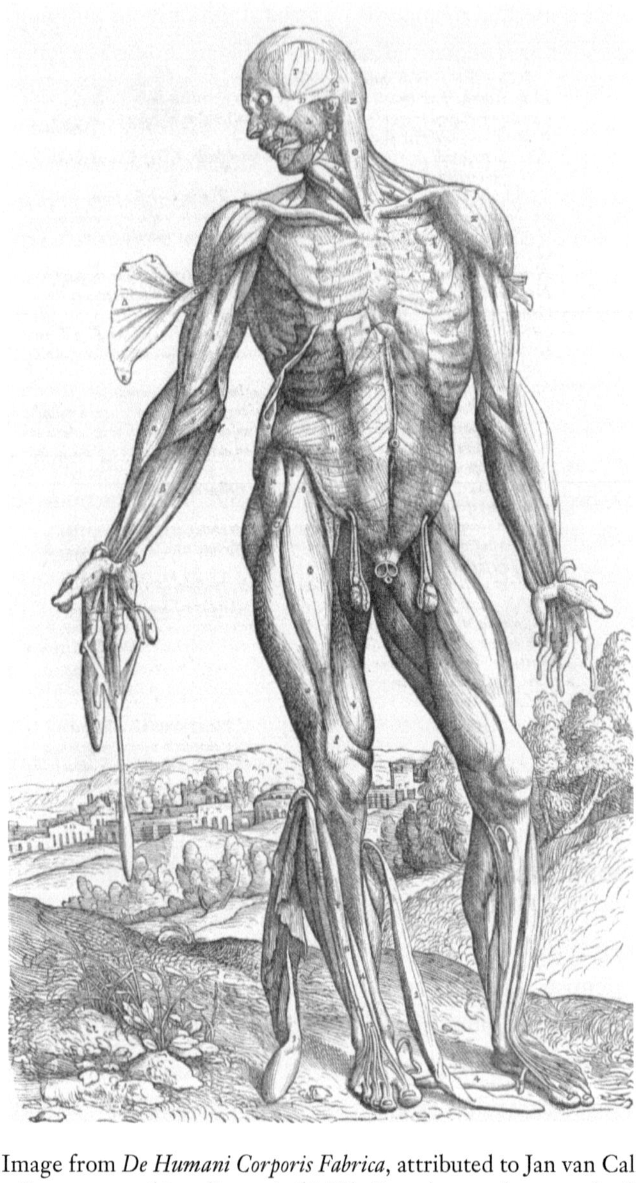

Figure 11. Image from *De Humani Corporis Fabrica*, attributed to Jan van Calkar, 1543. https://commons.wikimedia.org/wiki/File:Fourth_muscle_man,_by_Vesalius._Wellcome_L0001647.jpg.

does not detract from the originality of Vesalius's work. Not a single human body had been dissected for centuries. To arrive at a new theory about the human body, it first had to be mapped out. And that is what Vesalius did: scholarly medicine was transformed into an empirical discipline.

It was not unusual to attack Galen and other classical authors in the 16th century. The *Physics* of Aristotle, the *Almagest* of Ptolemy, and the *Donatio Constantini* document (see above) were also subjected to fierce scrutiny. The physician, alchemist, and theologian Paracelsus (1494–1541) went so far as to burn the medical textbooks of Galen and Avicenna publicly.[202] "Where the philosopher ends, the doctor begins," were his lofty words. But Paracelsus's approach was different from that of Vesalius. Although relying on an empirical method, Paracelsus combined mysticism, alchemy, astrology, and magic in his medical practices without testing their effects. Nevertheless, Paracelsianism—as the movement was called—had great appeal, with the use of chemical substances leading to new medical practices, such as we encounter with the Flemish Jan Baptist van Helmont (1579–1644). What connected Paracelsus and Vesalius was their rejection of Galen and their search for a new approach. We now associate the term "scientist" more with Vesalius than with Paracelsus, but in the 16th century that was far from clear.

We might wonder whether Vesalius's work should be categorized under medicine, or whether it could be better thought of as pertaining to anatomy, but then we would be overlooking the fact that medicine underwent a significant transformation during the 16th and 17th centuries. It was transformed from a discipline based primarily on book knowledge into an empirical field that studied all aspects of the human body. It is this field that was first referred to by the humanist physician Jean Fernel (1497–1558) as "physiology," a term he coined himself.[203] Physiology became one of the central subjects of medicine, and the study of anatomy was soon integrated into learned medicine by the likes of John Caius (1510–1573) in England, Conrad Gessner (1516–1565) in Switzerland, and Pieter Pauw (1564–1617) in the Netherlands.[204] Vesalius's primary follower was Fabricius of Acquapendente (1533–1619), who not only dissected human bodies but also addressed the question of how all those different parts of the body work.[205] Fabricius studied areas such as embryology and discovered the valves in the veins. He also designed the famous anatomical theater in Padua (1595), which attracted students and scholars from all over Europe, including the brilliant young man William Harvey.

William Harvey: From Empiricism to Theory and Back Again

Whereas Vesalius was mainly descriptive-analytical, with the English physician William Harvey (1578–1657) we find both empirical research and theory formation.[206] After graduating from Cambridge, Harvey joined Fabricius in Padua in 1599. But it wasn't until a few decades later, when Harvey was back in England, that he came up with his theory of blood circulation. It was this theory that provided a new picture of how the human body functioned, especially the heart, liver, and blood. In his *Exercitatio Anatomica de Motu Cordis et Sanguinis in Animalibus* (*An Anatomical Exercise on the Movement of the Heart and Blood in Living Beings*), usually abbreviated to *De Motu Cordis* (1628), Harvey asked how the blood moves through the body.[207] In his book he explains in detail how he had arrived at his discoveries and theory.

Harvey's first discovery occurred when estimating how much blood passes through the heart in a half hour. According to Galen's generally accepted theory, the liver was the source of the blood, which was then brought to all parts of the body, after which it did not return to the heart (see chapter 3.5). This meant that blood had to be continuously produced and supplied by the liver. To evaluate Galen's theory, Harvey began by estimating the following factors: the volume of the heart, the amount of blood expelled with each pumping movement of the heart, and the number of heartbeats per half hour. Harvey estimated the heart's volume to be about 1.5 imperial fluid ounces (28.4 ml), which is a major but deliberate underestimation. Each time the heart pumped, Harvey estimated that about one-eighth of that blood would be expelled, another deliberate underestimate. Furthermore, Harvey estimated the heart rate to be 1,000 beats per half hour, which is a pulse of fewer than 34 beats per minute—again an underestimate (resting heart rate is between 60 and 100 beats per minute). By multiplying these numbers, Harvey came to 10 pounds and 6 ounces of blood per half hour. By subsequently multiplying this number by 48 half-hour periods, he deduced that the liver produced 540 pounds of blood per day, making the amount of blood produced far greater than the weight of the entire body! Harvey then concluded that Galen's theory could not be correct.

Harvey argued that the blood had to flow back to the heart so that it could be recirculated without having to be produced by the liver. This meant that the blood must move in a cycle. Harvey immediately understood that this required connections between arteries (which run from the heart to organs) and veins (which run from organs to the heart) to allow circulation. But no such connections were detectable, even after dissecting a human body. So, Harvey came up with some experiments to test his theory of the circulatory system indirectly. He began by working with snakes and fish. He connected their veins and arteries and found that when veins were connected, the heart deflated, but when he did the same with arteries, the heart swelled up.

But Harvey's goal was to demonstrate the existence of the circulatory system in humans. He devised an experiment for this purpose without having to cut into human veins. He studied blood flow by means of a band he tied around a person's arm. By pulling the band tight, the blood flow was made to stop, as expected. But by loosening the band a bit, Harvey was able to allow the blood from the arteries to flow while stopping the blood from the veins. This is because the arteries lie deeper in the body than the veins. When only this arterial blood was allowed to pass through, the veins became visible and swollen. However, when the band was pulled very tight, the veins were not visible. From this, Harvey deduced that the blood was expelled from the heart through the arteries and that it flowed back through the veins (figure 12). And this meant that there had to be connections between veins and arteries. These connections are today

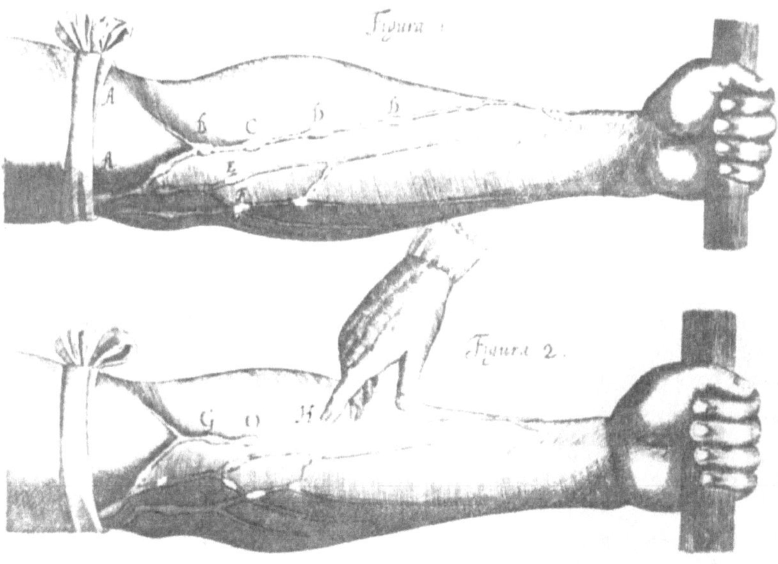

Figure 12. Image from Harvey's *Exercitatio Anatomica de Motu Cordis et Sanguinis in Animalibus*, 1628. From Henry E. Sigerist, (1965) *Große Ärzte* (Munich: J. F. Lehmans Verlag, 1958), plate 26, p. 120.

known as capillaries, but Harvey believed that veins and arteries absorbed blood through pores.

What we see at work here is an empirical cycle: Harvey derived a prediction from Galen's theory, which subsequently turned out to be incorrect. He then posited a new theory, the results of which he tested with further success, after which he further refined his theory of circulation by including the valves in the veins and explaining how they worked and what their function was. Then he tested the refined theory again, and so on. We have already encountered the empirical cycle in all other disciplines in this chapter, starting with 15th-century philology, linguistics, art theory, and musicology, followed by 16th-century astronomy and mechanics, which makes 17th-century medicine the youngest shoot off the trunk.

Now, the idea of circulation was not new: pulmonary circulation between the heart and the lungs had been described by Michael Servetus in the 16th century and a few centuries earlier by Ibn al-Nafis (see chapter 4.5). But Harvey had no knowledge of these texts. That may be understandable for al-Nafis's Arabic text, but less so for Servetus's work. However, Servetus published his theory as a theological work,[208] and it had been condemned as heretical by both Catholics and Protestants, after which Servetus was burned at the stake. As many of his books as possible were burned as well.

It is tempting to see Harvey as a "modern" scientist with his empirical cycle. But if you read his *De Motu Cordis*, it is immediately noticeable that Harvey's language is thoroughly Aristotelian. Harvey viewed the body as being moved by "vital forces" (see chapter 3.5). And, like Aristotle, he relied on the teleological aspects of the body: the "purpose" of the circulatory system was to transport vital blood to the periphery, after which it returns to the heart to be revived. Although Harvey rejected Galen's theory, he did not consider himself a supporter of the new philosophy of René Descartes according to which the world is subject to natural laws. But Descartes himself thought differently: in his *Discours de la méthode* (see above), he called Harvey's theory a support for his vision of considering the whole world—from the universe to the human body—as a mechanism. For Descartes, only the human mind was an exception, like a ghost in the machine.

The Microworld of Medicine: Evidence for Harvey's Theory

The invention of the microscope is attributed to the Dutch eyeglass makers Zacharias Janssen and his son Johannes Zachariassen. As far as we know, however, they did not conduct any physiological research with their discovery. One of the first to do this was Robert Hooke, who we already encountered in our discussion of Newton above and to whom we owe the first biological use of the word "cell."[209] The Italian Marcello Malpighi (1628–1694) used the microscope to derive a mechanical model of living beings.[210] What Harvey could not see with his own eyes in 1628, Malpighi succeeded in viewing in 1661: he was the first person to see the extremely fine-grained structure of blood vessels in frogs, confirming Harvey's theory of circulation. In *De Pulmonibus*, Malpighi describes the alveoli at the ends of the bronchial branches in the lungs, where air and blood are mixed, after which the blood flows back to the heart.

The Delft cloth merchant Antoni van Leeuwenhoek (1632–1724) managed to make the microscope even more powerful, allowing him to discover a whole new world.[211] Over the course of his life, he built 247 microscopes, some of which could magnify up to nearly 300 times. Unfortunately, van Leeuwenhoek took the secret of his microscope construction to his grave, and it was not until the 19th century that microscopes improved. Van Leeuwenhoek placed almost everything he could get his hands on under his microscope: wood, plants, insects, water droplets, bones, muscles, nerves, teeth, hair, blood, and even his own sperm, making one discovery after another with the greatest of ease. For example, he was the first to see and name corpuscles and the uric acid crystals responsible for gout, and to top it all off, the *dierkens* ("animalcules") he found in all sorts of liquids. This made van Leeuwenhoek the first person to see bacteria. When the British Royal Society was told of his discoveries, the society's secretary, Henry Oldenburg, learned Dutch in order to communicate with van Leeuwenhoek, who was not fluent in Latin. Many hundreds of publications followed in the Royal Society's *Philosophical Transactions.*

All these discoveries led to new theories, such as in the subdiscipline of reproductive medicine. Dutchmen like Reinier de Graaf and Jan Swammerdam, as well as the Dane Niels Stensen, plunged into this.[212] De Graaf, for example, discovered the "little balls" released during ovulation, which led him to deduce the existence of the egg, or ovum. It was also de Graaf who first suggested that the fetus was created out of the fusion of a sperm cell and an ovum. A fierce discussion immediately ensued between the animalculists and the ovists on the question as to whether the sperm or the ovum was most decisive in the formation of the fetus. Although these questions could not be answered until a few centuries later, by around 1660, traditional Aristotelian-Galenic medicine had given way to the interaction between empirical observations and theory. This did not mean that notions such as bodily fluids or humors or practices such as bloodletting disappeared all at once—that would not happen until the 19th century—but (early) modern medicine had embraced a new method. Classical medicine became subject to increasing ridicule, not only by medical doctors but also by playwrights such as Molière (1622–1673) in *Le malade imaginaire.*

Medicine in the Enlightenment and Beyond

Compared to the many 17th-century discoveries, 18th-century medicine in Europe was a bit lackluster. The most prominent physician in the Enlightenment was the Leiden professor Herman Boerhaave (see also above), whose fame stretched from China to South America.[213] In Boerhaave's time, Newtonianism as a philosophy and scientific approach was so predominant in the Netherlands that people believed that the greatest progress could be made through mathematical theorizing. The intricate interaction between theory and experimentation, as practiced by Harvey, suddenly seemed more distant than ever. Boerhaave and his followers believed that illness and health could be fully understood using the Newtonian concepts of force, weight, and pressure. Sickness and health were nothing but the presence or absence of hydrostatic equilibrium. It almost seemed like a return to Galen's theory of the bodily fluids (humorism) but with Newtonian underpinnings. That Boerhaave's medical theories could not be substantiated empirically did not prevent his *Institutiones Medicae* (1708) from becoming the most famous medical work of the Enlightenment.

Nevertheless, medicine's lag behind the other disciplines was striking. Around 1800, many recognized that medical science had few cures to offer for ailments, in spite of all the research that had been done on, for example, lung function in the interim. Important innovations were made, of course, such as the impressive classification of nature into binary taxonomies by the Swede Carl von Linné, better known as Linnaeus (1707–1778), in his *Systema Naturae* (1758). And then there was the fascinating discovery, made by the Italian Luigi Galvani in 1792 that nerves were powered by electricity. But these new insights did not lead to concrete medical results.

Medicine was and remained an outlier: while the empirical cycle had convincingly ensconced itself in the work of Harvey and a few of his 17th-century colleagues, it took more than 150 years before this cycle manifested itself again to the same degree in medicine. But then there was no stopping it: in the

19th century, medical practices started to be tested, improved, and tested again, sometimes successfully and sometimes not, but eventually with extremely impressive results—from the development of vaccination by Edward Jenner (1749–1823) and Louis Pasteur (1822–1895) to the discovery of the tuberculosis bacterium by Robert Koch (1843–1910); see below for further discussion. However, conducting and testing medical treatments does not equate to distilling these treatments into a set of theoretical principles that apply to the functioning of the human body as a whole. Even today, medicine is still the odd man out compared to most other sciences with respect to such principles. Some argue that medicine in the 21st century has even slipped to a new extreme, where therapies are investigated and successful ones applied, without any further search for deeper principles or theories.[214] That is disappointing, since theoretical principles are capable of producing new and unexpected patterns. *Evidence-based medicine*, which focuses on choosing a medical treatment based on the best available evidence, has mainly led to better testing of forms of treatment rather than to a theoretical foundation for those treatments.[215] It is in the new field of *biomedical science* that researchers investigate how the human body functions as a whole, from the molecular level to the organism. As happened centuries ago with anatomy, biomedical science will most likely be sooner or later integrated into medicine, if that is not already underway.

The Empirical Turn in Chinese Medicine: The Discovery of Inoculation

China, too, experienced an empirical turn in medicine accompanied by important discoveries. This appears to be at odds with the dominant view, in which all the great medical discoveries from the 16th century onward have taken place in Europe, while medical practices elsewhere were limited to "traditional medicine." According to this view, Europeans exported their medical knowledge to the rest of the world, leading to the emergence of a globalized medicine in the 20th century. This standard view is not entirely untrue, because Western medicine is practiced all over the world, albeit often alongside traditional forms of medicine. But clear misconceptions stand out in this standard view. For example, one of the most important medical discoveries took place not in Europe but in China: inoculation against smallpox. This inoculation method is described in detail in 16th- and 17-century Chinese sources, indicating a successful preventive inoculation practice.[216] Yet this discovery continues to be attributed to the Englishman Edward Jenner (1749–1823), who began to experiment with it at the end of the 18th century. Although the Chinese practice of inoculation had been known in Europe since 1700, it had long had a sketchy reputation and was thought to be unempirical and superstitious. Nevertheless, several regions of the world—the Ottoman Empire, Ethiopia, and India, in addition to China—had the smallpox virus (albeit unknown as such) fairly under control for several centuries when the first inoculation programs in Europe had yet to begin. In addition, the earlier principle of inoculation was the same as propagated by Jenner in 1796: the introduction of disease-carrying material (with reduced activity) into the body, resulting in immunity.[217] Smallpox was the deadliest disease the world has ever known: in the past 3,000 years, 1 in 10 people have died of the smallpox virus.[218]

However, we do not know which individual or group this important discovery of inoculation should be attributed to. According to Chinese tradition, Song chancellor Wang Dan (957–1017) lost his son to smallpox, prompting him to look for ways to spare his other relatives. He brought together doctors, sages, and magicians from all over the empire to arrive at recommendations on how to cure or prevent this disease, until a sacred person descended from the Buddhist Mount Emei and performed an inoculation with smallpox material, stopping the further spread of the disease.[219] Apart from this creation myth, there is no concrete evidence that inoculations actually took place in China prior to the 16th century.

It was not until the reign of the Longqing Emperor (1537–1572) in the Ming dynasty that inoculations against smallpox were mentioned.[220] The medical reports show that patients who had recovered from a mild bout of smallpox were chosen as donors to minimize the risk of a dangerous infection. The technique involved collecting and drying the crusts that formed after the smallpox vesicles had dried out. Three or four scabs were mashed into a powder and either placed in a cotton casing in a patient's nostril or blown into one of the nostrils through a pipe. In this way, the patient became infected with a less severe variant of the smallpox, leading him or her to develop only a few crusts and to then becoming immune. When the first symptoms appeared in the form of blisters, the patient was isolated from others until the symptoms disappeared. During the Qing dynasty, this method of inoculation was further refined, preserving pieces of crust in bottles and ritualizing the practice: the pipe needed to be of silver and the right nostril was reserved for boys, the left one for girls. This ritualization was probably one of the reasons why Europeans mistook Chinese inoculation for a form of superstition.

Origin of the Empirical Cycle in China: Medicine or Philology?

Was the Chinese practice of inoculation really the result of an empirical cycle? We lack sources on the first inoculation practices, but what we do know is that Chinese physicians had knowledge of the patterns of the disease as well as of some underlying principles. They understood that it was smallpox material that made someone ill, even though they did not know that it was a virus—although the 17th-century Wu Youxing did hypothesize the existence of "terribly small particles," *liqi*.[221] The Chinese also understood that those who survived the disease could not get smallpox again (the concept of immunity). In addition, it was understood that a small amount of pathogenic material from a mild variant could also lead to immunity (known today as the "weakening pattern"). These insights could lead to such a successful inoculation practice only through a cycle of empirical testing and hypothesis formation. The introduction of contaminated material must at least have been tested informally on the basis of a positive or negative result, after which the method was further improved using weakened material, until it developed into a relatively safe inoculation practice in the Qing dynasty. This is precisely what we call the empirical cycle.

But where did the Chinese Ming-Qing physicians get the knowledge of this interaction between experiment and theory? Was it invented in medicine itself or did the empirical cycle come from elsewhere? Could the empirical cycle come from the humanities in China as well? In European astronomy, physics, and medicine, we were able to reconstruct who influenced whom. For example, Kepler learned about the empirical cycle thanks to his philological training. In physics, Galileo experienced the empirical cycle owing to the musicological experiments of his father, Vincenzo. And in medicine, Vesalius knew of the empirical testing conducted by his philology teacher von Andernach as well as from art theory. This does not mean that Kepler, Galileo, and Vesalius could not have invented the empirical cycle independently, but rather that they were already familiar with this cycle from their humanist teachers or education. Can we also determine such personal influence between Chinese physicians and scholars from other areas of knowledge? That is no easy task, because it is unknown who applied the inoculation technique first.

If we focus on the 17th century, when the practice of inoculation was widespread in China, we can say with certainty that the physicians of the Ming and Qing dynasties must have been familiar with the work of the respected and widely read philologist Gu Yanwu. Besides being the founder of the Empirical

School of Textual Criticism (see above), Gu was a keen critic of the Chinese medicine of his day. He described many medical practices as a form of guesswork and eviscerated baseless medical prescriptions.[222] In his textual criticism, Gu successfully advocated a strict empirical approach, leading his students to succeed in exposing one classic text after another as a forgery (see above). However, there is no evidence of any direct collaboration between philologists and medical practitioners in China in the way we find in Europe. The question of the origin of the empirical transformation in the Chinese practice of inoculation therefore remains unanswered. We must of course keep open the possibility that this change did not come from learned medicine (or from other learned disciplines such as philology) but that it originated in folk medicine. After all, there was already an empirical tradition dating back thousands of years, especially in traditional herbal medicine (see chapter 2.5).

Inoculation Spreads from China to Europe via India, Turkey, and Africa

Chinese inoculation was a success story like no other. In the 17th and 18th centuries, we encounter the practice in many places in the world: initially in India, where the invention of inoculation has been claimed at times,[223] as well as in Turkey and Ethiopia. Europe, however, remained dormant for quite some time. In 1700, the English physician Joseph Lister wrote about the Chinese practice of inoculation in a letter to the Royal Society, but this did not lead to any action.[224] The philosopher Voltaire described the practice of inoculation in Turkey. In his *Lettres philosophiques* (*Philosophical Letters*, letter 11) from 1742, he describes the practice as originating from Turkey's neighboring country of Circassia, where smallpox had practically been eradicated. And that was not all: in the first half of the 18th century the method was found in Ethiopia, West Africa, and other parts of Africa.[225] In short, half the world was practicing inoculation, but not Europe.

So how did the idea of the inoculation reach Edward Jenner? Jenner's great contribution was that in 1796 he demonstrated that inoculation with cowpox also led to immunity against common smallpox. Furthermore, inoculating with cowpox was safer than using a mild form of smallpox, since cowpox is not fatal to humans.[226] The material used was therefore also referred to as a *vaccine* by Jenner, derived from *vacca*, the Latin word for "cow," and the practice itself came to be referred to as *vaccination*. Nevertheless, Jenner's method remained controversial for some time, both among physicians and theologians and among the public, but its success was hard to argue with. Thanks to his efforts, smallpox had largely disappeared from England by 1840. But Jenner did not come up with the idea of inoculation himself, nor did he reinvent it. Voltaire's works on Turkish inoculation techniques were fairly well known in England, but perhaps more famous was the courageous action of Lady Mary Wortley Montagu (1689–1762), whose brother had died of smallpox. Lady Montagu's husband served as a British diplomat in the Ottoman Empire from 1716 to 1718, where she observed inoculation practices up close. She now wanted to have her own daughter inoculated using this method. Back in England, experiments were conducted on six English prisoners, who would be released as their reward. All six survived the inoculation, after which in 1722 not only Lady Montagu's daughter but also the children of the English crown prince were inoculated. Slowly but surely, the practice of inoculation spread across the European mainland, first in Holland, where the practice began to be implemented in 1748, usually successfully, but sometimes with the alarming result of smallpox breaking out among those who were inoculated.[227] It is for this reason that inoculation remained a marginal phenomenon in Europe for decades, even though failure to inoculate resulted to many more deaths. In 1763, the practice was banned in France, prompting Voltaire to accuse the French parliament of being responsible for the deaths of thousands of children.[228] In the Americas, inoculation was mainly practiced by enslaved people who had brought the method with them from Africa.

Vaccination spread more widely only after Edward Jenner's successful experiments. Despite initial resistance to injecting humans with animal material, within three years of Jenner's initial experiments, more than 100,000 people had been vaccinated. As far as is known, none of these patients died from the vaccination. Louis Pasteur subsequently also found vaccines for cholera, anthrax, and rabies, and the discovery of other vaccines followed. Today, inoculation is perhaps the most imaginative form of preventive medicine. Although the possibility cannot be ruled out that it was invented independently in several locations in the world,[229] the oldest source for the technique is 16th-century China.

A Second Empirical Turn in China: From Li Zhizhen to Wang Qingren

The empirical turn in Chinese medicine was not restricted to the discovery of inoculation. A second empirical turn took place in the Ming-Qing period, but before this new medical science could get off the ground, it was overtaken by the arrival of European medical texts. Western anatomy and physiology turned out to be so much more advanced that their Chinese counterparts were nipped in the bud and quickly forgotten.

In a sense, this second empirical turn began in the 16th century with Li Zhizhen (1518–1593). In his immense *Compendium of Materia Medica* (*Bencao gangmu*), a work encompassing nearly 4,500 pages, he provided an overview of all known medicinal plants and herbs. But he also gave good recommendations on how to disinfect a sick person's bedding and clothing.[230] The discussion concerning infecting and disinfecting had barely begun in Europe at this time, but as we saw above with the practice of inoculation, the Chinese knew very well what contamination meant. Li was also an excellent observer. In his study of the properties of the medicinal plants, he recorded not only patterns but also exceptions, a habit dating back to Chinese antiquity (see the observation of anomalous sunspots, supernovas, and comets in Chinese astronomy in chapter 3.2). Li's observations were so detailed that the later botanist Pehr Osbeck (1723–1805), a student of Linnaeus, was able to immediately add 600 Chinese plants to Linnaeus's work when he visited China.

With Xu Dachun (1683–1771), in addition to an empirical approach, we also find a high degree of theory formation. According to Xu, every illness has a specific cause that cannot be explained by the condition of the body as a whole, as had been assumed in earlier Chinese medicine, but could be understood only when traced to a specific organ. Xu's vision had little to do with what is today regarded as "traditional" Chinese medicine, such as acupuncture or herbal medicine. When Chinese medicine became popular in 20th-century Europe, Xu was overlooked.[231]

The most convincing empirical approach is found in the anatomical studies of Wang Qingren (1768–1831).[232] Almost by chance, Wang discovered that older Chinese representations of the human body were incorrect. He started his observations when he was confronted with huge fields full of rotting corpses. Most of the bodies were in such a state of decomposition that their internal structure was visible without any dissection. It was easy for Wang to determine that these bodies looked very different on the inside from what he had learned in medical textbooks. What makes Wang special is that he put the value of his own observations above that of tradition or authority. He distanced himself from earlier Chinese medicine without relying on others and without access to European medical textbooks. In his *Correction of Medical Occupational Errors* (*Yilin gaicuo*) from 1830, Wang gives one striking description after another:

"The liver has four lobes; the wide surface is directed upwards and attached to the spine. The gallbladder is attached to the second lobe of the right half of the liver." And: "The pancreatic duct and the common duct are an inch to the left of the pylorus." It must be noted that Wang also made mistakes of his own, especially in interpreting the function of the aorta, which he mistook for a trachea. Wang recorded his observations not only textually but also visually in the form of numerous images of human organs (figure 13).

Unfortunately, Wang was not a great illustrator, and he did not employ an artist of the caliber of Jan van Calkar, who had captured the dissections of Vesalius with some of the best artwork of its time in Europe (see figure 11). Yet Wang's images and descriptions testify to a hitherto unknown medical empiricism in China. Thus, the Chinese empirical turn in anatomy begins with Wang's observations in 1830—several centuries later than in Europe, but in full glory nevertheless. Although Wang's work was reprinted with commentary several times, he did not create a new school of anatomy. After his death in 1831, European anatomical manuals poured into China in large numbers. These works were so much more detailed than Wang's studies that his empirical activities soon were drowned out. Wang's research may have been important for the acceptance of Western medicine in China, though. This time, foreign medicine was not seen as a "break from tradition" or "unsuitable" for China—as had happened several centuries earlier with the arrival of the Jesuits—but as a natural continuity with previous knowledge. This notion of continuity with the past was a sine qua non for Chinese scientists and scholars, as can be illustrated by the fate of an anatomical handbook by Pierre Dionis, previously translated by the Jesuits, from 1718: it was deemed unsuitable and placed under lock and key in the Imperial Library of the Kangxi Emperor.[233]

After Wang's death, Western medicine gradually became established as the basis for learned medicine in China. From the inception of the Republic of China in 1912, the new state's first goal was to advance Chinese medicine. There was the additional challenge of uniting the two medical traditions—Western and Chinese. It became common to say that Western medicine was good for acute conditions, while Chinese medicine was needed for chronic conditions.[234] However, the term "Chinese medicine" was not used to refer to the successful inoculation techniques from the 16th and 17th century, let alone the 19th-century empirical studies of Wang Qingren, but was restricted to the traditions of acupuncture, moxibustion, and food therapy. The pioneering work of the illustrious Xu Dachun and Wang Qingren was forgotten.

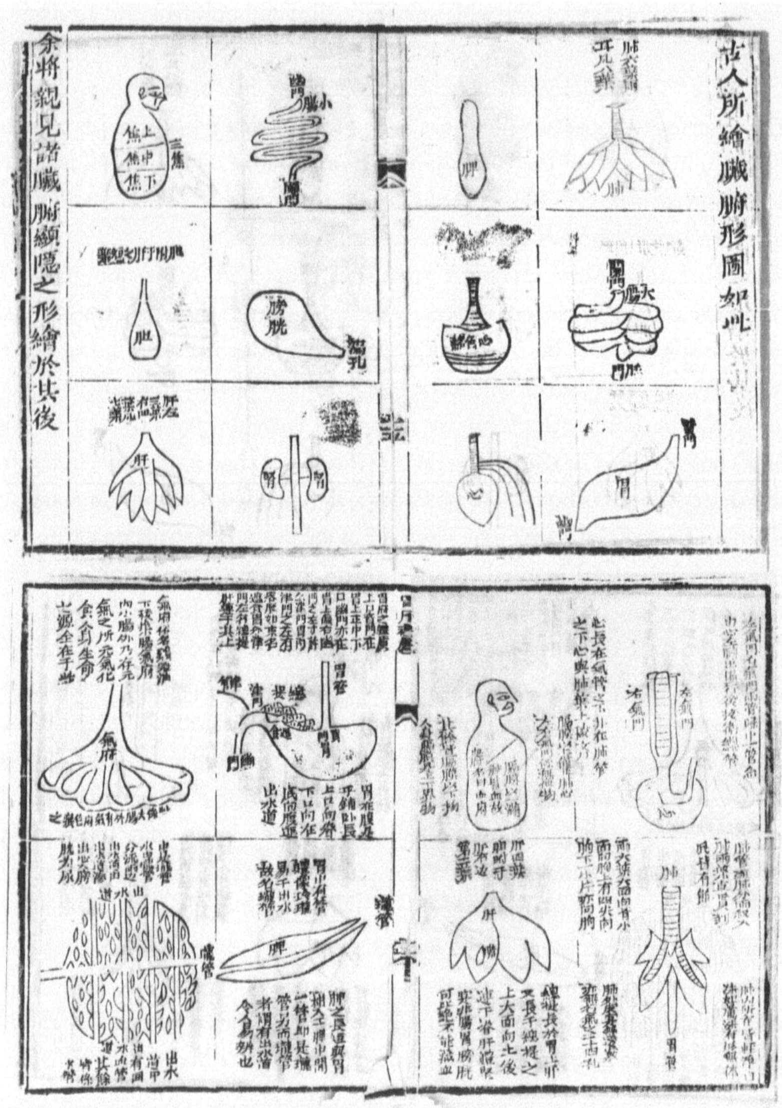

Figure 13. Images of human organs in Wang Qingren's *Correction of Medical Occupational Errors* (*Yilin gaicuo*), 1830.

Medicine Elsewhere in the World: The Mystery of Beneficial Herbs

Although Oceania and Mesoamerica are the only regions where no inoculation practices can be found until the arrival of Europeans, empirical medicine cannot be ruled out here either. We encounter traditional herbal medicine all across the world, and the benefits of local herbs and plants have been investigated by mainstream medicine since the beginning of the 20th century.[235] One of the first and most resounding results was the insight (from early antiquity) that a brew made of willow leaves has an analgesic effect. The active substance in these leaves led to the development of aspirin. New drugs are increasingly being "discovered" by investigating the traditional uses of beneficial herbs and plants, such as the kava plant used in Polynesian herbal medicine.[236] Meanwhile, the evaluation of traditional herbal treatments has become a subdiscipline in pharmacology.

Although it is not clear whether there was an empirical cycle in early modern herbal medicine, it is plausible, considering that monitoring the intake of an herb is a standard practice in traditional herbal medicine. The question is whether this test is accompanied by a cyclic interaction between empirical observations and hypotheses or with a single trial-and-error evaluation. In contrast to the sophisticated inoculation method, the effect of ingesting an herb can be determined quite easily. In contrast, the inoculation method required a much more complex cycle. In any case, herbal medicine is, if not cyclical, at least empirical, although not all of the remedies described in the past have been effective (see the discussion in chapter 2.5).

5.6 Jurisprudence: An Empirical Cycle in Legal Studies?

Legal Humanism

Unlike the medieval glossators, the first legal humanists, Andrea Alciato (1492–1550) and Guillaume Budé (1467–1540), were mainly active in France rather than in Italy.[237] Yet the kind of legal scholarship that these humanists had in mind was no different from that of their illustrious predecessors. The study of law remained an exegetical discipline with the main aim of explaining and applying Justinian's *Corpus iuris civilis*, using the commentary of earlier jurists, who often disagreed with each other. This historically informed method of interpretation made jurisprudence an extremely complex practice. As Michel de Montaigne (1533–1592), himself a lawyer, wrote in his renowned *Essais* (2.12), "I have heard of a lawyer who, when he encountered a sharp conflict between Bartolus and Baldus, would enter . . . in the margin of his book: 'a matter for a friend.'"

What the humanists succeeded in adding to jurisprudence was mainly the new philological-historical reconstruction method (see above). For jurisprudence this may seem like a peripheral contribution, but this technique could make all the difference when a reconstruction of an older legal text led to the rejection of an inaccurate copy.

The Emergence of Natural Law: Grotius

The greatest innovation in early modern legal studies was the development and elaboration of natural law. This kind of law assumes legal principles determined by nature, independent of place and time.[238] Natural law has a long history. We already encountered it with Aristotle, who argued that there were legal principles that applied to all peoples, as well as special laws that each people created for themselves (see chapter 3.7). With the rediscovery of Aristotle in 13th-century Latin Europe, medieval jurists and theologians divided this law into natural law (*ius naturale*) and positive law (*ius positivum*), a distinction found in the *Corpus juris civilis*. But it was only in the 16th and 17th centuries that natural law was given new impetus, especially in the Dutch Republic. This young state needed a justification for its aggressive trade policy, which regularly brought it into conflict with other nations, such as Portugal, which claimed an exclusive right to sail to the Indies. The Republic had broken through the Portuguese monopoly by force and now sought legal legitimacy. Since historical legal texts offered no solution, they had to rely on first principles, on reason. Ideally, the advocates of natural law minimized the role of legal history so as to base the law on fundamental principles from which the rules of law could be derived. This would make jurisprudence a real science, where reason was the measure of everything. The law would no longer depend on a king or on God but on human reason alone.

Hugo Grotius (1583–1645) is usually cited as the founder of this movement. Grotius was a child prodigy who at the age of 11 was already studying at Leiden University, where he came into contact with the renowned humanist Joseph Scaliger (see above). At the age of 15, Grotius obtained his doctorate from the University of Orléans, after which he published his first book at the age of 16, a learned edition of Martianus Capella's work on the seven liberal arts, *De nuptiis philologiae et Mercurii* (see chapter 4.2).[239] In the same year, in 1599, Grotius was appointed as a lawyer in The Hague, and two years later he was made an official historian of the States of Holland. After this meteoric rise, Grotius ran into political problems as a result of his activities in the Remonstrant community. This led to his banishment to Loevestein Castle and to his famous escape in a book chest, after which he worked in Paris and Sweden.

It was Grotius who sought to establish a new legal system that could regulate the relations between states on the basis of reason. Grotius put together a legal justification post factum for the Dutch forcefully breaking through the Portuguese trade route to the Indies. While Grotius's legitimatization was partisan, he tried to keep the underlying argumentation as pure as possible. Rather than invoking existing law, he reasoned that the sea could not be the object of private or state property. Ownership was possible only of things that were bounded, not of anything as immense as the sky or the ocean. According to nature, Grotius reasoned, all people were allowed to pursue their fortune, and trade was therefore free.[240] Grotius's most important work, *De Jure Belli ac Pacis* (*On the Law of War and Peace*, 1625) began with an explanation of this rational legal theory. This work provided the foundation for international law. Despite his great endeavor to appeal solely to pure principles, his elaboration is actually based on earlier law, especially Roman law, and is replete with references to historical (legal) sources. Moreover, his system does not provide derivations of laws from principles.

Nevertheless, Grotius considered his own work to be novel. He contended that previous attempts to turn jurisprudence into a science had failed because no distinction was made between natural and positive law. Grotius held that although natural law can be made into a science, because it is the same everywhere and at all times, positive law is too variable from place to place. This idea has an interesting analogy in astronomy: although human law depends on the arbitrary nature of time and place, the "higher" law follows nature and is just like the movement of the planets, unchanging and eternal.[241] Grotius even detached natural law from theology by positing that this law would also apply if God did not exist or if God did not concern himself with the world.[242] But Grotius immediately made it clear that he distanced himself from these two hypotheses.

Natural Law Flourishes: From Pufendorf to Wolff

Natural law rose to great heights in the late 17th and early 18th century, when it was developed into the major natural systems elaborated in the work of Thomas

Hobbes (1588–1679), Samuel von Pufendorf (1632–1694), and Christian Wolff (1679–1754).[243] These scholars continued on the path taken by Grotius and sought a system in which law was organized so that it could be constructed deductively from universal first principles.[244] Here we see an attempt to develop *predictive* principles in jurisprudence that could be used to derive the patterns (the legal rules) in a compelling way. Until then, jurisprudence had consisted in *restrictive* and *procedural* principles (see chapters 3.7 and 4.6), such as the restrictive principle stating that no commitment can oblige a person to do the impossible or the procedural principle that a more specific law takes precedence over a general one. These restrictive and procedural principles indicated the boundaries of what was possible but could not derive the legal rules themselves, let along predict them.

Wolff went the furthest in the pursuit of predictive principles in his *Jus Gentium Methodo Scientifica* (*The Law of Nations Treated according to the Scientific Method*) from 1749.[245] However, Wolff's ideal of a fully deductive system proved unfeasible: his abstract principles reproduce the social and political relations of his own time, which becomes evident when he discusses the position of women. Yet his attempt to break free of historical Roman law and replace it with axioms so evident that no one could deny them could be called heroic. But as with so many heroic attempts, this one failed miserably. In the second half of the 18th century, work on natural law waned, but it remained an open question as to whether there were higher, universal legal principles independent of time and place. For example, Immanuel Kant believed that the principle of the *categorical imperative* was universally valid, so that not only legal rules but also legal principles could be derived from them. According to the categorical imperative, for any moral action we should ask whether it can be elevated to a general rule. If it can, it is a universal principle. If it cannot, the action is morally deficient.[246] But with Kant we are dealing more with ethics than with concrete law.

Can natural law be tested empirically? In principle, it could, because legal principles assumed to be universal can be verified in different cultures. And it is only when such principles are valid everywhere that one can speak of a universal principle (see also the notion of *ius cogens* below). But this was beyond the field of vision of the 17th-century scholars of natural law: they restricted themselves to "civilized" cultures, which for them were limited to Europe.[247] Above all, they worked within a Cartesian framework, which started not from testable principles but from evident, clear principles that need not be tested. As a result, just as with Descartes himself, there was no empirical cycle. Reason had to suffice.

Positive Law: Noodt, Montesquieu, and the Search for the "Higher"

We do encounter the beginnings of an empirical cycle in positive law, which assumes that the laws and principles are in accordance with a people's nature and the natural conditions of a region. We find this idea elaborated in detail in the work of Gerhard Noodt (1647–1725), who proposed a number of factors that he believed could explain a people's legal system.[248] These factors include human needs, resources, natural conditions, education, religion, and political systems. Although Noodt's system seems far removed from natural law, he, like Grotius, stated that the highest power did not rest with the king nor was it given by God. Noodt's work was elaborated most extensively by Charles de Montesquieu (1689–1755) in his *De l'esprit des lois* (*The Spirit of Laws*) from 1748. In it, Montesquieu sets out his famous separation of powers: legislative, executive, and judicial. From our perspective on the search for deduction patterns, this work is important because Montesquieu derives concrete patterns from Noodt's factors in his work. For example, Montesquieu predicts a relationship between marriage law and climate: the colder the climate, the more monogamy there would be. Broadly speaking, the statement appears to be correct, albeit only statistically and not in absolute terms.

So, the ideal of reducing the legal system to as few principles as possible is found not only in natural law but also in the positive law of the Enlightenment thinkers. This pursuit of a minimal number of legal principles can actually be traced back to Tribonian and the earlier Roman scholars, as we saw in chapter 4.6. However, the early modern attempt at the reduction of principles did not last long. The law was radically reformed in around 1800, based not on principles but on pragmatic grounds: the Prussian Allgemeine Landrecht of 1794 seemed more removed than ever from a principle-based legal system (in terms of our notion of "principle"), although this is somewhat less true of the French Code civil of 1804.[249] Nevertheless, Grotius's ideas of maritime and war law and Noodt's and Montesquieu's factors did find a place in it.

Although natural law is marginal today, there is still a quest in jurisprudence for a "higher," universal form of law that transcends the legal rules of an individual state. We see this not only in legal practice, when judges invoke higher principles, but also in legal theory, particularly in international law, where principles that are valid in all cultures are said to belong to the so-called *ius cogens*, the "compelling law," more commonly referred to as the "peremptory norm." The principle of human dignity is an example of such a general principle.

The rules that follow from this principle, those prohibiting genocide, slavery, torture, and racial discrimination, are therefore part of the *ius cogens*.[250] While these rules are not time independent—slavery was generally accepted in many periods—today they are considered place independent. Even if a state has not explicitly consented to them, these legal rules are accorded general validity in institutions like the International Criminal Court in The Hague.

Chinese Jurisprudence: Ming, Qing, and the Chinese (People's) Republic

Western views of Chinese law long had one thing in common: they denied the existence of a full-fledged Chinese legal system. For example, Montesquieu argued that Chinese law was based on nothing more than fear, while the sociologist Max Weber (1864–1920) believed that China lacked any rational legal system. Joseph Needham (1900–1995) made a connection between the laws of jurisprudence and those of natural science, and since China did not have the concept of universal law, Needham said this had prevented the concept of natural law from emerging.[251] Broadly speaking, in the Western view, Chinese law served solely to punish crime and not to guarantee personal freedom or equality. But this is a judgment based on European principles. The question to ask here is, What were the principles of Chinese law in the Ming and Qing dynasties?

Law in China was first and foremost secular: the notion of a God-given law is nowhere to be found. Nevertheless, law was a moral basis for maintaining both social and cosmic order. Chinese law was thus seen as part of the cosmological unity, in which the harmony between heaven and earth was considered of utmost importance. By regulating and forbidding certain behaviors, the purpose of Chinese law was to bring everything under heaven into line with the cosmic order, guaranteeing the absolute authority of the emperor as the son of heaven. This is how the Chinese legal codes should be read.

The Ming dynasty (1368–1644) witnessed the creation of one of the most important legal codes in Chinese history, the Great Ming Code, which was published in five volumes during the Wanli reign (1573–1619).[252] The code starts with criminal laws, such as punishment for rioting, treason, and plotting rebellion, and then it goes on to deal with personal laws, such as correct procedures for appointing people to professional positions. This is followed by laws regarding income and revenues. There are severe penalties for failure to pay one's taxes, up to 100 blows with a heavy stick. The ritual laws are again accompanied by the specification of punishments, such as 50 blows with a light stick for failing to carry out imperial ancestor worship. An explanation of the military laws is followed by the criminal laws regarding the destruction of the imperial temples, mausoleums, and palaces. All who conspire to do so are to be sentenced to death by a thousand cuts (the "slow death penalty"). Finally, laws concerning public works are elaborated, again listing many punishments.

At first glance, this Ming code appears to fit Western stereotypes. How else should we interpret a code principally concerned with punishment? Recall from the previous chapter that in China, the law was primarily intended for those who strayed beyond the bounds of civilized behavior. For those who observed the right virtues, no law was needed. And here we find radically different view of "law" and "justice": in Europe it consisted in guarantees for personal freedom and, later, for equality, but in imperial China these guarantees were contained in the Confucian classics, such as the *Book of Rites*. There is therefore no point in searching legal codes for Chinese notions of European legal concepts. Nor will we find any legal empiricism or theory, let alone a cycle between the two.

However, at the end of the Qing dynasty (1644–1911), we do see an attempt to draft legal codes based on European models. It was only when the Chinese Republic was founded in 1912 that the Confucian classics were desecrated, more than 2,500 years after they had been put into place, and a system was sought that could do justice to the ambitions of the young republic. Chinese law was modeled on the German code, which had gained great prestige over the course of the 19th century and had previously served as a model in Japan. It is at this point that the history of Chinese jurisprudence became part of international jurisprudence.

Starting with the foundation of the People's Republic of China by Mao Zedong (1893–1976), the law was modeled on the Marxist-socialist system,[253] with the Cultural Revolution (1966–1976) as an extraordinarily cruel interlude. The Red Guards established by Mao persecuted anyone who did not follow the Maoist line. More than ever, China seemed to lack any rational legal system. But that is only an illusion: in Chinese history we find a surprising parallel with the Legalists of the Qin dynasty (221–206 BCE; see chapter 3.7). As with the Legalists, the Red Guards were keenly focused on achieving a new social order devoid of distinctions between social classes. As with its Legalistic predecessor, violators of Maoist ideology were deemed incorrigible and were relentlessly persecuted, with their eradication deemed the only solution.[254] It is unlikely that the Maoists themselves drew an analogy with the Legalists. According to the Maoist line, all dynasties prior to the advent of communism were considered part of the feudal period in their history.[255] The analogical use of rules of law from the Qin dynasty would therefore certainly not have met with Mao's approval. Furthermore, it is highly questionable whether the Red Guards had any knowledge of the Legalistic period from the distant Chinese past. Be that as it may, long-term analogies keep popping up in the impressively rich Chinese history of knowledge, as we have seen in other disciplines.

Ottoman Law and Xeer

The oldest surviving sources of Ottoman jurisprudence date back to the 14th century and are based on Islamic methodology of analogical reasoning, or *fiqh* (see chapter 4.6). The growing prestige of the Ottoman Empire attracted legal scholars to the Islamic world, and at the end of the 16th century, a recognizable Ottoman jurisprudence developed. In addition to traditional Islamic law, a secular law known as *Qanun* came into being. This secular law unified the diverse populations in the empire. Contrary to what has often been argued by Weber, among others—that Islamic law had become rigid and was applied arbitrarily—the Ottoman Empire actually had large collections of secular laws that were applied and enforced dynamically, depending on the situation and the population group.[256]

Starting in the 20th century, legal systems from around the world have become objects of study, especially among anthropologists. These include the legal system of the Trobriand Islands in Papua New Guinea, studied by Bronislaw Malinowski (1884–1942); that of the Barotse people in southern Africa, studied by Max Gluckman (1911–1975); and that of the Somalis, studied by Michael van Notten (1933–2002).[257] The questions asked include "What role does law play in social life?" and "To what extent does the law determine what happens in a society?" This anthropological research has led to the discovery of new, often unwritten legal systems, such as Xeer in Somalia. Xeer law is one of the few systems where the underlying legal principles require constant reinvention. The discovery of the underlying law starts from scratch for each case.[258] The Xeer legal system is probably pre-Islamic, but precise dating is difficult owing to the fact that the first oral sources were gathered only in the past century. Administration of justice traditionally takes place under a large tree, with judges chosen on the basis of their knowledge and wisdom. Judges are allowed to entertain their own legal principles, with multiple judges selected for each case, up to a maximum of 10. The aim is to reach consensus between the parties; there is no imprisonment. Xeer shows that the legal motivation for a judicial decision based on legal principles can be a continuous search, in which the principles are repeatedly questioned and even invented on the spot. This means that Xeer can be considered a non-prescriptive legal system, because there are no laws that Xeer judges fall back on. Their only guideline is to seek agreement between all parties and fellow judges. Of course, a judge is bound by local customs, and these customs are entirely normative. And judges bring their own, unwritten principles.

Legal anthropology has put the question of specific versus general principles of law back on the agenda. It has also made clear how difficult it is to compare legal practices in different cultures, especially when it comes to unwritten law. The principle of human dignity may seem to apply all over the world, but how this concept is interpreted varies widely. For example, the idea of equality (or equivalence) is not a universal principle: various legal systems assume that different laws apply to different social classes. The rules of law then depend on the position in the social hierarchy that a person occupies. This seems to contradict the aforementioned *ius cogens*, the peremptory norm that assumes the existence of universal legal principles. This contradiction again leads to the question of whether any universal principles exist at all or whether such universally assumed principles actually correspond to the principles of a dominant culture that imposes them on the rest of the world.[259]

Deduction Patterns in Jurisprudence?

If an empirical cycle is to be found anywhere in jurisprudence, it is in the positive law set in motion by Noodt and Montesquieu in the 18th century. They believed that in the many types of legal systems they could identify patterns that could be explained using underlying principles. This empirical approach was marginal in the 19th century, but it reemerged in the late 20th century as an approach called *empirical legal studies*.[260] In other regions it is more difficult to find a notion of an empirical cycle in jurisprudence, although we do see a different deduction pattern there: analogical thinking. In Ottoman law, as well as in current Islamic law, analogical thinking is a successful method for linking cases to principles, sometimes with new legal rules as a result. In addition, we ourselves have established an analogy between two legal systems in China—Legalism and Maoism—although it is unlikely that the historical actors themselves thought of them analogically.

5.7 Conclusion: Cyclic Interactions between Patterns and Principles

The Empirical Cycle Migrates from the Humanities to the Natural Sciences

In the course of the early modern period, the empirical cycle became embedded in all domains of knowledge. This happened first in the European humanities: in 15th-century philology, art theory, linguistics, and musicology, there came to be a cyclic interaction between theory and empirical observations, which led, among other things, to the discovery that the earth is older than the biblical date of Creation, the discovery of different forms of perspectival representations, and the discovery of new string laws in musicology. Over the course of the 16th century, the empirical cycle migrated from the humanities to the natural sciences, mathematics, and medicine, to reach jurisprudence in the 18th century.

The origin of the empirical cycle goes back further than the early modern period. We probably find it even in antiquity: theoretical foundations in Greek, Indian, and Chinese astronomy and musicology were also further refined, leading to a "process from misalignment to alignment" between theoretical principles and empirical patterns (see chapter 3). However, we must guard against equating this process with an empirical cycle. In an empirical cycle, a theory produces new predictions, which, after empirical testing, subsequently have repercussions on the theory itself. In a misalignment-to-alignment process, only a single convergence between principles and patterns is required, whether this convergence comes about through an empirical cycle or from, for example, a single experiment or logical inference. Nevertheless, it is plausible that the empirical cycle can also be found in some of the ancient domains of knowledge, especially in the increasingly refined fields of astronomy and musicology. But the cycle becomes dominant in the other domains of knowledge only when it is taken up by the early modern humanists. In the 15th-century humanistic studies, one discovery is made after another, whereas astronomy, mechanics, and medicine continue to build largely on existing ideas until the middle of the 16th century.

Humanistic discoveries had major implications for Europe, from the exposure of the *Donatio Constantini* document as a forgery, which would constitute one of the arguments for the Reformation, to a new vision of the earth's age. Because humanist scholars were the teachers of the later natural scientists, in the

17th century the empirical cycle would become the method of all domains of knowledge. The icons of the Scientific Revolution, such as Vesalius, Galileo, Kepler, and Newton, were themselves also active in the humanities. I have shown that the nature of these scholars' humanistic background (in philology, art theory, or musicology) codetermined their way of working in the natural sciences.

Interestingly, in the course of the 18th century, the success of the natural sciences led to a role reversal: the results obtained through the empirical cycle in the natural sciences now served as the example for the humanities. For example, Newtonianism was for some time a model for musicology, linguistics, and medicine. In an analogous way, the concepts of force, weight, and pressure were adopted and applied to conjugations in language, to harmonic intervals in music, and to physical balance in medicine, although without much success.

The Empirical Cycle as a Global Phenomenon?

We have seen that the empirical cycle was not put into practice in all domains of knowledge in China, but it was implemented in medicine and philology. In addition, the cycle can also be found in Indian, Ottoman, and Ethiopian medicine. This means that the empirical cycle is not just a local phenomenon but is more of a global phenomenon. We can at least speak of a general tendency in medicine.

Moreover, we have seen that, in addition to the empirical cycle, analogical thinking also constitutes a deduction pattern. We have found this deduction pattern in European astronomy (as in Kepler's comparison with the magnetic forces), in Chinese and Indian medicine, and in Ottoman law.[261]

The West versus the Rest: A Gap in Knowledge History?

Although the empirical cycle is a phenomenon on a global scale, at least in medicine, only in Europe did it gain a foothold in all domains of knowledge. Use of the cycle did not become widespread in other regions until the 20th century. But then it was no longer the Chinese, Indian, Arab, or African notions of knowledge that were the standard, but rather the Western and mostly colonial notions of disciplines. Colonial rule was accompanied by a scientific and scholarly expansion.[262]

However, the so-called Great Divergence[263] between knowledge in Europe and in the rest of the world is not merely a colonial phenomenon: Japan, for example, was not a colony of any country. While the phrase Great Divergence is most often used to refer to the growing rift between societies in terms of industrialization,[264] it can be generalized to denote the growing rift in the domain of knowledge. But there is one important difference: by the standard account, the world's civilizations were generally on par with each other until the 18th century; it is in the course of the 18th and especially in the 19th century that they slip out of balance, with the West pulling out ahead. However, this view appears to be too simplistic for our history of knowledge. This book has shown that the "Great Divergence" is far from unique. If we go back a thousand years to the 11th century, we see an equally large divergence between China and Europe in China's favor and an even bigger divergence between the Islamic world and Europe in favor of the former (see chapter 4). And if we go back even further, to the 5th century BCE, we encounter another "Great Divergence," this time between the Greek world and the rest of Europe (see chapter 3). The "Greatest Divergence" can probably be found in early antiquity, when the Fertile Crescent, with its agricultural revolution, created a virtually unfathomable gap with the rest of the world in the Fertile Crescent's favor. And yet all these gaps were bridged.

Every divergence proved temporary. Sooner or later, useful patterns from one civilization are adopted by other civilizations they come into contact with. Theoretical principles whose utility is less obvious often take longer to be adopted. However, as soon as it becomes clear that patterns considered important can be brought under control using foreign principles, it seems to be only a matter of time before the underlying principles are also adopted, albeit in combination with local principles, as we saw in medicine above.

CONCLUSION

The Origin, Growth, and Future of Knowledge

I began this book by asking where the quest for systematic knowledge started and how it developed in different regions and cultures. Although I did not set out to give a complete overview of systematic knowledge, I have tried to highlight the widest possible variety of knowledge disciplines from as many parts of the world as possible within the limitations of a single book.

The Long-Term Development of Knowledge

The search for systematic knowledge can be traced back to the Old Stone Age with the awareness of patterns as found on mammoth bones and in cave paintings in various places around the world (approx. 40,000 years ago). Since the invention of writing in early antiquity (ca. 3000 BCE) humanity has carried out a massive search for patterns in the surrounding nature and culture—linguistic, mathematical, historical, astronomical, magical, economic, legal, and medical patterns. In classical antiquity (from about 600 BCE) we see an awareness of underlying principles that generalize over a multitude of patterns. This search for principles was soon followed by a search for the precise relationships between patterns and principles, such as deductions, which took place mainly in the Hellenistic world (from about 300 BCE). The quest to reduce the number of principles was a constant throughout the postclassical period (starting at about 500 CE) and took place almost everywhere, but especially in the Islamic world: the fewer principles, the "better." We encounter the discovery of patterns in the relationships between principles and patterns starting in about the 15th and 16th centuries, and it ran through many regions and cultures (from Europe, Africa, and the Americas to Asia). This leads to an awareness of the cyclic interaction between empirical observations (patterns) and theory (principles). This interaction is known as the empirical cycle and has brought us to where we are today: testing, adapting, and improving our insights.

Sketched in this way, the long-term development of systematic knowledge looks rather abstract, but the patterns and principles found are far from it. Although many quests deadlocked or failed (see "Failed Knowledge" below), successful patterns and principles allow us to make concrete predictions. With the astronomical principles from ancient Greece, China, and India, we can calculate planetary positions to 1 degree; and this accuracy has increased over time, though not evenly. Using the early modern philological principles of Poliziano, we can reconstruct a text with great reliability on the basis of surviving copies; and here too reliability increased in the centuries that followed. And the inoculation pattern in 16th-century China, which rapidly spread to other regions, succeeded in preventing a number of viral diseases. Its relevance has only grown, and its application is more urgent than ever. Yet the inoculation pattern is nearly collapsing under the weight of its own success: as the experience of disease has receded, nonscientific arguments have gained traction. For this reason, the achievements of the search for patterns and principles must be repeatedly explained to the public.

Natural Science and the West Pushed Out of Their Central Position?

For most periods and regions, it was easy to decentralize the natural sciences: for example, in antiquity, linguistics and jurisprudence turned out to be more dominant than previously thought; the same was true of postclassical historiography and almost all the humanities in the (early) modern period. Even when in the West starting in the 18th century, the natural sciences finally surpassed the humanities as the dominant object of focus, it was mainly via the humanities that the most important epistemic tool—the empirical cycle—came to them.

I have also endeavored to move the West out of its central position. Yet, taking Europe out of the limelight is not an end in itself: my long-term goal is to understand what the history of knowledge looks like if we treat the various knowledge centers on an equal footing. Although such a polycentric history is still in its infancy, it puts forth a picture quite different from the one that has long been presented to us. My approach has shown, among other things, that major scholarly and scientific discoveries have taken place all over the world and in all disciplines, that knowledge centers as well as disciplines have regularly influenced each other, and that the so-called Great Divergence between the West and other regions (see chapter 5.7) is only one of many divergences—and one of the smallest in the history of knowledge.

Are Patterns, Principles, Deductions, and Deduction Patterns "Stages" in the Development of Knowledge?

While each era is dominated by a particular type of quest, we must guard against seeing this sequence of quests teleologically—as if *Homo sapiens* has developed cognitively over history in stages along the trajectory from child to adult, in the way described by the psychologist Jean Piaget (1896–1980).[1] This book is not about how children searched for patterns and principles but how adults sought to find them, and adult cognitive abilities have remained largely unchanged for the past 40,000 years, and certainly the past 6,000. Biologically and cognitively speaking, we are the same species as in the Stone Age (if not culturally, of course).

And thus we find patterns, principles, deductions, and deduction patterns in all eras, although they are not dominant in all eras. Yet we can detect the first impetus for principles as early as the Stone Age, and possibly even the empirical cycle, albeit implicitly. We see this, for example, in the incremental improvements to the hand ax that suggest perception of the principles underlying its function—such as principles concerning shape, weight, or texture—after which these assumptions were tested by cutting through various materials with it (cleaving patterns). The principles could be subsequently modified again, leading to new hand axes, which might or might not be better, and so on. Of course, these principles and patterns are implicit in the archeological evidence and therefore speculative. It is only in the 6th century BCE that we first encounter principles that are explicitly formulated (Thales, Panini; see chapter 3.1).

While I argue that patterns, principles, and their relationships are timeless, it is undeniable that one era is dominated by the quest for patterns, another by the quest for principles, and yet another by the quest to deduce patterns from principles, and even by the quest for patterns in these deductions. So, with respect to the dominance of the concepts in question, in a majority of regions there is a clear tendency in the development of human knowledge. Even though in biological terms, humans have hardly changed over the past 6,000 years, once knowledge such as patterns, principles, or deductions was acquired, this knowledge was regularly handed down to subsequent generations to build upon.

Can we explain this sequence in the dominance of concepts that we have found? Although my story is historical, there is a lot to be said for the fact that, logically speaking, no generalizations over patterns can be made until we first have some patterns. And relationships between patterns and principles cannot be established until we have both patterns and principles. And finally, patterns in relationships cannot be discovered until relationships have been established. But this logical order becomes interesting only if we can use it to understand the long-term history of human knowledge. And I believe that this order (in the dominance of concepts) has made the history of knowledge, with all its unexpected offshoots and dead ends, somewhat more comprehensible.

What about the Non-pattern-seeking Disciplines?

Not all knowledge activities we considered were pattern seeking. Already in antiquity, the anomalists of Pergamon rejected the search for patterns in texts and instead sought the best possible interpretation of a given text (see chapter 3.3). And today, although the empirical cycle is considered useful for empirical disciplines, it is assumed that this cycle does not hold for the interpretative, or hermeneutic, disciplines in the humanities. Are the non-pattern-seeking, interpretative disciplines an exception to our macrohistorical perspective consisting in patterns and principles, or do they lie beyond the scope of systematic knowledge?

It would be simplest to say that there are multiple measures of systematic knowledge and that there are multiple perspectives on what science and scholarship is. But before arriving at such an answer, we first need to consider what hermeneutic scholars do and how their activities relate to the empirical cycle. Then we can consider whether the premise is correct that the empirical cycle does not hold for the interpretative disciplines.

We have seen that in Europe the empirical cycle emerged from the early modern humanistic disciplines, such as 15th-century art theory, philology, linguistics, and musicology (see chapter 5.1). It was the natural scientists and physicians who adopted this empirical cycle from the humanities, mainly because these scientists (such as Vesalius, Galileo, and Kepler) had received a humanistic education themselves. In contrast, the 18th-century humanistic disciplines built upon the insights of the natural sciences, taking the Newtonian method as their example. It was only in the course of the 19th century that the humanities developed a methodology of its own. The hermeneutic method is explained in the work of Friedrich Schleiermacher (1768–1843), Wilhelm Dilthey (1833–1911), and especially Hans-Georg Gadamer (1900–2002), with one of the most important concepts being the *hermeneutic circle*. The notion of the hermeneutic circle differs somewhat in the works of these authors, so I will limit myself to the most influential interpretation, which is also the most recent: that of Gadamer.[2] According to him, the hermeneutic circle is an iterative process of developing a new understanding of historical or artistic reality by examining every specific detail of that reality. The hermeneutic circle leads from the specific to the general and back again. Any meaning assigned to a historical era or artistic work depends on its (historical) context, so in hermeneutics a word has no fixed meaning. By studying the vocabulary and historical sources of a certain period to the greatest extent possible, researchers can immerse themselves ever deeper in that period, in such a way that a so-called fusion of horizons eventually takes place, when points of intersection are found between one's own horizon (that of the researcher) and that of the matter being studied (text, culture) in order to gain a better understanding.

It may be clear where I want to go with this; namely, to defend the idea that the hermeneutic circle is not separate from the empirical cycle, but is actually an instance of it,[3] albeit a very special instance, in which the researcher becomes part of the subject being investigated.[4] First, consulting historical sources is itself an empirical activity (unlike thinking up new sources). Furthermore, the endeavor to get ever closer to a specific historical person using the hermeneutic circle is very similar to the empirical cycle: for example, a researcher studying a literary work from the 15th century will select a gateway into that period through historical sources from that time. This places the literary work and the author in the historical context, with the interpretation of the work again raising questions or problems that require further research into sources. The literary text (or work of art, musical composition, etc.) is then reinterpreted with the entire body of sources found, until the researcher has approached the historical person or historical work as closely as possible and can understand it (the fusion of horizons). Seen in this way, the hermeneutic circle is a form of empirical cycle, which, as in the other disciplines, never ceases but continues to approach ever closer to reality. The nature of humanistic material is of course different from that of the natural sciences, but if we consider historical sources to be empirical material, then the interpretative process is just as empirical and therefore verifiable[5]—with the difference that this assessment is performed within the researcher's own "horizon," making the researcher part of the object of study.

Here it is important to realize that as I interpret it, the empirical cycle (see chapter 5.1) is not only "empirical" but also "theoretical," despite the name. In an empirical cycle, observations and theory interact by comparing the two (which in the interpretative humanities is the historical context), after which the theoretical principles can be further refined or modified. Thus, a better name for the empirical cycle would perhaps be the *theoretical-empirical cycle*.[6]

The interpretative approach is not unique to the humanities. Newton also subjected his celestial mechanics to interpretation in order to understand why his system became unstable, leading him to introduce divine intervention (see chapter 5.3). And even today physicists grapple with the question of why the natural constants are precisely calibrated to make life possible (roughly the *anthropic principle*).[7] While natural scientists do not have to immerse themselves in a particular historical context, they do have to consider and compare multiple interpretations. And, indeed, many perspectives are also possible in the natural sciences.

This similarity between the humanities and natural sciences is often underestimated: the study of culture and the study of nature are usually described as opposing activities and approaches with different methods and values.[8] But this is mainly a Western view, one that has become increasingly outdated.[9] As the anthropologist Philippe Descola has shown in his book *Beyond Nature and Culture*, in many non-Western societies, the opposition between culture and nature is much less evident than in the West, if at all.[10] I was not introduced to Descola's work until quite late, but his book was a feast of recognition.

And What about the Unique?

Now I can already hear some humanities scholars object that while the hermeneutic circle may be cyclical and empirical, many humanities scholars do not look for patterns but instead endeavor to study and understand a *unique* event such as the French Revolution or the assassination of John F. Kennedy, to name just a couple of examples. I will not try to claim otherwise, for I myself have argued that there are both pattern-seeking and non-pattern-seeking activities dating from antiquity, and not only in the humanities, such as in Babylonian linguistics (chapter 2.1) and Greek philology in Pergamon (chapter 3.3), but also in other disciplines, such as the observation of unique phenomena like supernovas in Chinese astronomy (chapter 3.2). But how can the quest for the particular be reconciled with the quest for the general? They appear to be opposed, placing the unique outside the scope of our narrative.

But this is a sort of short-sightedness: patterns do *not* stand in opposition to unique events. Patterns themselves are made up of unique events (or unique objects, individuals, or phenomena). And such a unique entity, if part of a pattern, can also be linked to underlying principles in order to understand and interpret that event or individual. The underlying principles need not (yet) be known to the scholar or researcher. My starting point is that in principle, underlying generalizations can be found, although in the humanities they are many times more complex and unstable than in (some fields of) the natural sciences, making them more like tendencies.

Take the notion of style that we encountered in poetics (chapter 3.6), musicology (chapter 4.4), and art theory (chapter 5.1), for instance. A musical, literary, or artistic style such as baroque can be seen as a pattern: with some practice we quickly recognize the typical baroque style in music, literature, and the visual arts. Indeed, baroque paintings have certain features in common, which we could describe as a system of rules (as did the art historian Heinrich Wölfflin).[11] The same applies, for example, to the medieval musical style of the organum, as we saw in chapter 4.4. Indeed, every work of art shares a number of properties with a style, yet each work of art is unique. A scholarly researcher will not always be interested in similarities but may focus on differences, such as by describing the particular style of the medieval composer Perotinus. But ultimately, this "unique" style can only be described by comparing Perotinus with other medieval composers from his own period. In such cases, the researcher will refer to the common organum style patterns and to the underlying principles to the extent that they are known. And even when researchers are interested in studying a single piece of music, a single painting, or a single literary work, to interpret a piece of art, they still refer to the patterns relevant to the period in which the artist operated, which is often also the sort of contextualization that a hermeneutic scholar aims at. Obviously, the study of uniqueness is not immune to patterns and principles that constitute the context of the unique work or unique historical event. We thus conclude that the unique cannot exist without a pattern: the two notions are intimately connected—and with this observation these two notions have now crystallized (see the introduction).

Failed Knowledge

We have seen in this book that not all quests for patterns and principles succeed. Although unsuccessful quests rarely survive, I have nevertheless included evidence of those that have in this book. In the history of human knowledge, we have several examples of these failures:

- The search for a connection between grain prices and planetary motions: even after centuries of observations and searching for patterns, the Babylonians were unable to establish a connection.
- The endeavor to reduce the millions of legal rules of Roman law to a small number of principles. This only "succeeded" by disposing of great number of these laws and putting those remaining into a single, extensive law.
- The search for a grammar for all forms of the organum: centuries of attempts to capture the organum in a system of rules failed. The same goes for the motet, as well as for the later sonata, symphony, and so forth.
- The search for fewer axioms for Euclidean geometry: although the so-called parallel postulate was considered superfluous by mathematicians for centuries, no one has been able to prove the Euclidean theorems using fewer axioms.
- The search for universal principles for natural law: although the last word has yet to be said in this regard, we still cannot derive existing legal rules from them.
- The search for a deductive system of rules for medical diagnoses: the best we can do is establish *inductive* systems for medical diagnoses that generalize about previous diagnoses in a probabilistic way. This is also how modern medical expert systems work.

Despite these failures and many others, much can be learned from the quests: virtually every failure led to a new discovery in a different subfield. And meanwhile the development of knowledge continued. The Babylonians were not able to use the planetary motions to predict grain prices, but their centuries-long observations did lead to discoveries that included the Saros cycle of solar and lunar eclipses (see chapter 2.3). And while the axioms in Euclidean geometry could not be reduced in number, the attempt did result in a new, non-Euclidean geometry that allowed for other discoveries (see chapter 4.3). So the question is whether the category of "failed" knowledge makes any sense: there is almost always an interesting spin-off or an unexpected discovery. I will return to this at the end of the conclusion.

Is There Also Such a Thing as Non-knowledge?

Now that we have discovered in this book what constitutes systematic knowledge and in what historical course it was constructed and expanded, we can ask ourselves whether there is also such a thing as non-systematic knowledge, or simply non-knowledge. And here I do not refer to failed knowledge but ask whether there is a demarcation criterion that can distinguish knowledge from non-knowledge. If no such criterion existed, astrology, magic, and alchemy would be just as systematic as astronomy, medicine, and physics. And indeed we have seen that until the early modern period, disciplines such as astrology and astronomy were hardly distinguishable—astronomers such as Regiomontanus, Brahe, and Kepler were at the same time astrologers. But with the awareness of patterns in deductions, in particular of the empirical cycle, we have obtained a demarcation criterion. In astrology, magic, alchemy, the kabbalah, and other occult sciences, we see no improvement from bringing empirical observations and theory together. To be sure, an interaction can be found between observations and theory: for example, Regiomontanus improved the definition of astrological houses (see chapter 5.2), allowing him to make more explicit predictions, and Newton tried to assemble all alchemical writings to ascertain how to make gold from other metals (see chapter 5.3). But new experiments in these areas led neither to better results in astrological predictions or gold production nor to an adapted, improved theory.[12] The empirical cycle had no effect on the disciplines of astrology and alchemy, even after many generations, and even after many centuries.

So it appears that the discovery of patterns in deductions, in this case the empirical cycle, has yielded an interesting hypothesis: deduction patterns can serve as a local and temporal demarcation criterion between science and pseudoscience—or between scholarship and pseudoscholarship—a distinction that did not exist for centuries. Of course, our demarcation criterion is a *historically* derived criterion and not a *philosophical* criterion, as prevalent in the theory and philosophy of science. Sociologist of science Thomas Gieryn argued correctly that no formal-philosophical criterion could be used to distinguish science from nonscience,[13] but he was wrong to imply that there was no formal criterion at all: by exposing the underlying historical patterns in the history of knowledge, we can identify a criterion for the demarcation between science and nonscience and between humanities and non-humanities, in other words, between knowledge and non-knowledge. This is not to establish a historically independent *absolute* distinction between knowledge and non-knowledge but rather to understand how researchers distinguished between "knowledge" and "non-knowledge" at a given time and place.

Thus, no knowledge activity or discipline is independent of other disciplines. This does not mean that all disciplines are or need to be experimental, theoretical, or interpretative; it means that a successful way of linking patterns to principles in one group of disciplines is a challenge for another discipline or group of disciplines. And one of these challenges is the deduction pattern of the empirical cycle, which first served as an incentive but over time became a benchmark for disciplines.

Neglected or Forgotten Knowledge: Women in Knowledge History

This book has also shown that the contributions of women were much greater than long was believed. In earlier works on the general history of science—which, as mentioned, are usually limited to the natural sciences—practically no women appear on the scene until around 1900,[14] while it turns out that female physicists, mathematicians, and physicians, as well as philologists, linguists, and historians, were active at the highest level.[15] The fact that the vast majority of scientists and scholars were men makes the contributions of female scientists and scholars all the more remarkable.

For example, we have seen how Émilie du Châtelet developed the oldest known principle of energy conservation (see chapter 5.3). Yet her name is conspicuously absent in overview histories of science.[16] Another example is Maria Gaetana Agnesi, who fused the results and propositions from infinitesimal calculus into a whole, a sort of integration of calculus and algebra (see chapter 5.4). Whereas Descartes is regarded as a great innovator with a similar aim—the earlier fusion of geometry and algebra—Agnesi's contribution is seen merely as a pedagogical innovation. While in some sense it was, it is notable that many of Agnesi's contemporaries presented her work as a complete revision of traditional mathematics because of the way that it unified different parts of mathematics.[17] But in the 19th century her name disappears into oblivion. For this reason, further research into Agnesi's contribution and influence is in order.[18] It is important that scientists like Agnesi and du Châtelet, along with Laura Bassi and others, not only be studied individually—which has been done—but also in comparison to other scientists of their time, as I have sketched out in this book. It then becomes apparent that the history of knowledge will become an integrated history devoid of missing links only when both men and women are included (from both the humanities and the sciences).

In many cases we can no longer reconstruct the contribution and influence of female scholars and scientists. Was the 5th-century mathematician Hypatia (see chapter 3.4) a follower of her father, as is often stated, or did she also devise new propositions of her own? The surviving sources are inconclusive. And what about the 11th-century physician Trota of Salerno? Her very existence was long put in question, possibly because she was so exceptional. In contrast to her, history has been kinder to the abbess and physician Hildegard of Bingen (see chapter 4.5): thanks to her many surviving writings, her work cannot be ignored or downplayed.

Women were active not only in mathematics, physics, and medicine but also in the humanities. One of the great ignored women we have considered in this book is Ban Zhao, the Chinese historian and sister of Ban Gu from the 1st century CE (see chapter 3.3). She is accorded the honor only of finishing the work where her brother allegedly left off. It is time to check the veracity of this assumption by comparing writing styles. And why is the 9th-century founder of the University of Fez, Fatima al-Fihri (see chapter 4.1), shrouded in mystery? Was she merely the daughter of a wealthy Arab merchant, or was she intellectually active herself? And think of the 12th-century Byzantine historian Anna Comnena, who has a reputation of having produced only "strongly colored" history, as if her 12th-century male colleagues were not writing colored history themselves.[19] And of the many early Italian female philologists mentioned in chapter 5.1, until the 18th century they had little opportunity to develop their exceptional talent. Their fate was either seclusion or marriage. The philologists Anne Dacier and Anna Maria van Schurman were exceptions to this pattern, but an academic career was ruled out for them too—until Clotilde Tambroni broke this pattern at the end of the 18th-century with a professorship in Greek philology at the University of Bologna. Elsewhere in Europe, university chairs would remain reserved for men until the 20th century.

All these women deserve a place in the general history of science and humanities. I must also draw attention to the many female scientists and scholars I have omitted. For example, the influential natural philosopher Margaret Cavendish (1623–1673), the astronomer Maria Winckelmann (1670–1720), and the entomologist and artist Maria Sibylla Merian (1647–1717). And we mustn't forget the first computer programmer, Ada Lovelace (1815–1852), who wrote a program for the so-called *analytical engine*. Career opportunities for women did not improve until the 20th century, exemplified by Marie Skłodowska Curie (1867–1934). But despite her enormous international recognition, including two Nobel Prizes (in physics and chemistry), Marie Curie was not allowed to join the French Academy of Sciences. And she is an exception regardless: the insights of many brilliant 20th-century female researchers were long either ignored or suppressed, as is the case with Rosalind Franklin's (1920–1958) contribution to the discovery of DNA structure. While the situation has ameliorated in the 21st century, these improvements are relative. For example, in 2020, women accounted for barely over a third of (full) professors in the United States.

History of Knowledge versus Philosophy of Knowledge

Now that we have arrived at the end of this book, many new questions arise. For example, How does the relationship between patterns and principles relate to the practice of *modeling* in the various disciplines? In today's humanities, social sciences, and natural sciences, modeling is part of everyday practice.[20] The purpose of a model is to make part of the world more understandable by visualizing or simulating it or describing it mathematically. Models can be graphic, conceptual, mathematical, or computational. So, the notion of a model does not have a single, unambiguous definition, but in most cases it holds that a model "mediates" between phenomena and an underlying, sometimes still unknown, theory.[21] When a researcher models a phenomenon or pattern, it is not necessary to fully reduce it to underlying principles, as long as the patterns can be understood and described within the model. Sometimes a model is an approximation of the theory, and sometimes it is completely separate from the underlying theoretical principles. An intensive philosophical debate about the epistemic status of models has gone on for years.[22] However, most philosophers and researchers agree that models have taken on a life of their own in scientific and humanistic practice—even to the extent that research into the properties of physical, biological, linguistic, and archaeological models has become a subdiscipline unto itself. A multidisciplinary approach is indispensable for achieving a discipline-wide understanding of the notion of models, especially for understanding their relationship with underlying principles. For example, in many disciplinary practices, models are first developed for patterns, which are only later understood from the perspective of a broader theory, and thus from the perspective of principles. As a recurring meta-pattern between patterns and principles, in our terminology a model resembles a deduction pattern. But further research is needed to understand what it means when a model gains a more independent status that does not necessarily lead us back to deeper principles.

In addition, there is the question of how our history of knowledge relates to the problems in the philosophy of knowledge, better known as epistemology. This includes topics such as the problems of causality and induction.[23] How can we ascertain whether a given phenomenon or event is actually caused by another? And with regard to induction, how do we know whether the regularities we have found in the world actually correspond to reality or are accidental? After all, many other possible regularities could describe the same world, as we have seen at various points in this book, from astronomy to linguistics. My historical approach was not to answer these questions in a philosophical sense, let alone in an absolute sense, but to investigate how historical actors justified the patterns and principles they discovered at different times and places. The main questions in this book are therefore how people perceived patterns in history over time, what generalizations they made about these patterns, and whether they ever did so in a way that bridged the differences between cultures, periods, and disciplines. It is here that the historian can assist the philosopher. After all, philosophy cannot do without facts—and vice versa. We should therefore strive for an overarching field of history and philosophy of knowledge, similar to the existing history and philosophy of science.

The Future of Knowledge

My research has shown that every period is dominated by a certain type of quest. So, what should we think about the contemporary quest for patterns in *artificial intelligence* and *data science* in which little or no effort is made to explain these patterns with underlying principles?[24] As if history were repeating itself, the

Babylonian explosion of patterns has returned, with patterns derived on the basis of enormous amounts of data that are digitally available. Little or no effort is being made to seek out deeper explanations. Does this indicate a recurring dominance of that earlier quest from the distant past? It may seem a bit of a stretch to compare practices from early antiquity to those of contemporary science, but in many disciplines, classical insights are turning out to be more persistent than previously thought. Who would have expected that the seven exoplanets found to orbit the neighboring star Trappist-1 would do so in pure Pythagorean proportions?[25] At first such a system appeared to be excluded by the laws of celestial mechanics. But computer simulations have since shown that, over time, the orbits of planets can fall into a mutual harmonic resonance that corresponds to the time-honored idea of the harmony of the spheres (see chapter 3.2).[26] No one can escape the elegance of the Pythagorean idea of pure proportions between planetary orbits, but while Pythagoras was wrong insofar as our solar system is concerned, his intuition seems to hold true for another.

We can be inspired by ideas, concepts, or speculations from the past—and not just from ancient Greece but from all periods and from all regions, from Asia, the Americas, and Africa to Oceania. Many years ago I myself was inspired in my computational linguistic work by the 8th-century Persian linguist Sibawayh, who argued that a language can best be learned through large numbers of examples together with a generalization mechanism, rather than using rules of grammar (see chapter 4.4).[27] Our resulting *data-oriented parsing* model is based on Sibawayh's ideas, although I discovered this only later, which raised intriguing possibilities for a computer model that uses examples instead of rules to learn the regularities of language. Such a computer model not only appears to provide a deeper insight into children's acquisition of language; it has also proved to be important for concrete applications such as machine translation.[28]

So, the history of knowledge can be extremely fruitful for contemporary fields of study. It is important to realize this, considering that many researchers assume that only "recent" knowledge is relevant to contemporary knowledge practices. For example, the famous physicist Stephen Hawking wrote in his book, which appeared posthumously, "We spend a great deal of time studying history, which, let's face it, is mostly the history of stupidity."[29] My history of knowledge, however, shows that scientists and scholars alike make frequent and successful use of concepts, ideas, and methods from the past. Some of them were almost forgotten or were "failures." Yet they were picked up again and successfully applied in a new context. Think of the way Copernicus and others used Aristarchus's heliocentrism, the influence of the atomism of Lucretius on (early) modern disciplines, the impact of the oral Songhai histories on later historiography, or the influence of Chinese inoculation techniques on 18th- and 19th-century European medicine, not to forget the current revival of Babylonian, Pythagorean, and Sibawayhian ideas mentioned above.

The history of knowledge is a gold mine of ideas and practices that not only are important for understanding the past but that can also be inspiring and even decisive for the present. It is the historian's Herculean task to bring together knowledge practices from all periods and all parts of the world and make them accessible. Al-Masudi's words (see chapter 4.1) continue to resonate after more than a thousand years: "For any branch of knowledge to exist, it must be derived from history." The future depends on the past.

Acknowledgments

I am greatly indebted to those who read previous versions of my manuscript so carefully and who so generously provided suggestions and comments: I would like to thank the anonymous reviewers, as well as René Brouwer, Floris Cohen, Marjolein Degenaar, Marc de Wilde, Bart Karstens, Daniela Merolla, Floris Solleveld, Johan van Benthem, and Rienk Vermij. I am also grateful for the suggestions I received from many others with whom I have spoken about this project in recent years, too numerous to list here. It was a pleasure to work with my translator, Leston Buell. I am especially grateful for his original and inventive translations of some of my linguistic examples. Special thanks go to Carrie Watterson, whose excellent suggestions during her copy editing of the book led to significant improvements, both in style and content. Needless to say, any errors that remain are my sole responsibility.

By happy coincidence, I got involved in scientific research very early—perhaps too early. I published my first paper at the age of 15 after completing an astronomical summer school in Havelte, the Netherlands (IWAA, 1980), where young people participated in short-term research projects under the guidance of professional astronomers. My parents had sent me there in an attempt to rein in my boredom. They turned out to be right on target. Together with another summer school participant, I wrote a short article about an algorithmic procedure for detecting globular star clusters. It was a naive method that nevertheless produced some surprising results. For the first time I realized how much fun it was to write about a little discovery and then publish it.

The following year I attended the more international IAYC summer school in Ismailia, Egypt, where I participated in the Theory of Science group. We had both European and Egyptian instructors who involved us in their projects, one of the highlights of which was research on the step pyramids of Saqqara. It was my first acquaintance with the world outside of Europe, a revelation I still haven't gotten over. I would like to express my gratitude to the organizers and instructors at these summer schools.

When I began as a student at the University of Utrecht in 1983, in addition to the main subjects in physics and astronomy, I also did elective coursework in philosophy, musicology, history, and art history. After my first degree, I felt the need to leave the Netherlands and studied humanities at the University of La Sapienza in Rome, because there I could study all the humanities disciplines simultaneously, something that was not possible back home. However, I continued to prefer the more formal approach, such as formal stylistics and stemmatic philology. Ideally, I wanted to combine the sciences and the humanities. It was not until 1988, when I returned to the Netherlands, that I understood what I really wanted to pursue: artificial intelligence, in particular computational linguistics. In 1995 I obtained my PhD in computational linguistics from the University of Amsterdam. And in 1998 I published my first book, in which I aimed to make my work accessible to nonspecialists: *Beyond Grammar: An Experience-Based Theory of Language.*

After some wandering—from working as a consultant researcher at Xerox PARC (California) to being a professor of artificial intelligence at the University of St Andrews (Scotland)—I came back to the Netherlands full time in 2008. At the University of Amsterdam I became a professor of the digital humanities, a new field (at the time) that combines computer science with the humanities. I then realized that it was high time to write an overview of (the history of) the humanities, as part of the history of knowledge. My book *De vergeten wetenschappen* (The forgotten sciences) was published in 2010 and was translated into seven languages (see the preface). My desire to produce a more general history of knowledge has had to wait until the current book.

I could not have written this book without the support of multiple research institutions. Although I am critical of the ever-decreasing percentages of research proposals that get funded, I fully realize how much I am part of this competitive system and how much I owe to it. It has allowed me to immerse myself in the almost inexhaustible treasure house of human knowledge. The current work was supported by an NWO Open Competition grant (The Flow of Cognitive Goods: Towards a Post-disciplinary History of Knowledge), an NWO personal Vici grant (Integrating Cognition), an NSF/NWO Digging into Data grant (Legal Structures), and a personal grant from the Netherlands Institute for Advanced Study in the Humanities and Social Sciences (NIAS). I am also grateful to the Francqui Foundation for the International Francqui Chair at the University of Ghent that allowed me to work on the current English version.

The publication of a book often heralds the promise of a new line of research. That certainly was the case with my earlier work, but I doubt that I can live up to this promise again with the current book. Attempting to understand knowledge practices from a distant past or from another culture is indescribably fascinating. Slowly but surely you become initiated in the topic and then suddenly see a spectacular panorama before you. But sometimes you realize that it's time for something different. If there is one pattern in my own life, it is that nothing is interesting enough to devote an entire life to—except for my dearest Daniela, Livio, Ella, Luca, family, and friends.

Notes

Preface

1. The *Implementation Note NWO Strategy 2015–2018* ("Future-oriented humanities," p. 7) states, "And inversely, in their own way the humanities also contribute to the developments in those other fields of science, as Rens Bod in his book *De vergeten wetenschappen*, published in 2010, has convincingly demonstrated." And the KNAW writes in its 2012 report entitled *Outlines of a Renewal and Stimulation Program* (pp. 10–11), "However, Bod shows in his book *De vergeten wetenschappen* (2010) that in practice, the difference between the two fields of study is less fundamental than is often thought. He shows that, over the centuries, the boundary between the fields that we currently refer to as the natural sciences and the humanities was paper-thin and that humanities researchers have certainly made considerable contributions to the explanation of phenomena."

2. Shermer, "Scientia humanitatis," p. 80.

3. On one's amazement at the absence of great syntheses, see also Hakfoort, "The missing syntheses in the historiography of science."

4. Sarton, *Introduction to the History of Science.*

5. Although Sarton does include musicology and linguistics in his history, the other humanities disciplines, such as the study of literature and of art, are not included. According to him, the history of art sheds light on science only "from the outside" and does not contribute to scientific "progress"; see Sarton, *Introduction to the History of Science*, p. 1:5. Sarton has often been criticized for his strongly positivist attitude.

6. See classic works such as Dijksterhuis, *De mechanisering van het wereldbeeld*; Mason, *A History of the Sciences*; Dampier, *A History of Science and Its Relation to Philosophy and Religion*; and Gregory, *Natural Science in Western History.*

7. There are also books that address other aspects of knowledge, such as the history of knowledge institutions (McNeely and Wolverton, *Reinventing Knowledge*), of locations of knowledge practices (Jacob, *Lieux de savoir*), of the social aspects of knowledge (Burke, *A Social History of Knowledge*), or of the circulation of knowledge (Östling et al., *Circulation of Knowledge*).

Introduction

1. See, e.g., Tomlinson, *Culture and the Course of Human Evolution*, pp. 4–18. See also Geertz, *The Interpretation of Cultures*, p. 89.

2. For the definition of system, see, e.g., Backlund, "The definition of system."

3. For example, plants can learn from their experiences: experiments have shown that the sensitive plant (*Mimosa pudica*) can learn when it does and doesn't need to close its leaves in a given environment. See Mancuso et al., "Experience teaches plants."

4. For an overview, see McAllister, "The ontology of patterns in empirical data." See also Dennett, "Real patterns." And see also Dixon, "Analysis tool or research methodology: Is there an epistemology for patterns?"

5. Grenander, *Elements of Pattern Theory*; Psillos, "Regularities, natural patterns and laws of nature."

6. Cartwright, *How the Laws of Physics Lie.*

7. See https://hxwd.org/.

8. "The antithesis between explanation and description is quite illusory: we explain one thing *by* describing another," in Musgrave, *Essays on Realism and Rationalism*, p. 123.

9. See also the discussion in Burke, *What Is the History of Knowledge?*

10. See Raj, "Beyond postcolonialism"; Ganeri, "Polycentered history of science." See also Bala, *Asia, Europe, and the Emergence of Modern Science.*

11. Nicholas Jardine, "Uses and abuses of anachronism."

12. See Vansina, *Oral Tradition as History*, p. 27.

13. See Bod, *A New History of the Humanities*, pp. 253–254; and see the discussion in Armitage and Guldi, *The History Manifesto.* See also Tosh, *Historians on History.*

14. See the discussion in Kohler and Olesko, "Introduction"; Secord, "The big picture."

15. Huizinga, *Homo ludens.*

16. Braudel, *Civilisation matérielle.*

17. See, e.g., Nowak, *General Laws and Historical Generalizations*; and also Bod, "Who is afraid of patterns?"

18. Romein, *Historische lijnen en patronen.*

19. The anthropologist Mary Douglas referred to her colleagues' obsession with exceptions ("this does not apply to my tribe") as "Bongo-bongo-ism"; Richards, "Mary Tew Douglas." For a plea for making comparisons between different cultures, see Schipper, *Imagining Insiders.*

20. See Moretti, *Distant Reading.*

21. For an introduction to topic modeling, see Brett, "Topic modeling."

22. For useful links, see Clay Templeton, "Topic modeling in the humanities: An overview," Maryland Institute for Technology in the Humanities, August 1, 2011, http://mith.umd.edu/topic-modeling-in-the-humanities-an-overview/.

23. See Bod, Scha, and Sima'an, *Data Oriented Parsing*; Bod, "From exemplar to grammar." See also the tool developed by van Cranenburgh, "Rich Statistical Parsing and Literary Language."

24. See also the discussion in Lloyd, *Disciplines in the Making.*

25. See https://books.google.com/ngrams. See also Phillips, "Francis Bacon and the Germans."

26. Pliny, *Natural History*, book 36.

27. Nicholas Jardine, mentioned above, cites the application of modern disciplinary designations as one of the many enlightening uses of anachronisms; see Jardine, "Uses and abuses of anachronism."

28. Where I discuss the humanistic disciplines in this book, there is some overlap with my previous book, *A New History of the Humanities*, although my treatment of the Stone Age and early antiquity, which were beyond the scope of my previous book, is new. I revisit my treatment of the humanities because my insights into the history of knowledge have been progressively enriched since 2013. For example, the notion of "principle" in the current book differs from that in my previous book, as explained above.

CHAPTER ONE: The Awareness of Patterns

1. Joordens et al., "Homo erectus at Trinil."

2. See, e.g., Harari, *Sapiens.*

3. See Lanzarote-Guiral, "The recognition of cave art in the Iberian Peninsula."

4. Pike et al., "U-series dating."
5. Vogelsang et al., "New excavations of Middle Stone Age Deposits."
6. The cave paintings in Sulawesi are even older than their European counterparts; see Aubert et al., "Pleistocene cave art from Sulawesi."
7. Eir, *A Very Brief History of Eternity*, p. 10.
8. For a discussion about this, see North, *Cosmos*, pp. 5–6.
9. For a discussion, see Marshack, *The Roots of Civilization*, pp. 148–150.
10. See Rappenglück, "The Pleiades in the 'Salle des Taureaux.'"
11. See Wenke, *Patterns in Prehistory*, pp. 258–269. For a recent overview, see Zohary, Hopf, and Weiss, *Domestication of Plants in the Old World*, pp. 1–6.
12. Druzhkova et al., "Ancient DNA analysis affirms the canid."
13. Zohary, Hopf, and Weiss, *Domestication of Plants in the Old World*, p. 19.
14. Wenke, *Patterns in Prehistory*, p. 131.
15. This has also been observed in animals, where an accidental discovery is passed on by a bird or ape to the entire group and to subsequent generations.
16. See Roebroeks and Villa, "On the earliest evidence for habitual use of fire in Europe."
17. See Goudsblom, *Fire and Civilization*.
18. See for example Wiessner, "Embers of society."
19. See Liebenberg, *The Art of Tracking*, p. 29.
20. See, e.g., Insoll, *Oxford Handbook*.
21. See Bailey and Geary, "Hominid brain evolution."
22. See, e.g., Wenke, *Patterns in Prehistory*. See also Diamond, *Guns, Germs, and Steel*, pp. 114–115.
23. See Pringle, "The science of inequality."
24. Flannery and Marcus, *The Creation of Inequality*, pp. 66–67.
25. Salen and Zimmerman, *Rules of Play*, chapter 18.
26. Matthew 25:29, New Revised Standard Version. See also Robert Merton, "The Matthew effect in science."
27. McGee, *On Food and Cooking*, pp. 33–39.
28. Brenda Fowler, *Iceman*.
29. McClellan and Dorn, *Science and Technology in World History*, pp. 20–21.
30. Taylor, *Celestial Geometry*.
31. See North, *Cosmos*, p. 11.
32. Although the interpretation of this painting as a landscape is widely accepted, opposing views can be found; see, for example, Meece, "A bird's eye view."
33. Li et al., "The earliest writing?"
34. Haarmann, *Geschichte der Schrift*, p. 20.
35. Hayes, *A Manual of Sumerian Grammar and Texts*, p. 266.
36. For an overview, see Hock and Joseph, *Language History*.
37. For example, chimpanzees practice self-medication, and they have acquired knowledge of dozens of medicinal plants and herbs; see Michael Huffman, *The Study of Primate Self-Medication*.

CHAPTER TWO: The Explosion of Patterns and the Awareness of Principles

1. This is certainly an exaggeration. And yet there are cases where a single linguist developed multiple theories that superseded one another.
2. In my book *A New History of the Humanities* (2013), I ignored the Babylonians, stating that linguistics began with the Indian linguist Panini in around 600 BCE. But now that I have

been able to examine Babylonian clay tablets in detail, I have revised my position. What can be said is that no description of language "as a whole" (that is, of phonology, morphology, syntax, semantics, and pragmatics together) can be found before Panini.

3. Jacobson, "Very ancient linguistics."

4. Huber, "On the Old Babylonian understanding of grammar."

5. See, for example, the discussion in Frank, Bod, and Christiansen, "How hierarchical is language use?"

6. For an overview of all known linguistic clay tablets (including OBGT), see Landsberger et al., *Materialien zum Sumerischen Lexikon IV.*

7. It has been suggested that the Paleolithic dashes on the mammoth bones contain some deeper knowledge about numbers, but these speculations are extremely controversial. For a critical discussion, see Rudman, *How Mathematics Happened.*

8. For an overview, including an index of the mathematical clay tablets, see Robson, *Mathematics in Ancient Iraq.*

9. See Ifrah, *The Universal History of Numbers.*

10. Aaboe, *Episodes from the Early History of Mathematics*, p. 30.

11. Neugebauer and Sachs, *Mathematical Cuneiform Texts*, pp. 38–41. See also Bruins, "On Plimpton 322."

12. See Neugebauer, *The Exact Sciences in Antiquity.*

13. See Mansfield and Wildberger, "Plimpton 322 is Babylonian exact sexagesimal trigonometry." See also the reaction to this paper by Lamb, "Don't fall for Babylonian trigonometry hype.

14. Robson, "Words and pictures."

15. Fowler and Robson, "Square root approximations."

16. Robins and Shute, *The Rhind Mathematical Papyrus.*

17. Staal, "Greek and Vedic geometry."

18. Walker, "Notes on the Venus tablet of Ammisaduqa." See also North, *Cosmos*, p. 41.

19. MUL.APIN (ca. 1000 BCE) also contains a catalog of stars and constellations. Zodiac signs are among the oldest known patterns observed in the sky and probably date back to the Stone Age, but they are first described in MUL.APIN, which contains 18 constellations. We now know that constellations are random collections of stars that suggest figures if we draw lines to connect them. Accordingly, they constitute patterns of random similarities between star configurations and figures on earth.

20. See, for example, Brack-Bernsen, "The 'days in excess' from MUL.APIN."

21. Hunger, *State Archives of Assyria.*

22. See Osita, *A Day in the Life of God.* See also Rochberg, *The Heavenly Writing.*

23. Grasshoff, "Globalization of ancient knowledge."

24. Grasshoff, "Globalization of ancient knowledge."

25. Norriss, *Cosmology*, p. 46.

26. See North, *Cosmos*, p. 24.

27. Tripathi, "Astrology in India."

28. See Witzel, "Autochthonous Aryans?"

29. Kaufholz, *Sonne, Mond und Sterne.*

30. Gurney and Kramer, "Two Fragments of Sumerian Laws."

31. Wilcke, "Der Kodex Urnamma (CU)."

32. For a legal analysis of Hammurabi's laws, see Driver and Miles, *The Babylonian Laws.* See also Petschow, "Zur Systematik und Gesetzestechnik im Codex Hammurabi."

33. See Karstens et al., "Reference structures of national constitutions."

34. Ascalone, *Mesopotamia.*
35. Klein and Sharlach, "A collection of model court cases."
36. VerSteeg, *Law in Ancient Egypt*, pp. 20–24.
37. Heeβel, "Diagnosis, divination and disease."
38. See Geller, *Ancient Babylonian Medicine*, p. 25.
39. Stevens, "Gynaecology from ancient Egypt."
40. Allen, *The Art of Medicine in Ancient Egypt*, p. 70. See also Ghalioungui, *The House of Life*, p. 38.
41. Nunn, *Ancient Egyptian Medicine*, p. 28.
42. Nunn, *Ancient Egyptian Medicine*, pp. 131–132.
43. Magic has meant different things and has been held in different regards in different times. I use "magic" as a neutral term for a specific but widespread activity: contacting the supernatural to influence reality. See also Hanegraaff, *Esotericism and the Academy*, pp. 14–15.
44. Geller, *Ancient Babylonian Medicine*, 2010.
45. See Rochberg, "Empiricism in Babylonian omen texts."
46. Herodotus, *The History*, book 2.
47. Michalowski, "History as charter."
48. Baines, "On the evolution, purpose and forms of Egyptian annals."
49. Wilkinson, "Hydraulic landscapes and irrigation systems of Sumer."
50. Anthony, *The Horse, the Wheel, and Language.*
51. See Nissen, Damerow, and Englund, *Archaic Bookkeeping*, p. 51.
52. See Slotsky, *The Bourse of Babylon.*
53. See, for example, Sasson, *Civilizations of the Ancient Near East*, p. 4:2305: "In fact, the whole of its 'science' consists in the enumeration and classification of all natural and cultural entities."

CHAPTER THREE: The Explosion of Principles and the Awareness of Deduction

1. See Kirk and Raven, *The Presocratic Philosophers*, p. 3.
2. Diogenes Laertius, *Lives of Eminent Philosophers.*
3. Aristotle, *Metaphysics*, 983 b6 8–11. See also Diogenes Laertius, *Lives of Eminent Philosophers.*
4. Herodotus, *History*, 1.73–74.
5. See O'Grady, *Thales of Miletus.*
6. Panini's grammar is incorporated in the *Ashtadhyayi* (Eight books).
7. Line 1.4.2 in Panini's *Ashtadhyayi.*
8. Conche, *Anaximandre.*
9. Lindberg, "The Greeks and the Cosmos," p. 29.
10. For an overview of commentaries on Panini, see Staal, *A Reader on Sanskrit Grammarians.*
11. Dionysius Thrax, *Téchne grammatiké.*
12. Householder, *The Syntax of Apollonius Dyscolus*, p. 2; see also Blank, *Ancient Philosophy and Grammar.*
13. Law, *The History of Linguistics in Europe*, p. 42.
14. Plato, *The Republic*, 529c7–d5. My translation.
15. No works by Pythagoras have survived to this day—if he wrote anything at all—but many insights are attributed to him and his followers. See Diogenes Laertius, *Lives of Eminent Philosophers*, book 8.
16. See, for example, Horky, *Plato and Pythagoreanism.*

17. This question is apocryphal and was attributed to Plato by the Neoplatonist Simplicius (ca. 490–560 CE), who allegedly posed the question to his contemporaries in the 4th century BCE; see Lloyd, *Early Greek Science*, p. 84.

18. To be precise, Callippus was a student of Polemarchus, who in turn was a student of Eudoxus.

19. Mendell, "Reflections on Eudoxus, Callippus and their curves."

20. For an overview, see Shields, *The Oxford Handbook of Aristotle*.

21. For a description of the introduction of this new mathematical notion in ancient astronomy, see Neugebauer, *A History of Ancient Mathematical Astronomy*, p. 264.

22. Toomer, "Hipparchus and Babylonian astronomy."

23. Jones, *A Portable Cosmos*.

24. *Almagest* was the title of the Arabic translation of the original Greek work *Hē mathēmatikē syntaxis*, or *The Mathematical Order*. It has remained known as the *Almagest* after its 12th-century translation from Arabic into Latin (see chapter 4). See Toomer, *Ptolemy's Almagest*.

25. See Goldstein, "The Arabic version of Ptolemy's planetary hypotheses."

26. Feraboli, *Claudio Tolomeo*.

27. Even though ellipses are themselves an approximation of the actual orbits of the planets, albeit a better approximation than circles; see chapter 5.3.

28. Gottschalk, *Heraclides of Pontus*, p. 69.

29. See von Erhardt and von Erhardt-Siebold, "Archimedes' sand-reckoner."

30. There were Pythagorean ideas that both the sun and the earth moved around a central fire. See also the beginning of this chapter.

31. Even the founder of the history of Chinese science, Joseph Needham, dips his toes into mathematical astronomy. See Needham, *Science and Civilization in China*.

32. According to tradition, the works of Confucius were destroyed during the great book burning of 213 BCE, but Confucian scholars were able to reproduce them effortlessly thanks to their legendary memory.

33. See Hetherington, *Cosmology*, pp. 25–37.

34. Cullen, "Understanding the planets in ancient China."

35. See Cullen, "The first complete Chinese theory of the moon."

36. See, for example, the discussion in McClellan and Dorn, *Science and Technology in World History*, p. 133: "Unlike the Greeks, they [the Chinese] did not develop explanatory models for planetary motion."

37. Cullen, "The first complete Chinese theory of the moon," p. 36.

38. Sivin, *Cosmos and Computation in Early Chinese Mathematical Astronomy*.

39. North, *Cosmos*, p. 141.

40. D. Pingree et al., "The Paitamahasiddhanta of the Visnudharmottapurana."

41. See the discussion in Duke, "The equant in India."

42. See http://en.wikipedia.org/wiki/List_of_intervals for links to sound clips of consonant and dissonant intervals.

43. See Draffkorn Kilmer, "The discovery of an ancient Mesopotamian theory of music." See also M. L. West, "The Babylonian musical notation and the Hurrian melodic texts."

44. See Riedweg, *Pythagoras*.

45. Gibson, *Aristoxenus of Tarentum and the Birth of Musicology*.

46. See Bod, *A New History of the Humanities*, p. 39.

47. Huffman, *Aristoxenus of Tarentum*.

48. Sengupta, *Foundations of Indian Musicology*, p. 104.

49. Kaufmann, *Musical References in the Chinese Classics*, p. 37.
50. McClain and Hung, "Chinese cyclic cunings in late antiquity."
51. Herodotus, *The History*, 1.5.
52. Thucydides, *The Peloponnesian War*, 1.22.
53. See Bebbington, *Patterns in History*.
54. Anthony Clark, *Ban Gu's History of Early China*.
55. Barbara Bennet Peterson, *Notable Women of China*.
56. Barwick, *Problems of the Sprachlehre and Rhetorik*, p. 21.
57. Casper de Jonge, *Between Grammar and Rhetoric*.
58. Casper de Jonge, *Between Grammar and Rhetoric*, p. 283.
59. Yang, *Dragon-Carving and the Literary Mind*.
60. Pliny the Elder, *Natural History*, 35.103.
61. Pliny the Elder, *Natural History*, 34.55.
62. Agrawala, *On the Sadanga Canons of Painting*.
63. Sirén, *The Chinese on the Art of Painting*, p. 219.
64. Callanan, *Die Sprachbeschreibung at Aristophanes von Byzanz*.
65. Schironi, *Aristarco di Samotracia negli etimologici bizantini*.
66. See Fehling, "Varro und die grammatische Lehre von der Analogie und der Flexion."
67. Broggiato, *Cratete di Mallo*.
68. Garcea, *Caesar's De Analogia*.
69. See Diogenes Laertius, *Lives of Eminent Philosophers*, book 8.
70. Dante, *La Divina Commedia, Paradiso*, canto 13, 101–102.
71. Diogenes Laertius, *Lives of Eminent Philosophers*, book 8.84.
72. In reality, we find the oldest known proof that $\sqrt{2}$ cannot be written as a ratio of two integers in Euclid's *Elements* (book 10, theorem 117).
73. However, it should be emphasized that very little is known with certainty about when Plato's Academy was first established; see D. H. Fowler, *The Mathematics or Plato's Academy*.
74. For a translation, see the 13 books of Euclid's *Elements*, translated and with commentary by Thomas Heath.
75. Proclus, *Commentary on the First Book on Euclid's Elements*, p. 56.
76. Incidentally, the Greeks made a distinction between geometry and arithmetic (see below), but the bulk of Greek mathematics concerns quantities, rather than numbers.
77. See, e.g., Katz, *A History of Mathematics*, p. 63.
78. See Kneale and Kneale, *The Development of Logic*, p. 24.
79. Aristotle, *Metaphysics*, book 4.
80. See Bobzien, "Stoic logic."
81. See Dijksterhuis, *Archimedes*, part 1.
82. Rashed and Houzel, *Les "Arithmétiques" de Diophante*.
83. Watts, *Hypatia*.
84. Waithe, *Ancient Women Philosophers*, p. 175.
85. Graham, *Later Mohist Logic, Ethics and Science*. See also Johnston, *The Mozi*.
86. Shen, Crossley, and Lun, *The Nine Chapters on the Mathematical Art*.
87. For an analysis of Chinese algorithmic argumentation, see Chemla, *The History of Mathematical Proof*, pp. 462–471. So, preoccupation with argumentation is not uniquely Greek, as is often claimed.
88. Zha, "Research on Tsu Ch'ung-Chih's approximate method for π."

89. These principles correspond to A74 and A75 in Johnston, *The Mozi*.

90. For a comparison between Mohistic and classical Greek logic, see Zhang and Liu, "Some thoughts on Mohist logic."

91. See Joseph, *The Crest of the Peacock*, p. 314; see also Plofker, *Mathematics in India*, pp. 53–57.

92. Van Nooten, "Binary numbers in Indian."

93. See North, *Cosmos*, p. 18.

94. Boyer, *A History of Mathematics*, p. 209. See also Plofker, *Mathematics in India*, p. 235.

95. Hamilton, *Indian Philosophy*, pp. 4–7.

96. Lloyd, *Hippocratic Writings*, pp. 260–271. Incidentally, several versions of the theory of the humors were used, with different numbers of humors.

97. Arikha, *Passions and Tempers*, p. 7.

98. Lloyd, *Hippocratic Writings*, p. 262.

99. Von Staden, *Herophilos*.

100. Ivan, *Erasistrato*.

101. From an artistic standpoint, the surviving manuscripts of the *De materia medica* are unsurpassed; see, for instance, the 7th-century manuscript in the Biblioteca Nazionale di Napoli: Dioscurides Neapolitanus, Codex ex Vindobonensis Graecus 1.

102. Wujastyk, *The Roots of Ayurveda*.

103. Almast, "History and evolution of the Indian method of rhinoplasty."

104. See Unschuld, *Huang Di nei jing su wen*. See also Sivin, "Science and medicine in imperial China."

105. Martens, *Huang Di nei jing su wen*, chapter 29.

106. Perez, *Introduction au Shanghanlun*.

107. For the notion of experiment, see Hacking, *Representing and Intervening*, pp. 220–232.

108. Plutarch, *Quaestiones convivales*, 641C5.

109. Nicomachus, *Enchiridion harmonices*, 2nd century CE.

110. Barnes, *The Presocratic Philosophers*, p. 313.

111. Aristotle, *History of Animals*, 561a4–21.

112. For an overview, see Russo, *The Forgotten Revolution*.

113. Geus, *Eratosthenes von Kyrene*.

114. Vitruvius, *Ten Books on Architecture*, 9.9–12.

115. See Renn, "From the History of Science to the History of Knowledge—and Back."

116. Van Leeuwen, *The Aristotelian Mechanics*.

117. Aristotle, *Mechanical Problems*, 850b2.

118. By the way, quantitative formulations were not invented by the Greeks; we encountered these sorts of formulations before with the Babylonians; see chapter 2.

119. Aristotle, *Mechanical Problems*, 848b4–6.

120. See the discussion in Cohen, *How Modern Science Came into the World*, pp. 11–13.

121. See http://www.archimedespalimpsest.net/.

122. Laird, "Archimedes among the Humanists."

123. Section B25b in the Mohistic Canon. See also Needham, *Science and Civilisation in China*, p. 4:22. And see Renn and Schemmel, "Mechanics in the Mohist canon and its European counterpart," pp. 24–31.

124. Major, *Heaven and Earth in Early Han Thought*.

125. The similarity between this principle ("a very small weight [can] support a very large thing") with the statement attributed to Archimedes: "Give me a position to stand, and I will move the world" is striking.

126. Needham, *Science and Civilisation in China*, p. 4:37.

127. Weld and De Kleer, *Readings in Qualitative Reasoning about Physical Systems.*

128. See also Miao and Baichun, "The development of knowledge on levers in ancient China."

129. Guerra, "Weights and measures in pre-Columbian America."

130. See Aristotle, *Physics.*

131. See Needham, *Science and Civilisation in China*, p. 4:56.

132. Carawan, *Rhetoric and the Law of Draco*, 1998.

133. See, e.g., Maris, *Legal Philosophy*, chapter 1.

134. Cicero, *On the Laws*, book 2.

135. See Brouwer, "Ulpian's appeal to nature"; see also Brouwer, *Law and Philosophy in the Late Roman Republic.*

136. See Frier, *The Rise of the Roman Jurists.*

137. See Proverbs 18:17: "In a lawsuit the first to speak seems right, until someone comes forward and cross-examines."

138. Thanks to René Brouwer for this suggestion.

139. Strangely enough, Roman law still has the reputation of concerning itself solely with concrete cases rather than with principles—see, e.g., Menski, *Comparative Law in Global Context*, p. 140. It is correct that Roman legal scholars rarely searched for universal principles of natural law, but as I have shown in this section, they were committed to establishing principles of legal practice. In this respect, Roman jurists were looking for what later philosophers of law called a balance between "principles" and "rules"; see, e.g., Dworkin, *A Matter of Principle.*

140. Tolsa, "Ptolemy's law court analogy and Alexandrian philosophy"; see also Lehoux, *What Did the Romans Know?*, p. 127.

141. See Kane, *History of Dharmaśāstra.*

142. Head and Wang, *Law Codes in Dynastic China.*

CHAPTER FOUR: The Reduction of Principles

1. Fehling, *Herodotus and His "Sources."*

2. The three-volume *Encyclopedia of the History of Arabic Science*, edited by Roshdi Rashed attempts to provide an overview of the Arabic fields of knowledge as a whole, but the only field it covers in the humanities is musicology, mainly because that field was then classified as one of the mathematical sciences.

3. See Ibn Ali ibn Hajr al-Asqalani, *Al-Nukat ala Kitab ibn al-Salah*, p. 263.

4. Guillaume, *The Life of Muhammad.*

5. Robinson, *Islamic Historiography*, p. 49; see also Donner, *Narratives of Islamic Origins.*

6. Yar-Shater, *The History of al-Ṭabarī.*

7. Makdisi, *The Rise of Humanism.*

8. Al-Masudi, *The Meadows of Gold*, section 989.

9. For an in-depth analysis, see Cohen, *How Modern Science Came into the World.*

10. Although hardly any reference was made to Herodotus and Thucydides, they were well known by Islamic historians. Herodotus and Thucydides, for example, feature prominently on al-Sijistani's list of pre-Islamic scholars, the *Siwan al-Hikma* (Vessel of

wisdom); see Rosenthal, *The Classical Heritage in Islam*, pp. 36–37; Robinson, *Islamic Historiography*, p. 49.

11. Al Biruni, *Indica*, p. 1:3.
12. Al-Masudi, *The Meadows of Gold.*
13. This book is also known as *Chronology of Ancient Nations*, after the translation by Edward Sachau, W. H. Allen, 1879.
14. See Ibn Khaldun, *The Muqaddimah.*
15. See Joseph and Najmabadi, *Encyclopedia of Women & Islamic Cultures*, p. 314. It has now become a fairly common practice to apply the term "university" not only to institutions within Europe but also to those elsewhere if they independently award degrees for different levels of study.
16. See Twitchett, *The Writing of Official History under the T'ang.*
17. Ng and Wang, *Mirroring the Past*, p. 113.
18. Liu, *Shitong tongshi.*
19. See Pollman, *Saint Augustine the Algerian.*
20. Augustine, *De doctrina christiana*, book 2.18.28.
21. See Bod, *A New History of the Humanities*, chapter 3, for more details.
22. *Decem libri historiae*; see Heinzelmann, *Gregory of Tours.*
23. See Higham, *(Re-)Reading Bede.*
24. See Blackburn and Holford-Strevens, "Calendars and chronology."
25. For a list of Latin works translated from Arabic and Greek during the 12th-century wave of translations, see, e.g., Grant, *A Source Book in Medieval Science*, pp. 35–41.
26. See Stein, *Kalhana's Rajatarangini.*
27. *Kebra nagast.*
28. See Neugebauer, *The Astronomical Tables of al-Khwarizmi.* See also Hogendijk, "Al-Khwārizmī's table of the 'Sine of the hours.'"
29. See Sabra, "Configuring the universe."
30. Ragep, *Nasir al-Din Tusi's Memoir.*
31. See Roberts and Kennedy, "The planetary theory of Ibn al-Shatir."
32. Al-Biruni, *Indica.*
33. Martianus Capella, *The nuptiis philologiae et Mercurii*, 857.1–3, p. 333. See also Eastwood, *Ordering the Heavens*, pp. 238–239.
34. Lefèvre, Renn, and Schoepflin, *The Power of Images*, p. 208.
35. Heilbron, *The Sun in the Church*, p. 35.
36. See Nancy Marie Brown, *The Abacus and the Cross.*
37. Burnett, "The coherence of the Arabic–Latin translation program," pp. 275–281.
38. Rosen, "Alfonsine tables and Copernicus."
39. Heil and Ritter, *Corpus Dionysiacum*
40. Walter Eugene Clark, *The Âryabhaṭîya.*
41. Neugebauer and Pingree, *The Pancasiddhantika of Varamihira.*
42. Almeida, John, and Zadorozhnyy, "Keralese mathematics."
43. Bonnet-Bidaud, Praderie, and Whitfield, "The Dunhuang Chinese sky."
44. See Sivin, *Science in Ancient China*, part 2, pp. 71–72.
45. Grube, *Der Dresdner Mayan Calendar.*
46. Aveni, *Skywatchers of Ancient Mexico*, pp. 173–199.
47. Iwaniszewski, "Ancient cosmologies."
48. See, for example, Forster's travel report, *Observations Made during a Voyage round the World.*
49. Chadwick and Paviour-Smith, *The Great Canoes in the Sky*, chapter 5.

50. Lewis, "Voyaging stars."
51. See Harris et al., "A review of Maori astronomy."
52. Fischer, "Preliminary evidence for cosmogonic texts."
53. Smith, "The remarkable Ibn al-Haytham."
54. Rosenfeld, *A History of Non-Euclidean Geometry*, pp. 65–73.
55. Rashed, *Encyclopedia of the History of Arabic Science*, p. 469.
56. Rashed, *Al-Khwarizmi.*
57. Rashed and Armstrong, *The Development of Arabic Mathematics*, p. 11.
58. Hughes, *Robert of Chester's Latin Translation of Al-Khwarizmi's Al-Jabr.*
59. Devlin, *The Man of Numbers.*
60. Sigler, *Fibonacci's Liber Abaci.*
61. Singh, "The so-called Fibonacci numbers."
62. Clagett, *The Science of Mechanics in the Middle Ages*, pp. 332–333.
63. Manekin, *The Logic of Gersonides.*
64. See Kaplan, *The Nothing That Is.*
65. Proposition 18.30 in *Brahmasphuta-siddhanta*, GRETIL (Göttingen Register of Electronic Texts in Indian Languages), http://gretil.sub.uni-goettingen.de/gretil/1_sanskr/6_sastra/8_jyot/brsphutu.htm.
66. Proposition 18.33 in *Brahmasphuta siddhanta.*
67. See Plofker, *Mathematics in India*, pp. 7–57.
68. Rajagopal and Rangachari, "On medieval Keralese mathematics." See also Sarma, *A History of the Kerala School of Hindu Astronomy.*
69. See Roy, "Discovery of the series formula for π."
70. Libbrecht, *Chinese Mathematics in the Thirteenth Century*, p. 362.
71. Shuchun, Zaixin, and Jinhai, *Zhu Shijie.*
72. For an introduction to ethnomathematics, see Ascher, *Ethnomathematics.*
73. Sibawayh's *Kitab* has not been translated into English, but there are French and German translations, such as Gustav Jahn, *Sībawaihis Buch über die Grammatik übersetzt und erklärt.* I myself encountered a 10th-century copy of the *Kitab* in 2009 in the Biblioteca Ambrosiana in Milan.
74. For an in-depth study into the influence of Greek on Arabic linguistics, see Versteegh, *Greek Elements in Arabic Linguistic Thinking.*
75. Owens, *Early Arabic Grammar Theory*, p. 13.
76. Versteegh, *Greek Elements in Arabic Linguistic Thinking*, p. 11.
77. Al Biruni, *Indica*, p. 135.
78. For Aristotle's influence on early European linguistics, see Law, *The History of Linguistics in Europe*, p. 171.
79. Aristotle, *Metaphysics*, 4.1025b27.
80. Bursill-Hall, *Speculative Grammars of the Middle Ages.*
81. Seuren, *Western Linguistics*, p. 32.
82. Eco, *The Search for the Perfect Language*, p. 44.
83. Roger Bacon, *Opus maius*, part 3, 1267. Bacon's famous statement about universal grammar reads, "Grammatica una et eadem est secundum substantiam in omnibus linguis, licet accidentaliter varietur" (In substance, grammar is the same in all languages, although it may vary in random ways).
84. Covington, *Syntactic Theory in the High Middle Ages.*
85. Itkonen, *Universal History of Linguistics*, pp. 70–78, 335.
86. See Rau, *Bhartṛharis Vākyapadīya I & II.*
87. Wriggins, *Xuanzang.*

88. Dieter Torkewitz, *Das älteste Dokument zur Entstehung der abendländischen Mehrstimmigkeit.*
89. Erickson, "Musica enchiriadis, Scholia enchiriadis."
90. See Christensen, *The Cambridge History of Western Music Theory*, p. 480.
91. Atkinson, "Johannes Afflighemensis as a historian of fashion."
92. *Ad organum faciendum*, in Huff, *Musical Theorists in Translation.*
93. See Taschow, *Nicole Oresme and the Frühling der Moderne.*
94. See Chabrier, "Musical Science."
95. See Habib, *The Music of the Arabs*, pp. 17–20.
96. Kilpatrick, *Making the Great Book of Songs.*
97. See Bertrand, *La musique carnatique.*
98. See Picken and Nickson, *Music from the Tang Court.*
99. Cho, *The Discovery of Musical Equal Temperament*, pp. 172–175.
100. Augustine, *De doctrina christiana*, book 2.
101. Aquinas, *Summa theologica*, part 1, question 1, article 10.
102. See Harland, *Literary Theory from Plato to Barthes*, p. 25.
103. O'Daly, *Days Linked by Song*, p. 347.
104. Fulgentius, *The Exposition of the Content of Virgil.*
105. Anglade, *Leys d'amors.*
106. "Caedite eos! Novit andim Dominus qui sunt eius," attributed to the Cistercian abbot and crusader Arnaud Amalric by the German monk Caesarius of Heisterbach in the *Dialogus miraculorum* (written between 1219 and 1223).
107. Occitan had enormous prestige in the Middle Ages: courtly literature and troubadour songs originate from the Occitan courts, and Dante even considered writing his *Divina Commedia* in Occitan.
108. Of course, the Quran itself is not poetry. The Quran even takes a pronounced anti-poetry stance. But to their credit, Arabs nevertheless continued to compose poetry.
109. Dahiyat, *Avicenna's Commentary on the Poetics of Aristotle.*
110. Cantarino, *Arabic Poetics in the Golden Age.*
111. Ouyang, *Literary Criticism in Medieval Arabic-Islamic Culture*, p. 181.
112. Van Gelder, *Beyond the Line*, pp. 37–42.
113. Ramaswami Shastri, *A Short History of the Purva Mimamsa Shastra.*
114. See Kracke, *Civil Service in Early Song China.*
115. Kirkpatrick, "China's first systematic account of rhetoric," p. 150.
116. Kirkpatrick, "China's first systematic account of rhetoric."
117. See McGinnis, *Avicenna.* See also Goehl, *Avicenna and seine Darstellung der Arzneiwirkungen.*
118. Dante, *Divina Commedia, Inferno*, 4.132.
119. Savage-Smith, "Medicine," 917.
120. Meyerhof, "New light on universal Ibn Isḥâq and his period."
121. John West, "Ibn al-Nafis."
122. Badri, *Abu Zayd al-Balkhi's Sustenance of the Soul.*
123. The phenomenon that the merits of female scientists are minimized and attributed to male colleagues is known as the Matilda effect. See Rossiter, "The Matthew/Matilda effect in science." About the Salernitan school, see Jacquart and Bagliani, *La Scuola Medica Salernitana.*
124. See the biography by Maddocks, *Hildegard of Bingen.*
125. Hildegard of Bingen, *Causae et curae.*
126. Hildegard of Bingen, *Scivias* 1, vision 2.12.

127. Glick, Livesey, and Wallis, *Medieval Science, Technology, and Medicine*, p. 178.
128. Campbell and Colton, *The Surgery of Theodoric.*
129. Mensa i Valls, *Arnau de Vilanova.*
130. Ahmad and Qadeer, *Unani.*
131. See Hinrichs and Barnes, *Chinese Medicine and Healing*, pp. 87–88.
132. See Unschuld, *Medicine in China*, pp. 154–156.
133. In *A New History of the Humanities*, I deal with the Byzantine humanities in more detail, especially philology and history.
134. For a discussion about the Byzantine natural sciences, see "Viewpoint: Science and Orthodox Christianity: Hippopedes and Callippopedes," *Centaurus*, 107, 2016, pp. 541–596.
135. See Liebs, *Die Jurisprudenz im spätantiken Italien.*
136. See Lokin, "Codificatie in Constantinopel?," 41.
137. Leppin, "Die Gesetzgebung Iustinian."
138. There are few monographs of the brilliant Tribonian. An exception is Honoré's *Tribonian.*
139. James, *Visigothic Spain*, p. 132.
140. Drew, *The Laws of the Salian Franks.*
141. Rota, *Lo stato and diritto nella concezione di Irnerio.*
142. Maffei, *La "Lectura super digesto veteri" di Cino da Pistoia.*
143. Benedetto, *Bartolo da Sassoferrato.* See also Cecil Nathan Sidney Woolf's classic *Bartolus of Sassoferrato.*
144. See Lowry, *Early Islamic Legal Theory.*
145. Esposito, *Islam*, p. 78.
146. Bauer, *Bausteine des Fiqh*, p. 67.
147. For an introduction to the Tang code, see Johnson, *The T'ang Code.*
148. Johnson, *The T'ang Code.*

CHAPTER FIVE: The Discovery of Patterns in Deductions

1. I take this term from de Groot, *Methodologie.*
2. Although the "Scientific Revolution" has been under debate for several years now, it remains a dominant concept in much historiography of knowledge; see Shapin, *The Scientific Revolution.*
3. For an introduction to Petrarch's philological work, see Billanovich, *Petrarch and Primo Umanesimo.* See also Barolini and Storey, *Petrarch and the Textual Origins of Interpretation.*
4. See Reynolds and Wilson, *Scribes and Scholars*, p. 129.
5. See Billanovich, *Tradizione and fortuna di Livio tra Medioevo and Umanesimo.*
6. Salutati, *Epistolario*, p. 4:216.
7. Buck, "The 'studia humanitatis' im italienischen Humanismus," pp. 11–24.
8. Makdisi, *The Rise of Humanism.*
9. See Fubini and Caroti, *Poggio Bracciolini.* See also Greenblatt, *The Swerve.*
10. See Bowersock, *Lorenzo Valla, On the Donation of Constantine.* See also Zinkeisen, "The Donation of Constantine as applied by the Roman Church."
11. For a comparison between Cusanus's *De Concordantia Catholica* and Valla's *De Falso Credita*, see Fubini, "Humanism and truth."
12. For a translation of, see Bowersock, *On the Donation of Constantine*, 2007.
13. Nauta, "Lorenzo Valla and Quattrocento skepticism."
14. Poliziano, *Miscellanea.*

15. Poliziano, *Miscellanea*, 1.39; see also Grafton, *Defenders of the Text*, p. 56.

16. See Metzger, *Der Text des Neuen Testamentsa*, pp. 100–102; see also H. J. de Jonge, "The character of Erasmus' translation of the New Testament."

17. In this book I do not do justice to Erasmus's extremely multifaceted work but limit myself to mentioning his contribution to the development of the empirical cycle in philology. For the works of Erasmus, see Hoffmann and Tracy, *Collected Works of Erasmus.*

18. See also van Miert, *The Emancipation of Biblical Philology.*

19. For a thorough biography of J. J. Scaliger, see Grafton, *Joseph Scaliger.*

20. Nothaft, *Dating the Passion*, pp. 2–9.

21. According to the Hebrew Bible, see Grafton, "Joseph Scaliger and historical chronology."

22. Elrington, *The Whole Works of the Most Rev. James Ussher*, p. 489.

23. See Popkin, *Isaac La Peyrère*, p. 26.

24. See Jorink and van Miert, *Isaac Vossius.*

25. See Piet Steenbakkers, "Spinoza in the history of biblical scholarship."

26. See Israel, *Radical Enlightenment.*

27. As Jorink states in *Het boeck der natuere*, p. 429: "There is a rarely recognized line that runs from Scaliger, through Saumaise and Isaac Vossius, to Spinoza." See also van Miert, *The Emancipation of Biblical Philology* 8.

28. See also Aalders, *Gevecht met de tijd.*

29. See Manuel, *Isaac Newton, Historian.*

30. See Rademaker, *Leven en werk van Gerardus Joannes Vossius*, p. 186.

31. See Wyles and Hall, *Women Classical Scholars*, p. 35.

32. See Wyles and Hall, *Women Classical Scholars*, p. 78.

33. Tosi, *I carmi greci di Clotilde Tambroni.*

34. Cavazza, "Donne e scienza nell'Italia del settecento."

35. For a further explanation of Karl Lachmann's philology, see Glenn Most, "Karl Lachmann (1793–1851)."

36. See Engelstein, *Sibling Action.*

37. Baur, *Versuch über die Historik des jungen Ranke.*

38. Bod, "A comparative framework."

39. Baxter, *A Handbook of Old Chinese Phonology*, pp. 154–155.

40. Willard Peterson, "The life of Ku Yen-Wu."

41. Wang, "Beyond East and West."

42. Elman, "Philology and its enemies."

43. The Remonstrants formed a movement that split from the Dutch Reformed Church in the early 17th century. Although the Dutch States-General issued an edict tolerating both parties, a number of Remonstrants had to flee the Dutch Republic.

44. Goodrich and Fang, *Dictionary of Ming Biography*, p. 811.

45. The Timbuktu manuscripts were jeopardized in 2011 when the city was overrun by extremists. A large number of manuscripts were moved to Bamako in time, but many other manuscripts are being kept in home libraries. See also Jeppie, "Timbuktu scholarship."

46. See Shillington, *Encyclopedia of African History*, p. 640.

47. John Hunwick, *Timbuktu and the Songhay Empire.*

48. Milton, *Paradise Lost*, book 11, line 399.

49. See the analysis in Tolmacheva, *The Pate Chronicle.*

50. Ely and Johansen, *Handlist of Manuscripts.*

51. For a discussion, see Ki-Zerbo, *General History of Africa I*, pp. 1–6. See also Minicka, "Towards a conceptualization."

52. See chapter 3 in Goody, *The Interface between the Written and the Oral.* This book also discusses the accuracy of oral tradition in history, starting with the Vedas, through Homer, and up to contemporary traditions.
53. See Ki-Zerbo, *General History of Africa I.*
54. Thanks to Daniela Merolla for this suggestion.
55. Dowson, *The History of India.*
56. *The Akbarnama of Abul Fazl.*
57. Beach, Koch, and Thackston, *The King of the World.*
58. Hill Boone, *Stories in Red and Black.*
59. See the stories in Tedlock, *2000 Years of Mayan Literature.*
60. Tedlock, *Popol Vuh.*
61. Fischer, *Island at the End of the World.*
62. Patota, *L. B. Alberti.*
63. Tavoni, "The 15th-century controversy on the language spoken by the ancient Romans."
64. See Breva-Claramonte, *Sanctius' Theory of Language.*
65. See Lakoff, review of Herbert Brekle, *Grammaire générale et rassonnée*, p. 356.
66. Muller, "Early stages of language comparison."
67. Burke, *A Social History of Knowledge*, pp. 164–165.
68. As Jorink states in *Het boeck der natuere*, p. 307.
69. See Grafton, *Leon Battista.*
70. See Alberti, *On Painting.*
71. Kemp, *The Science of Art*, pp. 17–20.
72. Lindberg, *Roger Bacon and the Origins of Perspectiva.*
73. Andersen, *The Geometry of an Art*, p. 40.
74. Da Vinci, *A Treatise on Painting.*
75. See Dürer, *De Symmetria Partium.*
76. So, Ptolemy tried to bring the theory closer to practice. This was also his endeavor in astronomy (see 4.2) and even in the study of light. For an analysis of Ptolemy's program, see Cohen, *How Modern Science Came into the World.*
77. Ramis de Pareia, *Musica Practica.*
78. Young, *The Practica Musicae of Franchinus Gafurius*, pp. xxii–xxiii.
79. Zarlino, *Istituzioni armoniche.*
80. See Cohen, *Quantifying Music*; see also Palisca, "Was Galileo's father an experimental scientist?"
81. See Macrobius, *Commentary on the Dream of Scipio*, 2.1.15; and Boethius, *De Institutione Musica*, 1.x–xi.
82. See also D. P. Walker, "Some aspects of the musical theory of Vincenzo Galilei and Galileo Galilei."
83. See also Gouk, "Music and the emergence of experimental science in early modern Europe."
84. Mersenne, *Harmonie universelle.*
85. The proportions remain the same considering that a string n times long has a vibration frequency n times as small.
86. For an overview, see Cohen, *Quantifying Music.*
87. See Cohen, "Music as science and as art."
88. See Fabris, "Galileo and music."
89. See Lam, "Chinese music: History and theory."
90. Lam, "Ming Music and Music History."

91. See Powers, "The Background of the South Indian Raga System"; and also Powers, "The structure of musical meaning."

92. Gray, *African Music*; see also Scherzinger, "African Music and the History of Time."

93. Owen Wright, *Demetrius Cantemir.*

94. See for instance Marie Boas Hall when she writes, "Scientists were ready to adopt the methods of humanism," and "scientists were mostly scholars, physicians or magicians." Marie Boas Hall, *The Scientific Renaissance, 1450–1630*, pp. 18 and 19.

95. Zinner, *Regiomontanus.*

96. Holden, *A History of Horoscopic Astrology*, pp. 157–159.

97. See Zimmer, *Regiomontanus*, p. 121.

98. For this question, see Cohen, *How Modern Science Came into the World.*

99. There is a great amount of literature on Copernicus, see, e.g., Gingerich, *The Book Nobody Read.* See also Kuhn, *The Copernican Revolution.*

100. Mosley, *Bearing the Heavens.*

101. Hashimoto, "Longomontanus's 'Astronomia Danica' in China."

102. Bialas, *Johannes Kepler.* See also Gingerich, *The Eye of Heaven.*

103. See the chapter on Kepler's philology and chronology in Grafton, *Defenders of the Text.*

104. See Kepler's collected works, https://kepler.badw.de/.

105. See Kepler, *Vom wahren Geburtsjahr Christi.*

106. See, e.g., Grafton, "Kepler as reader."

107. See https://kepler.badw.de/.

108. Ferguson, *Tycho & Kepler.*

109. For Kepler's theological motives, see Barker and Goldstein, "Theological foundations of Kepler's astronomy."

110. Roller, *The "De Magnete" of William Gilbert.*

111. Nellen, *Ismaël Boulliau.*

112. Henry, "Seth Ward."

113. For the role of Gassendi, see Fisher, *Pierre Gassendi's Philosophy and Science.*

114. There is a vast literature on Galileo; see, e.g., Biagioli, *Galileo Courtier*; Sharratt, *Galileo.*

115. Van Helden, *The Invention of the Telescope.*

116. Van Helden, *Galileo Galilei, Sidereus Nuncius.*

117. Galileo, *Dialogue concerning the Two Chief World Systems.*

118. Duncan, *Johannes Kepler—Mysterium Cosmographicum*, p. 82.

119. Although Galileo worked systematically in his observations, there does not seem to be a cyclic interaction between theory and experimentation (or testing). But we cannot be certain about this.

120. Grafton, "Kepler as reader."

121. North, *Cosmos*, pp. 149, 153–157.

122. Pingyi, "Scientific dispute in the imperial court.

123. See Krisciunas and Paksoy, *The Legacy of Ulugh Beg.*

124. See Ágoston and Masters, *Encyclopedia of the Ottoman Empire*, p. 552.

125. Crew and de Salvio, *Dialogues concerning Two New Sciences.*

126. Or, as Galileo put it, that the consecutive distances traveled are proportional to the consecutive odd numbers, which mathematically has the same result (because the consecutive odd numbers 1, 3, 5, 7, 9, 11 . . . are equal to the differences between the squares of the consecutive numbers 0, 1, 4, 9, 16, 25, 36 . . .).

127. The symbol *v* stands for velocity (*velocitas*), *a* for acceleration (*acceleratio*), and *t* for time (*tempus*).

128. Galileo's experiments have been replicated based on his descriptions by Settle, "An experiment in the history of science"; see also Drake, *Galileo*.

129. Cohen, *Quantifying Music*, pp. 78–84.

130. See Palisca, *Humanism in Italian Renaissance Musical Thought*; see also Moyer, *Musica Scientia*.

131. Descartes, *A Discourse on Method*.

132. This mechanistic world was elaborated by Descartes in his *Principia Philosophiae* from 1644.

133. See van Ruler, *The Crisis of Causality*.

134. Bacon's works are available online: Spedding, Ellis, and Heath, *The Works of Francis Bacon*, http://onlinebooks.library.upenn.edu/webbin/metabook?id=worksfbacon.

135. See Vermij, *Christiaan Huygens*; see also Andriesse, *Huygens*.

136. Huygens's laws of collision did not appear until after his death, under the title *De Motu Corporum ex Percussione* (*The Motion of Colliding Bodies*) as part of the *Opera Posthuma* (1703).

137. Yoder, *Unrolling Time*, p. 5 ; Huygens, *Oeuvres complètes*.

138. See Huygens, *Oeuvres complètes*, pp. 21:653–842.

139. There is an untold variety of works written on Newton. I will just mention Westfall, *Never at Rest*; White, *Isaac Newton*.

140. See Feingold, *Before Newton*.

141. Halley's letter no. 289 to Newton, in Newton, *Correspondence*.

142. On Newton's obsessive nature, see Westfall, *Never at Rest*, p. 404.

143. Newton, *Principia*, book 1.

144. See Terrall, *The Man Who Flattened the Earth*.

145. See Lancaster-Brown, *Halley and His Comet*.

146. "The Planets move one and the same way in Orbs concentrick, some inconsiderable Irregularities excepted, which may have arisen from the mutual Actions of Comets and Planets upon one another, and which will be apt to increase, till this System wants a Reformation." Newton, *Opticks*, query 31, 1704.

147. Antommarchi, *Mémoires du docteur F. Antommarchi*, p. 282.

148. See van Helden, "Willem Jacob 's Gravesande, 1688–1742," pp. 450–453.

149. Hagengruber, *Émilie du Chatelet between Leibniz and Newton*.

150. Mitford, *Voltaire in Love*.

151. See La Vopa, *The Labor of the Mind*. On Poullain de la Barre, see Stuurman, *François Poulain de la Barre and the Invention of Modern Equality*.

152. From the 1713 edition of the *Principia*.

153. A substantial portion of Newton's unpublished work (329 manuscripts, a third of them in the field of alchemy) was put up for auction at Sotheby's in 1936 and has been extensively studied ever since. See Keynes, "Newton, the man." For a historical review, see Schilt, "Created in our image."

154. See Christianson, *Isaac Newton*, p. 144.

155. Recall that the square root in Vincenzo Galilei's string law is the inverse of a square. See McGuire and Rattansi, "Newton and the 'Pipes of Pan.'"

156. See Walker, "Some aspects of the musical theory of Vincenzo Galilei and Galileo Galilei."

157. See Manuel, *Isaac Newton, Historian*; see also Schilt, "Created in our image."

158. For heresy around 1700 in England and Scotland, see Hunter, "Aikenhead the atheist."

159. See White, *Isaac Newton.*

160. See Westfall, *Never at Rest*, pp. 530–531.

161. See Karstens, "Bopp the builder."

162. See Bod, *A New History of the Humanities*, chapter 5.

163. Needham, *Science and Civilization in China*, p. 4:55.

164. See, e.g., Lu, *A History of Chinese Science and Technology*; see also Schäfer, *The Crafting of the 10,000 Things.*

165. See also Cullen, "The science/technology interface in seventeenth-century China."

166. See, e.g., Diamond, "Peeling the Chinese onion."

167. Thanks to Karine Chemla, who pointed this out to me. See also Elman, *Science in China, 1600–1900*, p. 64.

168. Buringh and van Zanden, "Charting the 'Rise of the West.'"

169. See Blaydes and Chaney, "The feudal revolution and Europe's rise."

170. See also the discussion in Schemmel, "The transmission of scientific knowledge from Europe to China."

171. See, e.g., Soll, *The Reckoning.*

172. See Giusti, *Luca Pacioli.*

173. Amir-Moez, "A paper of Omar Khayyam."

174. See Bottazzini, "La 'grande arte,'" p. 72.

175. Godard, "François Viète."

176. Droste, *Simon Stevin.*

177. For an annotated translation of Descartes's *Géométrie*, along with the original French text, see Smith and Latham, *The Geometry of René Descartes.*

178. Smith and Latham, *The Geometry of René Descartes*, p. 48.

179. Dopper, "A Life of Learning in Leiden."

180. Bardi, *The Calculus Wars*; see also Boyer, *The History of the Calculus.*

181. See Baron, *The Origins of the Infinitesimal Calculus.*

182. See Berggren, "Innovation and tradition in Sharaf al-Din al-Tusi's Muadalat."

183. See Whiteside, *The Mathematical Papers of Isaac Newton*, p. 32.

184. See Iliffe and Smith, *The Cambridge Companion to Newton*, p. 593.

185. See Stigler, *The History of Statistics.*

186. Calinger, *Leonhard Euler.*

187. See also Mancosu, *Philosophy of Mathematics.*

188. Agnesi, *Analytical Institutions.*

189. See Ogilvie, *Women in Science*; see also Mazzotti, *The World of Maria Gaetana Agnesi*; Paula Findlen, "Calculations of faith."

190. See, e.g., Katz, *A History of Mathematics*, pp. 565–567.

191. Kramer, "Maria Agnesi."

192. For an overview, see Martzloff, *A History of Chinese Mathematics.*

193. Needham, *Science and Civilisation in China*, volume 3.

194. Zhihui, "Scholars' recreation of two traditions."

195. See, e.g., El-Rouayheb, "Sunni Muslim scholars on the status of logic"; El-Rouayheb, "Was there a revival of logical studies?"

196. Katz, "Ideas of calculus in Islam and India." For an overview, see Plofker, *Mathematics in India.* For a wider discussion, see Raj, "Beyond postcolonialism."

197. See Eckart, "Die Medizin der Renaissance."

198. O'Malley, *Andreas Vesalius of Brussels.*

199. In his later work *De Statua* (About sculpture, 1462), Alberti goes a step further and argues that a realistic representation of reality requires in-depth knowledge of anatomy.

200. See Bod, *A New History of the Humanities*, chapter 4.

201. For a translation, see Vesalius, *On the Fabric of the Human Body.*

202. Weeks, *Paracelsus.*

203. Mazliak, *Jean Fernel.*

204. Many of these scholars remained just as much philologists as physiologists; see Grafton, "Philological and artisanal knowledge making."

205. Muccillo, "Fabrici d'Acquapendente."

206. See Thomas Wright, *Circulation.*

207. Harvey, *On the Motion of the Heart and Blood in Animals*, https://www.bartleby.com/38/3/.

208. See Bainton, *Hunted Heretic.*

209. Lisa Jardine, *The Curious Life of Robert Hooke.*

210. Meli, *Mechanism, Experiment, Disease.*

211. See Snyder, *Eye of the Beholder.*

212. See the contributions in Houtzager, *Reinier de Graaf.*

213. Kooijmans, *Het orakel.*

214. Miettinen, "The modern scientific physician."

215. Doi, *Understanding Evidence in Health Care.*

216. For a bibliography on Chinese smallpox vaccination, see Unschuld and Jinsheng, *Chinese Traditional Healing.*

217. See Gross and Sepkowitz, "The myth of the medical breakthrough."

218. Henderson, *Smallpox.*

219. See Needham, Gwei-Djen, and Sivin, *Science and Civilisation in China*, p. 6:154.

220. Wan, "Douzhen xinfa" [Smallpox patterns based on personal experiences], from 1549; and also Yu Chang, "Yuyi cao" [Notes on my judgment], from 1643; see also Needham, Gwei-Djen, and Sivin, *Science and Civilisation in China*, vol. 6.

221. Unschuld, *What Is Medicine?*, pp. 140–141.

222. See Leung, "Organized medicine in Ming-Qing China," p. 150.

223. Holwell, *An Account of the Manner of Inoculating for the Small Pox.*

224. See Needham, Gwei-Djen, and Sivin, *Science and Civilisation in China*, p. 6:145.

225. Pankhurst, *An Introduction to the Medical History of Ethiopia*, pp. 26–28.

226. Henderson, *Smallpox*, p. 45.

227. Janssens, "Matthieu Maty and the adoption of inoculation for smallpox in Holland."

228. Voltaire to A. M. Dalmilaville, June 1763, in *Lettres inédites.*

229. Boylston, "The origins of inoculation."

230. Li, *Compendium of Materia Medica.*

231. Unschuld, *What Is Medicine?*, p. 143.

232. See Fu, "A forgotten reformer of anatomy in China"; see also Andrews, "Wang Qingren and the history of Chinese anatomy."

233. Asen, "'Manchu anatomy.'"

234. Hinrichs and Barnes, *Chinese Medicine and Healing*, p. 262.

235. See, e.g., Ji, Li, and Zhang, "Natural products and drug discovery."

236. Norton, "Herbal medicines in Hawaii."

237. See Montheit, "Guillaume Budé, Andrea Alciato, Pierre de l'Estoile."

238. For an overview and introduction, see Passerin d'Entreves, *Natural Law.*
239. See Nellen, *Hugo de Groot.*
240. This summary reasoning by Grotius comes from van den Bergh, *Geleerd recht*, p. 78.
241. From van den Bergh, *Geleerd recht*, pp. 78–79.
242. Grotius, *Prolegomena*, 11.
243. See Hochstrasser, *Natural Law Theories.*
244. See Dufour, "L'influence de la méthodologie," p. 33.
245. Hochstrasser, *Natural Law Theories* pp. 40–71.
246. Kant, *Ausgabe der Preußischen Akademie der Wissenschaften*, AA IV, p. 420.
247. With a few exceptions, such as Grotius's attempt to establish the rights of non-Western peoples on the basis of natural law, especially the passages on the rights of the indigenous people in the Indies in *De Jure Praedae* (On the law of prize and booty, 1604). Thanks to Marc de Wilde, personal communication.
248. See van den Bergh, *Die holländische elegante Schule.*
249. See Wieacker, *Privatrechtsgeschichte der Neuzeit*, pp. 339–347.
250. See Weatherall, *Jus Cogens*, pp. 41–54.
251. See Yonglin, *The Mandate of Heaven and the Great Ming Code*, p. 5.
252. *Da Ming lu*, translated by Jiang Yonglin in *The Mandate of Heaven and the Great Ming Code.*
253. See Menski, *Comparative Law in Global Context*, p. 566.
254. See Chen, *An Introduction to the Legal System of the People's Republic of China.*
255. See Feuerwerker, *History in Communist China.*
256. See Gerber, *State, Society, and Law in Islam*; see also Uriel, *Studies in Old Ottoman Criminal Law.*
257. For an overview, see Moore, *Law and Anthropology.*
258. Van Notten, *The Law of the Somalis*; see also Abdile, "Customary dispute resolution in Somalia."
259. See the discussion in Donald Brown, "Human universals and their implications."
260. See Ho and Kramer, "The empirical revolution in law." For a concrete example of a search for patterns and principles in empirical legal studies, see Karstens et al., "Reference structures of national constitutions."
261. For a description of analogy in the natural sciences, see Achinstein, "Models, analogies and theories."
262. My use of the phrase "the West versus the rest" here is borrowed from Stuart Hall's "The West and the Rest." The dichotomy between the West and the "Third World" has been effectively critiqued by Homi Bhabha in *Nation and Narration.*
263. Pomeranz, *The Great Divergence.*
264. See Pomeranz, *The Great Divergence.*

Conclusion

1. See Piaget, *The Child's Conception of the World.*
2. Gadamer, *Wahrheit und Methode.*
3. Thanks to Stefani Engelstein, who asked me the right question on this topic during her visit to the Vossius Center in Amsterdam in 2017. My view of the hermeneutic circle has since been subject to growing insight and is no longer in line with what I previously wrote on this subject in *A New History of the Humanities* pp. 333–334.
4. Surprisingly, this is similar to the notion of *participant observation* in cultural anthropology, but I will leave this out of consideration for the current book.

5. See also Kwa, *Kernthema's in de wetenschapsfilosofie*, p. 97, where he advocates that the practice of the hermeneutic researcher "comes functionally close to being a 'test.'"

6. De Groot, *Methodologie.*

7. Barrow and Tipler, *The Anthropic Cosmological Principle.*

8. See, e.g., Dilthey, *Einleitung in die Geisteswissenschaften*; Windelband, *Geschichte und Naturwissenschaft*; Snow, *The Two Cultures and the Scientific Revolution.*

9. On similar virtues in the humanities and natural sciences, see in particular van Dongen and Paul, *Epistemic Virtues.*

10. Descola, *Beyond Nature and Culture.*

11. Wölfflin, *Principles of Art History.*

12. This does not alter the fact that astrological and alchemical investigations have yielded other insights; see, e.g., Principe, *The Secrets of Alchemy.*

13. Gieryn, "Boundary-work."

14. See, e.g., McClellan and Dorn, *Science and Technology in World History*; Gregory, *Natural Science in Western History*; Fara, *Science.*

15. One place where women are included in overview histories is in books that focus specifically on female (natural) scientists, such as Abir-Am and Outram, *Uneasy Careers and Intimate Lives*; Schiebinger, *The Mind Has No Sex?*; Ogilvie, *Women in Science*; Fara, *Pandora's Breeches*; Fara, *A Lab of One's Own.* One of the few overviews that focuses on female humanities scholars is Wyles and Hall, *Women Classical Scholars.*

16. See McClellan and Dorn, *Science and Technology in World History*; Gregory, *Natural Science in Western History*; Fara, *Science.*

17. For references and quotes, see Findlen, *Calculations of Faith*, pp. 248–291.

18. This further research has only accelerated in recent years; see, e.g., Roero, "M. G. Agnesi, R. Rampinelli and the Riccati family."

19. See Bod, *A New History of the Humanities*, p. 91.

20. See Jeffrey Koperski, "Models," *Internet Encyclopedia of Philosophy*, http://www.iep.utm.edu/models/. For modeling in the humanities, see Bod, "Modeling in the humanities."

21. See Morgan and Morrison, *Models as Mediators.*

22. See Bailer-Jones, *Scientific Models in Philosophy of Science*; Cartwright, *The Dappled World.*

23. See also the discussion in Bod, review of Peter Burke, *What Is the History of Knowledge?*

24. See Dhar, "Data science and prediction."

25. See Chang, "The harmony that keeps Trappist-1's 7 Earth-size worlds from colliding."

26. Tamayo et al., "Convergent migration renders TRAPPIST-1 long-lived."

27. See Bod, *Beyond Grammar.*

28. Bod, Scha, and Sima'an, *Data-Oriented Parsing.* For language learning based on examples, see Bod, "From exemplar to grammar."

29. Hawking, *Brief Answers to the Big Questions.*

Bibliography

Aaboe, Asger. *Episodes from the Early History of Mathematics*. Random House, 1998.

Aalders, Gerard. *Gevecht met de tijd: Hoe de aarde in 4 eeuwen 4 miljard jaar ouder werd*. Uitgeverij Aspekt, 2014.

Abdile, Mahdi. "Customary dispute resolution in Somalia." *African Conflict & Peacebuilding Review*, 2(1), 2012, pp. 87–110.

Abir-Am, Pnina, and Dorinda Outram (eds.). *Uneasy Careers and Intimate Lives: Women in Science, 1787–1979*. Rutgers University Press, 1987.

Achinstein, Peter. "Models, analogies and theories." *Philosophy of Science*, 31, 1964, pp. 328–349.

Agnesi, Maria Gaetana. *Analytical Institutions*. Translated by John Colson. John Hellins, 1801.

Ágoston, Gábor, and Bruce Masters. *Encyclopedia of the Ottoman Empire*. Infobase, 2009.

Agrawala, Prithvi. *On the Sadanga Canons of Painting*. Prithivi Prakashan, 1981.

Ahmad, Jamil, and Hakim Ashhar Qadeer. *Unani: The Science of Graeco-Arabic Medicine*. Luster Press, 1998.

The Akbarnama of Abul Fazl. Translated by Henry Beveridge. 3 volumes. Asiatic Society, 1902–1939, reprint 2010.

Al-Biruni, *Chronology of Ancient Nations*. Translated by Edward Sachau, W. H. Allen, 1879.

Al-Biruni. *Indica*. Translated by Edward Sachau. In *Alberuni's India*. Trübner, 1888.

Alberti, Leon Battista. *On Painting*. Translated by Cecil Grayson, with an introduction and notes by Martin Kemp. Penguin Classics, 1991.

Allen, James. *The Art of Medicine in Ancient Egypt*. Metropolitan Museum of Art, 2005.

Almast, S. C. "History and evolution of the Indian method of rhinoplasty." In *Transactions of the Fourth International Congress of Plastic and Reconstructive Surgery*. Edited by G. Sanvenero-Rosselli. Excerpta Medica Foundation, 1967, pp. 19–25.

Al-Masudi. *The Meadows of Gold: The Abbasids*. Translated by Paul Lunde and Caroline Stone. Kegan Paul, 1989.

Almeida, D., J. John, and A. Zadorozhnyy. "Keralese mathematics: Its possible transmission to Europe and the consequential educational implications." *Journal of Natural Geometry*, 20, 2001, pp. 77–104.

Amir-Moez, Ali. "A paper of Omar Khayyam." *Scripta Mathematica*, 26(4), 1963, pp. 323–337.

Andersen, Kirsti. *The Geometry of an Art: The History of the Mathematical Theory of Perspective from Alberti to Monge*. Springer, 2007.

Andrews, Bridie. "Wang Qingren and the history of Chinese anatomy." *Journal of Chinese Medicine*, 35, 1991, pp. 30–36.

Andriesse, Cees. *Huygens: The Man behind the Principle*. Cambridge University Press, 2005.

Anglade, Joseph (ed.). *Las leys d'amors*. 4 volumes. E. Privat, 1919–1920.

Anthony, David. *The Horse, the Wheel, and Language: How Bronze Age Riders from the Eurasian Steppes Shaped the Modern World*. Princeton University Press, 2007.

Antommarchi, Francesco. *Mémoires du docteur F. Antommarchi, ou Les derniers momens de Napoléon*. Vol. 1. Barrois L'Ainé, 1825.

Aquinas, Thomas. *Summa Theologica*. Translated as *The Summa Theologiæ of St. Thomas Aquinas*, by Fathers of the English Dominican Province. Benzinger Brothers, 1911.

Arikha, Nougat. *Passions and Tempers: A History of the Humors*. Harper Perennial, 2007.

Aristotle. *History of Animals*. Translated by D'Arcy Wentworth Thompson. Clarendon Press, 1910.

Aristotle. *Mechanical Problems*. Translated by Joyce van Leeuwen. In *The Aristotelian Mechanics: Text and Diagrams*. Springer, 2016.

Aristotle. *Metaphysics*. Translated by Hugh Tredennick. Harvard University Press, 1933.

Aristotle. *Physics*. Translated by Richard Hope. In *Aristotle's Physics: With an Analytical Index of Technical Terms*. University of Nebraska Press, 1961.

Armitage, David, and Jo Guldi. *The History Manifesto*. Cambridge University Press, 2014.

Ascalone, Enrico. *Mesopotamia: Assyrians, Sumerians, Babylonians*. University of California Press, 2007.

Ascher, Marcia. *Ethnomathematics: A Multicultural View of Mathematical Ideas*. Brooks/Cole, 1991.

Asen, Daniel. "'Manchu anatomy': Anatomical knowledge and the Jesuits in seventeenth- and eighteenth-century China." *Social History of Medicine*, 22(1), 2009, pp. 23–44.

Atkinson, Charles. "Johannes Afflighemensis as a historian of mode." *Dobszay Festschrift*. Special issue, *Spolia Berolinensia* 7, 1995, pp. 1–10.

Aubert, M., et al. "Pleistocene cave art from Sulawesi, Indonesia." *Nature*, 514, 2014, pp. 223–227.

Augustine. *De doctrina christiana*. Edited and translated by R. P. H. Green. Clarendon Press, 1995.

Aveni, Anthony. *Skywatchers of Ancient Mexico*. University of Texas Press, 1980.

Backlund, Alexander. "The definition of system." *Kybernetes*, 29(4), 2000, pp. 444–451.

Bacon, Roger. *Opus maius*. Translated by Robert Belle Burke. 1928, reprint Kessinger, 2002.

Badri, Malik. *Abu Zayd al-Balkhi's Sustenance of the Soul: The Cognitive Behavior Therapy of a Ninth Century Physician*. International Institute of Islamic Thought, 2013.

Bailer-Jones, Daniela. *Scientific Models in Philosophy of Science*. University of Pittsburgh Press, 2009.

Bailey, Drew, and David Geary. "Hominid brain evolution: Testing climatic, ecological, and social competition models." *Human Nature*, 20, 2009, pp. 67–79.

Baines, John. "On the evolution, purpose and forms of Egyptian annals." In *Zeichen aus dem Sand*. Edited by Eva-Maria Engel, Vera Müller, and Ulrich Hartung. Wiesbaden, 2008, pp. 19–40.

Bainton, Roland. *Hunted Heretic: The Life and Death of Michael Servetus, 1511–1553*. Beacon Press, 1953.

Bala, Arun (ed.). *Asia, Europe, and the Emergence of Modern Science: Knowledge Crossing Boundaries*. Palgrave Macmillan, 2012.

Bardi, Jason. *The Calculus Wars: Newton, Leibniz, and the Greatest Mathematical Clash of All Time*. Basic Books, 2007.

Barker, Peter, and Bernard Goldstein. "Theological foundations of Kepler's astronomy." *Osiris*, 16, 2001, pp. 88–113.

Barnes, Jonathan. *The Presocratic Philosophers*. Routledge, 2002.

Barolini, Teodolinda, and H. Wayne Storey. *Petrarch and the Textual Origins of Interpretation*. Brill, 2007.

Baron, Margaret. *The Origins of the Infinitesimal Calculus*. Dover, 1969.

Barrow, John, and Frank Tipler. *The Anthropic Cosmological Principle*. Oxford University Press, 1986.

Barwick, Karl. *Problems of the Sprachlehre and Rhetorik*. Akademie-Verlag, 1957.

Bauer, Wolfgang Johann. *Bausteine des Fiqh: Kernbereiche der 'Uṣūl al-Fiqh; Quellen und Methodik der Ergründung islamischer Beurteilungen*. Peter Lang Edition, 2013.

Baur, Siegfried. *Versuch über die Historik des jungen Ranke*. Duncker & Humblot, 1998.

Baxter, William. *A Handbook of Old Chinese Phonology*. Mouton de Gruyter, 1996.

Beach, Milo, Ebba Koch, and Wheeler Thackston. *The King of the World: An Imperial Manuscript for the Royal Library Windsor Castle*. Thames & Hudson, 1997.

Bebbington, David. *Patterns in History*. Regent College, 1990.

Benedetto, Maria Ada. *Bartolo da Sassoferrato*. Novissimo Digesto Italiano, 1958.

Berggren, J. L. "Innovation and tradition in Sharaf al-Din al-Tusi's Muadalat." *Journal of the American Oriental Society*, 110(2), 1990, pp. 304–309.

Bertolini Meli, Domenico. *Mechanism, Experiment, Disease: Marcello Malpighi and Seventeenth-Century Anatomy*. Johns Hopkins University Press, 2011.

Bertrand, Daniel. *La musique carnatique*. Éditions du Makar, 2001.

Bhabha, Homi (ed.). *Nation and Narration*. Routledge, 1990.

Biagioli, Mario. *Galileo Courtier: The Practice of Science in the Culture of Absolutism*. University of Chicago Press, 1994.

Bialas, Volker. *Johannes Kepler*. Beck, 2004.

Biela, Adam. *Analogy in Science*. Peter Lang, 1991.

Billanovich, Giuseppe. *Petrarca e il Primo Umanesimo*. Antenore, 1996.

Billanovich, Giuseppe. *Tradizione and fortuna di Livio tra Medioevo and Umanesimo*. Antenore, 1981.

Blackburn, Bonnie, and Leofranc Holford-Strevens. "Calendars and chronology." In *The Oxford Companion to the Year: An Exploration of Calendar Customs and Time Checking*. Oxford University Press, 1999, pp. 659–937.

Blank, David. *Ancient Philosophy and Grammar: The Syntax of Apollonius Dyscolus*. Scholars Press, 1982.

Blaydes, Lisa, and Eric Chaney. "The feudal revolution and Europe's rise: Political divergence of the Christian West and the Muslim world before 1500 CE." *American Political Science Review*, 107(1), 2013, pp. 16–34.

Bobzien, Susanne. "Stoic logic." In *The Cambridge History of Hellenistic Philosophy*. Edited by K. Algra, J. Barnes, J. Mansfeld, and M. Schofield. Cambridge University Press, 1999, pp. 92–157.

Bod, Rens. *Beyond Grammar: An Experience-Based Theory of Language*. CSLI, 1998.

Bod, Rens. "A comparative framework for studying the histories of the humanities and science." *Isis*, 106(2), 2015, pp. 367–377.

Bod, Rens. "From exemplar to grammar: A probabilistic analogy-based model of language learning." *Cognitive Science*, 33(5), 2009, pp. 752–793.

Bod, Rens. *A New History of the Humanities: The Search for Patterns and Principles from Antiquity to the Present*. Oxford University Press, 2013.

Bod, Rens. "Who is afraid of patterns? The particular versus the universal and the meaning of humanities 3.0." *BMGN—Low Countries Historical Review*, 128(4), 2013, pp. 171–180.

Bod, Rens. Review of Peter Burke, *What Is the History of Knowledge?*, Polity Press, 2016. *History of Humanities*, 1(2), 2016, pp. 426–428.

Bod, Rens. "Modeling in the humanities: Linking patterns to principles." *Historical Social Research*, supplement 31, 2018, pp. 78–95.

Bod, Rens, Jeroen van Dongen, Sjang ten Hagen, Bart Karstens and Emma Mojet. "The flow of cognitive goods: A historiographical framework for the study of epistemic transfer." *Isis*, 110(3), 2019, pp. 483–496.

Bod, Rens, Remko Scha, and Khalil Sima'an (eds.). *Data-Oriented Parsing*. University of Chicago Press, 2003.

Boethius, Anicius Manlius Severinus. *De Institutione Musica*. Edited by Claude Palisca. Translated as *Fundamentals of Music* by Calvin Bower. Yale University Press, 1989.

Bonnet-Bidaud, Jean-Marc, Francoise Praderie, and Susan Whitfield. "The Dunhuang Chinese sky: A comprehensive study of the oldest known star atlas." *Journal of Astronomical History and Heritage*, 12(1), 2009, pp. 39–59.

Bottazzini, Umberto. "La 'grande arte': l'algebra del rinascimento." In Paolo Rossi (ed.), *Storia della scienza moderna e contemporanea*. Vol. 1: *Dalla rivoluzione scientifica all'età dei lumi*. Feltrinelli, 2000, pp. 59–84.

Bowersock, Glen. *On the Donation of Constantine: Lorenzo Valla*. Harvard University Press, 2007.

Boyer, Carl. *A History of Mathematics*. 2nd edition. Wiley, 1989.

Boyer, Carl. *The History of the Calculus and Its Conceptual Development*. Dover, 1959.

Boylston, Arthur. "The origins of inoculation." *Journal of the Royal Society of Medicine*, 105(7), 2012, pp. 309–313.

Brahmagupta. *Brahmasphuta-siddhanta*. GRETIL (Göttingen Register of Electronic Texts in Indian Languages). http://gretil.sub.uni-goettingen.de/gretil/1_sanskr/6_sastra/8_jyot/brsphutu.htm.

Brack-Bernsen, Lis. "The 'days in excess' from MUL.APIN: On the 'first intercalation' and 'water clock'; Schemes from MUL.APIN." *Centaurus*, 47(1), 2005, pp. 1–29.

Braudel, Fernand. *Civilisation matérielle, économie et capitalisme (XVe et XVIIIe siècles)*. Vol. 1. Armand Colin, 1967.

Brett, Megan. "Topic modeling: A basic introduction." *Journal of Digital Humanities*, 2(1), 2012, pp. 12–16.

Breva-Claramonte, Manuel. *Sanctius' Theory of Language: A Contribution to the History of Renaissance Linguistics*. John Benjamins, 1983.

Broggiato, Maria (ed.). *Cratete di Mallo: I frammenti*. Agorà Edizioni, 2001.

Brouwer, René. *Law and Philosophy in the Late Roman Republic*. Cambridge University Press, 2021.

Brouwer, René. "Ulpian's appeal to nature: Roman law as universal law." *Tijdschrift voor Rechtsgeschiedenis*, 83(1–2), 2015, pp. 60–76.

Brown, Donald, "Human universals and their implications." In *Being Humans: Anthropological Universality and Particularity in Transdisciplinary Perspectives*. Edited by N. Roughley. Walter de Gruyter, 2000, pp. 156–174.

Brown, Nancy Marie. *The Abacus and the Cross: The Story of the Pope Who Brought the Light of Science to the Dark Ages*. Basic Books, 2010.

Bruins, Evert. "On Plimpton 322, Pythagorean numbers in Babylonian mathematics." *Proceedings Koninklijke Nederlandse Akademie van Wetenschappen*, 52, 1949, pp. 629–632.

Buck, August. "The 'studia humanitatis' im italienischen Humanismus." In *Humanism im Bildungswesen des 15. und 16. Jahrhunderts*. Edited by Wolfgang Reinhard. Acta Humaniora Weinheim, 1984, pp. 11–24.

Buringh, Eltjo, and Jan Luiten van Zanden. "Charting the 'rise of the West': Manuscripts and printed books in Europe, a long-term perspective from the sixth through eighteenth centuries." *Journal of Economic History*, 69(2), 2009, pp. 409–445.

Burke, Peter. *A Social History of Knowledge*. 2 volumes. Polity Press, 2000–2012.

Burke, Peter. *What Is the History of Knowledge?* Polity Press, 2016.

Burnett, Charles. "The coherence of the Arabic–Latin translation program in Toledo in the twelfth century." *Science in Context*, 14, 2001, pp. 275–281.

Bursill-Hall, G. L. *Speculative Grammars of the Middle Ages: The Doctrine of the Partes Orationis of the Modistae*. Mouton, 1971.

Calinger, Ronald. *Leonhard Euler: Mathematical Genius in the Enlightenment*. Princeton University Press, 2016.

Callanan, Christopher. *Die Sprachbeschreibung at Aristophanes von Byzanz*. Vandenhoeck und Ruprecht, 1987.

Campbell, Eldridge, and James Colton (eds.). *The Surgery of Theodoric*. 2 volumes. Appleton-Century-Crofts, 1955, 1960.

Cantarino, Vicente. *Arabic Poetics in the Golden Age*. Brill, 1975.

Carawan, Edwin. *Rhetoric and the Law of Draco*. Oxford University Press, 1998.

Cartwright, Nancy. *How the Laws of Physics Lie*. Oxford University Press, 1983.

Cartwright, Nancy. *The Dappled World*. Cambridge University Press, 1999.

Cavazza, Marta. "Donne e Scienza nell'Italia del Settecento." In *Actes de la VIII Trobada d'historia de la ciencia i de la tecnica*. SCHCT, 2006, pp. 15–27.

Chabrier, Jean-Claude. "Musical Science." In *Encyclopedia of the History of Arabic Science*. Edited by Roshdi Rashed. Vol. 2. Routledge, p. 581.

Chadwick, Stephen Robert, and Martin Paviour-Smith. *The Great Canoes in the Sky: Starlore and Astronomy of the South Pacific*. Spinger, 2017.

Chang, Kenneth. "The harmony that keeps Trappist-1's 7 Earth-size worlds from colliding." *New York Times*, May 10, 2017.

Chemla, Karine (ed.). *The History of Mathematical Proof in Ancient Traditions*. Cambridge University Press, 2012.

Chen, Albert. *An Introduction to the Legal System of the People's Republic of China*. Butterworths Asia, 1998.

Cho, Gene. *The Discovery of Musical Equal Temperament in China and Europe in the Sixteenth Century*. Edwin Mellen Press, 2003.

Christensen, Thomas (ed.). *The Cambridge History of Western Music Theory*. Cambridge University Press, 2002.

Christianson, Gale. *Isaac Newton*. Oxford University Press, 2005.

Cicero, Marcus Tullius. *On the Laws*. Translated by David Fott. Cornell University Press, 2014.

Clagett, Marshall. *The Science of Mechanics in the Middle Ages*. University of Wisconsin Press, 1959, pp. 332–333.

Clark, Anthony. *Ban Gu's History of Early China*. Cambria Press, 2008.

Clark, Walter Eugene. *The Âryabhaṭîya or Âryabhaṭa: An Ancient Indian Work on Mathematics and Astronomy*. University of Chicago Press, 1930.

Cohen, H. Floris. *How Modern Science Came into the World: Four Civilizations, One 17th-Century Breakthrough*. Amsterdam University Press, 2010.

Cohen, H. Floris. "Music as science and as art: The 16th/17th-century destruction of cosmic harmony." In *The Making of the Humanities*. Edited by Rens Bod, Jaap Maat, and Thijs Weststeijn. Vol. 1: *Early Modern Europe*. Amsterdam University Press, 2010, pp. 59–71.

Cohen, H. Floris. *Quantifying Music: The Science of Music at the First Stage of the Scientific Revolution, 1580–1650*. Reidel, 1984.

Conche, Marcel. *Anaximandre: Fragments et témoignages*. Presses Universitaires de France, 1991.

Covington, Michael. *Syntactic Theory in the High Middle Ages: Modistic Models of Sentence Structure*. PhD thesis, Yale University, 1982.

Crew, Henry, and Alfonso de Salvio. *Dialogues concerning Two New Sciences*. McMillan, 1914.

Cullen, Christopher. "The first complete Chinese theory of the moon: The innovations of Liu Hong c. AD 200." *Journal of the History of Astronomy*, 35, 2000, pp. 21–39.

Cullen, Christopher. "The science/technology interface in seventeenth-century China: Song Yingxing 宋應星 on 'qi' 氣 and the 'wu xing' 五行." *Bulletin of the School of Oriental and African Studies*, 53(2), 1990, pp. 295–318.

Cullen, Christopher. "Understanding the planets in ancient China: Prediction and divination in the Wu Xing Zhan." *Early Science and Medicine*, 16, 2011, pp. 218–251.

da Vinci, Leonardo. *A Treatise on Painting*. Translated by John Francis Rigaud. Kessinger, 2004.

Dahiyat, Ismail. *Avicenna's Commentary on the Poetics of Aristotle: A Critical Study with an Annotated Translation of the Text*. Brill, 1974.

Dampier, William. *A History of Science and Its Relation to Philosophy and Religion*. Cambridge University Press, 1966.

Dante Alighieri. *La Divina Commedia*. Biblioteca Universale Rizzoli, 2007.

de Groot, Adriaan. *Methodologie: Grondslagen van onderzoek en denken in de gedragswetenschappen*. Mouton, 1961. Translated into English as *Methodology: Foundations of Inference and Research in the Behavioral Sciences*. Mouton 1969.

de Jonge, Casper. *Between Grammar and Rhetoric: Dionysius of Halicarnassus on Language, Linguistics, and Literature*. Brill, 2008.

de Jonge, H. J. "The character of Erasmus' translation of the New Testament." *Journal of Medieval and Renaissance Studies*, 14, 1984, pp. 81–88.

Dennett, Daniel. "Real patterns." *Journal of Philosophy*, 88, 1991, pp. 27–51.

Descartes, René. *A Discourse on Method: Meditations and Principles*. Translated by John Veitch. Orion, 2004.

Descartes, René. *Principles of Philosophy*. Translated by John Cottingham. In *The Philosophical Writings of Descartes*, Vol. 1. Cambridge University Press, 1985, pp. 177–292.

Descola, Philippe. *Beyond Nature and Culture*. Translated by Janet Lloyd. University of Chicago Press, 2013.

Devlin, Keith. *The Man of Numbers: Fibonacci's Arithmetic Revolution*. Walker Books, 2012.

Dhar, Vasant. "Data science and prediction." *Communications of the ACM*, 56(12), 2013, pp. 64–73.

Diamond, Jared. *Guns, Germs, and Steel*. Random House, 1997.

Diamond, Jared. "Peeling the Chinese onion." *Nature*, 391, January 29, 1998, pp. 433–434.

Dijksterhuis, Eduard. *Archimedes*. Noordhoff, 1938.

Dijksterhuis, Eduard. *De mechanisering van het wereldbeeld*. Meulenhoff, 1950.

Dilthey, Wilhelm. *Einleitung in die Geisteswissenschaften: Versuch einer Grundlegung für das Studium der Gesellschaft und der Geschichte*. Teubner, 1959 [1883].

Diogenes Laertius. *Lives of Eminent Philosophers*. Book 1. Translated by R. D. Hicks. Harvard University Press, 1989.

Dionysius Thrax. *Téchne grammatiké*. Translated as "The Tekhne Grammatike of Dionysius Thrax translated into English" by J. Alan Kemp. *Historiographia Linguistica*, 13(2/3), 1986, pp. 343–363.

Dixon, Dan. "Analysis tool or research methodology: Is there an epistemology for patterns?" In *Understanding Digital Humanities*. Edited by David Berry. Palgrave Macmillan, 2012, pp. 191–209.

Doi, Suhail. *Understanding Evidence in Health Care: Using Clinical Epidemiology*. Palgrave Macmillan, 2011.

Donner, Fred. *Narratives of Islamic Origins: The Beginnings of Islamic Historical Writing*. Darwin Press, 1998.

Dopper, Jantien Gea. *A Life of Learning in Leiden: The Mathematician Frans van Schooten (1615–1660)*. PhD thesis, Utrecht University, 2014.

Dowson, John (ed.). *The History of India, as Told by Its Own Historians: The Muhammadan Period; The Posthumous Papers of the Late Sir H. M. Elliot*. Calcutta, 1959.

Draffkorn Kilmer, Anne. "The discovery of an ancient Mesopotamian theory of music." *Proceedings of the American Philosophical Society*, 115(2), 1971, pp. 131–149.

Drake, Stillman. *Galileo: Pioneer Scientist*. University of Toronto Press, 1990.

Drew, Katherine. *The Laws of the Salian Franks (Pactus Legis Salicae)*. University of Pennsylvania Press, 1991.

Driver, G., and John Miles. *The Babylonian Laws*. Wipf and Stock, 2007.

Droste, F. G. *Simon Stevin: Wetenschapper in oorlogstijd 1548–1620*. Aspekt, 2007.

Druzhkova, Anna et al. "Ancient DNA analysis affirms the canid from Altai as a primitive dog." *PLoS ONE*, 8(3), 2013, p. e57754.

Dufour, A. "L'influence de la méthodologie des sciences physiques et mathematiques sur les fondateurs de l'École du droit naturel moderne." *Grotiana*, 1(1), 1980, pp. 33–52.

Duke, Dennis. "The equant in India: The mathematical basis of ancient Indian planetary models." *Archive for History of Exact Sciences*, 59, 2005, pp. 563–576.

Duncan, A. M. *Johannes Kepler—Mysterium Cosmographicum: The Secret of the Universe*. Abaris 1981.

Dürer, Albrecht. *De Symmetria Partium in Rectis Formis Humanorum Corporum / Underweysung Der Messung*. Translated by Silvio Levy and Walter Strauss, with an introduction by David Price. Octavo, 2003.

Dworkin, Ronald. *A Matter of Principle*. Harvard University Press, 1985.

Eastwood, Bruce. *Ordering the Heavens: Roman Astronomy and Cosmology in the Carolingian Renaissance*. Brill, 2007.

Eckart, Wolfgang. "Die Medizin der Renaissance." In *Geschichte der Medizin*. Springer, 1994, pp. 107–133.

Eco, Umberto. *The Search for the Perfect Language*. Wiley-Blackwell, 1995.

Eir, Carlos. *A Very Brief History of Eternity*. Princeton University Press, 2010.

Elman, Benjamin. "Philology and its enemies: Changing views of late imperial Chinese classicism." *Colloquium "Images of Philology."* 2006.

Elman, Benjamin. *Science in China, 1600–1900: Essays by Benjamin A. Elman*. Edited by Ho Yi Kai. World Scientific, 2015.

Elrington, Charles Richard. *The Whole Works of the Most Rev. James Ussher, DD*. Vol. 11. Hodges and Smith, 1847.

El-Rouayheb, Khaled. "Sunni Muslim scholars on the status of logic, 1500–1800." *Islamic Law and Society*, 11, 2004, pp. 213–232.

El-Rouayheb, Khaled. "Was there a revival of logical studies in eighteenth-century Egypt?" *Die Welt des Islams*, 45(1), 2005, pp. 1–19.

Ely, Sidi Amar Ould, and Julian Johansen (eds.). *Handlist of Manuscripts in the Centre de Documentation et de Recherches Historiques Ahmed Baba, Timbuktu*. Vol. 1: *Handlist of Islamic Manuscripts Series V: African Collections, Mali*. Al-Furqan Islamic Heritage Foundation, 1995.

Engelstein, Stefani. *Sibling Action: The Genealogical Structure of Modernity*. Columbia University Press, 2017.

Erickson, Raymond. "Musica enchiriadis, Scholia enchiriadis." In *The New Grove Dictionary of Music and Musicians*. 2001. https://www.oxfordmusiconline.com/grovemusic.

Esposito, John. *Islam: The Straight Path*. Oxford University Press, 1988.

Euclid. *Elements*. Translated and commentary by Thomas Heath. 3 volumes. Dover, 1956.

Fabris, Dinko. "Galileo and music: A family affair." In *The Inspiration of Astronomical Phenomena VI, ASP Conference Series*. Vol. 441. Astronomical Society of the Pacific, 2011, pp. 57–72.

Fara, Patricia. *A Lab of One's Own: Science and Suffrage in the First World War*. Oxford University Press, 2017.

Fara, Patricia. *Pandora's Breeches: Women, Science & Power in the Enlightenment*. Pimlico, 2004.

Fara, Patricia. *Science: A Four Thousand Year History*. Oxford University Press, 2009.

Fehling, Detlev. *Herodotus and His "Sources": Citation, Invention, and Narrative Art*. Arca Classical and Medieval Texts, Papers, and Monographs 21. Francis Cairns, 1989.

Fehling, Detlev. "Varro und die grammatische Lehre von der Analogie und der Flexion." *Glotta*, 35, 1956, pp. 214–270.

Feingold, Mordechai (ed.). *Before Newton: The Life and Times of Isaac Barrow*. Cambridge University Press, 1990.

Feraboli, Simonetta (ed.). *Claudio Tolomeo: Le previsioni astrologiche (Tetrabiblos)*. Fondazione Valla/Mondadori, 1985.

Ferguson, Kitty. *Tycho & Kepler: The Unlikely Partnership That Forever Changed Our Understanding of the Heavens*. Walker, 2002.

Feuerwerker, Albert. *History in Communist China*. MIT Press, 1968.

Findlen, Paula. "Calculations of faith: Mathematics, philosophy, and sanctity in 18th-century Italy (new work on Maria Gaetana Agnesi)." *Historia Mathematica*, 38(2), 2011, pp. 248–291.

Fischer, Steven. *Island at the End of the World*. Reaktion Books, 2005.

Fischer, Steven. "Preliminary evidence for cosmogonic texts in Rapanui's rongorongo inscriptions." *Journal of the Polynesian Society*, 104, 1995, pp. 303–321.

Fisher, Saul. *Pierre Gassendi's Philosophy and Science*. Brill, 2005.

Flannery, Kent, and Joyce Marcus. *The Creation of Inequality: How Our Prehistoric Ancestors Set the Stage for Monarchy, Slavery, and Empire*. Harvard University Press, 2012.

Forster, Johann Reinhold. *Observations Made during a Voyage round the World*, 1778. Edited by Nicholas Thomas, Harriet Guest, and Michael Dettelbach. University of Hawai'i Press, 1996.

Fowler, Brenda. *Iceman: Uncovering the Life and Times of a Prehistoric Man Found in an Alpine Glacier*. Random House, 2000.

Fowler, D. H. *The Mathematics or Plato's Academy: A New Reconstruction*. Clarendon Press, 1987.

Fowler, David, and Eleanor Robson. "Square root approximations in Old Babylonian mathematics: YBC 7289 in Context." *Historia Mathematica*, 25(4), 1998, pp. 366–378.

Frank, Stefan, Rens Bod, and Morten Christiansen. "How hierarchical is language use?" *Proceedings of the Royal Society of London B: Biological Sciences*, 2012, rspb20121741.

Frier, Bruce. *The Rise of the Roman Jurists*. Princeton University Press, 1985.

Fu, Louis. "A forgotten reformer of anatomy in China: Wang Ch'ing-Jen." *ANZ Journal of Surgery*, 78(12), 2008, pp. 1052–1058.

Fubini, Riccardo. "Humanism and truth: Valla writes against the Donation of Constantine." *Journal of the History of Ideas*, 57, 1996, pp. 79–86.

Fubini, Riccardo, and Stefano Caroti (eds.). *Poggio Bracciolini 1380–1980, Nel VI centenario della nascita*. Sansoni, 1982.
Fulgentius. *The Exposition of the Content of Virgil*. Translated by O. B. Hardison. In *Classical and Medieval Literary Criticism: Translations and Interpretations*. Edited by Alex Preminger, O. B. Hardison, and Kevin Kerrane. F. Ungar, 1974, pp. 333–339.
Gadamer, Hans-Georg. *Wahrheit und Methode: Grundzüge einer philosophischen Hermeneutik*. Mohr, 1960.
Galilei, Galileo. *Dialogue concerning the Two Chief World Systems: Ptolemaic and Copernican*. Translated by Stillman Drake. Modern Library, 2001.
Ganeri, Jonardon. "Polycentered history of science." In *Encyclopaedia of the History of Science, Technology, and Medicine in Non-Western Cultures*. Edited by Helaine Selin. Springer, 2013, pp. 3556–3562.
Garcea, Alessandro. *Caesar's De Analogia: Edition, Translation and Commentary*. Oxford University Press, 2012.
Garofalo, Ivan. *Erasistrato: I frammenti*. Giardini, 1988.
Geertz, Clifford. *The Interpretation of Cultures*. Basic Books, 1973.
Geller, Markham. *Ancient Babylonian Medicine: Theory and Practice*. Wiley-Blackwell, 2010.
Gerber, Haim. *State, Society, and Law in Islam: Ottoman Law in Comparative Perspective*. State University of New York Press, 1994.
Geus, Klaus. *Eratosthenes von Kyrene: Studien zur hellenistenen Kultur- und Wissenschaftsgeschichte*. Beck, 2002.
Ghalioungui, Paul. *The House of Life: Magic and Medical Science in Ancient Egypt*. B. M. Israel, 1973.
Gibson, Sophie. *Aristoxenus of Tarentum and the Birth of Musicology*. Routledge, 2005.
Gieryn, Thomas. "Boundary-work and the demarcation of science from non-science: Strains and interests in professional ideologies of scientists." *American Sociological Review*, 48(6), 1983, pp. 781–795.
Gingerich, Owen. *The Book Nobody Read: Chasing the Revolutions of Nicolaus Copernicus*. Walker, 2004.
Gingerich, Owen. *The Eye of Heaven: Ptolemy, Copernicus, Kepler*. American Institute of Physics, 1993.
Giusti, Enrico (ed.). *Luca Pacioli: Summa de arithmetica, geometria, proportioni et proportionalità*. Abrizzi, 1994.
Glick, Thomas, Steven Livesey, and Faith Wallis (eds.). *Medieval Science, Technology, and Medicine: An Encyclopedia*. Routledge, 2014.
Godard, Gaston. "François Viète (1540–1603), père de l'algèbre moderne." *Recherches vendéennes*, 9, 2002, pp. 297–346.
Goehl, Konrad. *Avicenna and seine Darstellung der Arzneiwirkungen*. Deutscher Wissenschafts-Verlag, 2014.
Goldstein, Bernard. "The Arabic version of Ptolemy's planetary hypotheses." *Transactions of the American Philosophical Society*, 57(4), 1967, pp. 9–12.
Goodrich, Carrington, and Chaoying Fang (eds.). *Dictionary of Ming Biography*. Columbia University Press, 1976.
Goody, Jack. *The Interface between the Written and the Oral*. Cambridge University Press, 1987.
Gottschalk, Hans. *Heraclides of Pontus*. Oxford University Press, 1980.
Goudsblom, Johan. *Fire and Civilization*. Penguin Books, 1995.
Gouk, Penelope. "Music and the emergence of experimental science in early modern Europe." *Sound Effects: An Interdisciplinary Journal of Sound and Sound Experience*, 2(1), 2012, pp. 6–21.
Grafton, Anthony. *Defenders of the Text*. Harvard University Press, 1991.

Grafton, Anthony. *Joseph Scaliger: A Study in the History of Classical Scholarship*. 2 volumes. Oxford University Press, 1983, 1993.
Grafton, Anthony. "Joseph Scaliger and historical chronology: The rise and fall of a discipline." *History and Theory*, 14(2), 1975, pp. 156–185.
Grafton, Anthony. "Kepler as reader." *Journal of the History of Ideas*, 53(4), 1992, pp. 561–572.
Grafton, Anthony. *Leon Battista Alberti: Master Builder of the Renaissance*. Harvard University Press, 2002
Grafton, Anthony. "Philological and artisanal knowledge making in Renaissance natural history: A study in cultures of knowledge." *History of Humanities*, 3(1), 2018, pp. 39–55.
Graham, Angus Charles. *Later Mohist Logic, Ethics and Science*. Chinese University Press, 1978.
Grant, Edward. *A Source Book in Medieval Science*. Harvard University Press, 1974.
Grasshoff, Gerd. "Globalization of ancient knowledge: From Babylonian observations to scientific regularities." In *The Globalization of Knowledge in History*. Edited by Jürgen Renn. epubli GmbH, 2012, pp. 175–190.
Gray, John. *African Music: A Bibliographical Guide to the Traditional, Popular, Art and Liturgical Musics of Sub-Saharan Africa*. Greenwood Press, 1991.
Greenblatt, Stephen. *The Swerve: How the World Became Modern*. W. W. Norton, 2011.
Gregory, Frederick. *Natural Science in Western History*. Wadsworth, 2007.
Grenander, Ulf. *Elements of Pattern Theory*. Johns Hopkins University Press, 1996.
Gross, Cary, and Kent Sepkowitz. "The myth of the medical breakthrough: Smallpox, vaccination, and Jenner reconsidered." *International Journal of Infectious Diseases*, 3(1), 1998, pp. 54–60.
Grotius, Hugo. *On the Law of Prize and Booty*. With an introduction by Martine Julia van Ittersum. Liberty Fund, 2006.
Grotius, Hugo. *Prolegomena to the Law of War and Peace*. Translated and introduced by Edward Dumbauld. Library of Liberal Arts, No. 65.
Grube, Nikolai. *Der Dresdner Mayan Calendar: Der vollständige Codex*. Verlag Herder, 2012.
Guerra, Francisco. "Weights and measures in pre-Columbian America." *Journal of the History of Medicine and Allied Sciences*, 15(4), 1960, pp. 342–345.
Guillaume, Alfred. *The Life of Muhammad: A Translation of Isḥaq's "Sirat Rasul Allah."* Oxford University Press, 1955, reprint 2004.
Gurney, O. R., and Samuel Kramer. "Two Fragments of Sumerian Laws." *Assyriological Studies*, 16, 1965, pp. 13–19.
Haarmann, Harald. *Geschichte der Schrift*. C. H. Beck, 2002.
Habib, Touma. *The Music of the Arabs*. Amadeus Press, 2003.
Hacking, Ian. *Representing and Intervening*. Cambridge University Press, 1983.
Hagengruber, Ruth (ed.). *Émilie du Chatelet between Leibniz and Newton*. Springer, 2011.
Hakfoort, Casper. "The missing syntheses in the historiography of science." *History of Science*, 29(2), 1991, pp. 207–216.
Hall, Marie Boas. *The Scientific Renaissance, 1450–1630*. 3rd edition. Dover, 1994.
Hall, Stuart. "The West and the Rest: Discourse and Power." In *The Indigenous Experience: Global Perspectives*. Edited by Roger Maaka and Chris Andersen. Canadian Scholars Press, 1992, pp. 165–173.
Hamilton, Sue. *Indian Philosophy: A Very Short Introduction*. Oxford University Press, 2001.
Hanegraaff, Wouter. *Esotericism and the Academy: Rejected Knowledge in Western Culture*. Cambridge University Press, 2012.
Harari, Yuval. *Sapiens: A Brief History of Humankind*. Harper, 2015.
Harland, Richard, *Literary Theory from Plato to Barthes*. Palgrave, 1999.

Harris, Pauline, Rangi Matamua, Takirirangi Smith, Hoturoa Kerr, and Toa Waaka. "A review of Maori astronomy in Aotearoa-New Zealand." *Journal of Astronomical History and Heritage*, 16(3), 2013, pp. 325–336.

Harvey, William. *On the Motion of the Heart and Blood in Animals*. Translated by Robert Willis. Harvard Classics. Vol. 38, part 3. New York: P. F. Collier & Son, 1909–14; New York: Bartleby.com, 2001. https://www.bartleby.com/38/3.

Hashimoto, Keizo. "Longomontanus's 'Astronomia Danica' in China." *Journal for the History of Astronomy*, 18(2), 1987, pp. 95–110.

Hawking, Stephen. *Brief Answers to the Big Questions*. Bantam Books, 2018.

Hayes, John. *A Manual of Sumerian Grammar and Texts, Aids and Research Tools in Ancient Near Eastern Studies* 5. Undena, 1990.

Head, John Warren, and Yanping Wang. *Law Codes in Dynastic China: A Synopsis of Chinese Legal History in the Thirty Centuries from Zhou to Qing*. Carolina Academic Press, 2005.

Heeβel, Nils. "Diagnosis, divination and disease: Towards an understanding of the rationale behind the Babylonian Diagnostic Handbook." In *Magic and Rationality in Ancient Near Eastern and Graeco-Roman Medicine*. Edited by H. Horstmanshoff, M. Stol, and C. van Tilburg. Brill, 2004, pp. 97–116.

Heil, Gunter, and Adolf Martin Ritter (eds.). *Corpus Dionysiacum: Pseudo-Dionysius Areopagita, De Coelesti Hierarchia, De Ecclesiastica Hierarchia, De Mystica Theologia, Epistulae*. Walter De Gruyter, 1991.

Heilbron, John. *The Sun in the Church*. Harvard University Press, 1999.

Heinzelmann, Martin. *Gregory of Tours: History and Society in the Sixth Century*. Cambridge University Press, 2001.

Henderson, Donald. *Smallpox: The Death of a Disease*. Prometheus Books, 2009.

Henry, John. "Seth Ward." In *Dictionary of National Biography*. Oxford University Press, 2004, https://www.oxforddnb.com/.

Herodotus. *The History*. Translated by David Grene. University of Chicago Press, 1987.

Hetherington, Norriss S. (ed.). *Cosmology: Historical, Literary, Philosophical, Religious, and Scientific Perspectives*. Taylor & Francis, 1993.

Higham, N. J. *(Re-)Reading Bede: The Historia Ecclesiastica in Context*. Routledge, 2006.

Hildegard of Bingen. *Causae et curae*. Translated by Manfred Pawlik and Patrick Madigan. Edited by Mary Palmquist and John Kulas. Liturgical Press, 1994.

Hildegard of Bingen. *Scivias*. Translated by Bruce Hozeski. In *Hildegard von Bingen's Mystical Visions*. Bear, 1995.

Hill Boone, Elizabeth. *Stories in Red and Black: Pictorial Histories of the Aztec and Mixtec*. University of Texas Press, 2000.

Hinrichs, T. J., and Linda Barnes (eds.). *Chinese Medicine and Healing*. Harvard University Press, 2013.

Ho, Daniel, and Larry Kramer. "The empirical revolution in law." *Stanford Law Review*, 65, 2013, pp. 1195–1205.

Hochstrasser, T. J. *Natural Law Theories in the Early Enlightenment*. Cambridge University Press, 2000.

Hock, Hans, and Brian Joseph. *Language History, Language Change and Language Relationship*. Mouton de Gruyter, 1996.

Hoffmann, Manfred, and James Tracy (eds.). *Collected Works of Erasmus*. 78 volumes. University of Toronto Press, 1974–2011.

Hogendijk, Jan. "Al-Khwārizmī's table of the 'Sine of the hours' and the underlying sine table." *Historia Scientiarum*, 42, 1991, pp. 1–12.

Holden, James. *A History of Horoscopic Astrology*. American Federation of Astrologers, 1996.
Holwell, John. *An Account of the Manner of Inoculating for the Small Pox in the East Indies*. T. Becket & P. A. de Hondt, 1767.
Honoré, Tony. *Tribonian*. Cornell University Press, 1978.
Horky, Phillip. *Plato and Pythagoreanism*. Oxford University Press, 2013.
Householder, Fred. *The Syntax of Apollonius Dyscolus*. John Benjamins, 1981.
Houtzager, H. L. (ed.). *Reinier de Graaf 1641–1673*. Erasmus, 1991.
Huang Di nei jing. Translated by Guus Martens. La Douze VOF, 2011.
Huber, Peter. "On the Old Babylonian understanding of grammar: A reexamination of OBGT VI–X." *Journal of Cuneiform Studies*, 59, 2007, pp. 1–17.
Huff, Jay. *Musical Theorists in Translation*. Vol. 8. Institute of Mediaeval Music Brooklyn, 1963.
Huffman, Carl (ed.). *Aristoxenus of Tarentum: Texts and Discussions*. Transaction, 2012.
Huffman, Michael. *The Study of Primate Self-Medication*. Primate Research Institute, Kyoto, 2005.
Hughes, Barnabas (ed.). *Robert of Chester's Latin Translation of Al-Khwarizmi's Al-Jabr*. New critical edition. Franz Steiner, 1989.
Huizinga, Johan. *Homo ludens: Proeve ener bepaling van het spelelement der cultuur*. Wolters-Noordhoff, 1938.
Hunger, Hermann (ed.). *State Archives of Assyria: Astrological Reports to Assyrian Kings*. Vol. 8. Helsinki University Press, 1992.
Hunter, Michael. "Aikenhead the atheist: The context and consequences of articulate irreligion in the late seventeenth century." In *Atheism from the Reformation to the Enlightenment*. Edited by Michael Hunter and David Wootton. Clarendon Press, 1992, pp. 221–254.
Hunwick, John (ed.). *Timbuktu and the Songhay Empire: Al-Sadi's "Tarikh al-Sudan" Down to 1613 and Other Contemporary Documents*. Brill, 1999.
Huygens, Christiaan. *Oeuvres complètes*. Edited by D. Bierens de Haan (vols. 1–5), J. Bosscha (vols. 6–10), D. J. Korteweg (vols. 11–15), A. A. Nijland (vol. 15), and J. A. Vollgraf (vols. 16–22). Martinus Nijhoff, 1888–1950.
Ibn Ali ibn Hajr al-Asqalani, Ahmad. *Al-Nukat ala Kitab ibn al-Salah*. Vol. 1. Maktabah al-Furqan, 2003.
Ibn Khaldun. *The Muqaddimah: An Introduction to History*. Translated by Franz Rosenthal. Princeton University Press, 1989.
Ifrah, Georges. *The Universal History of Numbers: From Prehistory to the Invention of the Computer*. John Wiley and Sons, 2000.
Iliffe, Rob, and George Smith (eds.). *The Cambridge Companion to Newton*. 2nd edition. Cambridge University Press, 2016.
Insoll, Timothy. *The Oxford Handbook of the Archaeology of Ritual and Religion*. Oxford University Press, 2011.
Israel, Jonathan. *Radical Enlightenment*. Oxford University Press, 2002.
Itkonen, Esa. *Universal History of Linguistics*. John Benjamins, 1991.
Iwaniszewski, Stanisław. "Ancient cosmologies: Understanding ancient skywatchers and their worldviews." *Journal of Cosmology*, 9, 2010, pp. 2121–2129.
Jacob, Christian (ed.). *Lieux de savoir*. Albin Michel, 2007–2011.
Jacobson, Thorkild. "Very ancient linguistics: Babylonian grammatical texts." In *Studies in the History of Linguistics: Traditions and Paradigms*. Edited by Dell Hymes. Indiana University Press, 1974, pp. 41–62.
Jacquart, Danielle, and Agostino Paravicini Bagliani (eds.). *La Scuola Medica Salernitana: Gli autori e i testi*. Edizione nazionale la Scuola Medica Salernitana. Edizioni del Galluzzo, 2007.

James, Edward (ed.). *Visigothic Spain: A New Reconstruction.* Clarendon Press, 1980.
Janssens, Uta. "Matthieu Maty and the adoption of inoculation for smallpox in Holland." *Bulletin of the History of Medicine*, 55(2), 1981, pp. 246–256.
Jardine, Lisa. *The Curious Life of Robert Hooke: The Man Who Measured London.* Harper Perennial, 2005.
Jardine, Nicholas. "Uses and abuses of anachronism in the history of the sciences." *History of Science*, 38, 2000, pp. 251–270.
Jeppie, Shamil. "Timbuktu scholarship: But what did they read?" *History of Humanities*, 1(2), 2016, pp. 213–229.
Ji, Hong-Fang, Xue-Juan Li, and Hong-Yu Zhang. "Natural products and drug discovery: Can thousands of years of ancient medical knowledge lead us to new and powerful drug combinations in the fight against cancer and dementia?" *EMBO Reports*, 10(3), 2009, pp. 194–200.
Johnson, Wallace. *The T'ang Code.* Vol. 1: *General Principles.* Princeton University Press, 1979.
Johnston, Ian. *The Mozi: A Complete Translation.* Chinese University Press, 2010.
Jones, Alexander. *A Portable Cosmos.* Oxford University Press, 2017.
Joordens, Josephine, et al. "Homo erectus at Trinil on Java used shells for tool production and engraving." *Nature*, 518(7538), 2015, p. 228.
Jorink, Eric. *Het boeck der natuere: Nederlandse geleerden en de wonderen van Gods schepping 1575–1715.* Primavera Pers, 2007.
Jorink, Eric, and Dirk van Miert (eds.). *Isaac Vossius (1618–1689): Between Science and Scholarship.* Brill, 2012.
Joseph, George. *The Crest of the Peacock: Non-European Roots of Mathematics.* 3rd edition. Princeton University Press, 2010.
Joseph, Suad, and Afsaneh Najmabadi. *Encyclopedia of Women & Islamic Cultures: Economics, Education, Mobility and Space.* Brill, 2003.
Juynboll, Gautier. *Encyclopedia or Canonical Hadith.* Brill, 2007.
Kane, Pandurang Vaman. *History of Dharmaśāstra: Ancient and Medieval Religious and Civil Law in India.* 8 vols. Bhandarkar Oriental Research Insitute, 1958, reprint 2006.
Kant, Immanuel. *Kant's gesammelte Schriften.* In *Ausgabe der Preußischen Akademie der Wissenschaften.* Reimer, 1900.
Kaplan, Robert. *The Nothing That Is: A Natural History of Zero.* Oxford University Press, 1999.
Karstens, Bart. "Bopp the builder: Discipline formation as hybridization; The case of comparative linguistics." In *The Making of the Humanities.* Edited by Rens Bod, Jaap Maat, and Thijs Weststeijn. Vol. 2: *From Early Modern to Modern Disciplines.* Amsterdam University Press, 2012, pp. 104–127.
Karstens, Bart, Marijn Koolen, Giuseppe Dari Mattiacci, Rens Bod, and Tom Ginsburg. "Reference structures of national constitutions." In *Proceedings NAiL 2014: 2nd International Workshop on "Network Analysis in Law."* Edited by R. Winkels and N. Lettieri. 2014, pp. 1–12.
Katz, Victor. *A History of Mathematics.* 2nd edition. Addison-Wesley, 1998.
Katz, Victor. "Ideas of calculus in Islam and India." *Mathematics Magazine*, 68, 1995, pp. 163–174.
Kaufholz, Ute. *Sonne, Mond und Sterne: Das Geheimnis der Himmelsscheibe.* Anderbeck, 2004.
Kaufmann, Walter. *Musical References in the Chinese Classics.* Detroit Monographs in Musicology. No. 5. Harmonie Park, 1976.
Kebra nagast. Translated by E. A. Wallis Budge. London, 1932.
Kemp, Martin. *The Science of Art: Optical Themes in Western Art from Brunelleschi to Seurat.* Yale University Press, 1990.

Kepler, Johannes. *Vom wahren Geburtsjahr Christi*. Translated by Otto Schönberger and Eva Schönberger. Itinera Classica, 10. Verlag Marie Leidorf, 2016.

Keynes, John Maynard. "Newton, the man." In *Proceedings of the Royal Society Newton Tercentenary Celebrations*. Cambridge University Press, 1947, pp. 363–374.

Kilpatrick, Hilary. *Making the Great Book of Songs: Compilation and the Author's Craft in Abu l-Faraj al-Isbahani's "Kitab al-Aghani."* Routledge, 2002.

Kirk, Geoffrey, and John Raven. *The Presocratic Philosophers*. 2nd edition. Cambridge University Press, 1983.

Kirkpatrick, Andy. "China's first systematic account of rhetoric: An introduction to Chen Kui's *Wen ze*." *Rhetorica*, 23(2), 2005, pp.103–152.

Ki-Zerbo, Joseph, et al. (eds.). *General History of Africa I: Methodology and African Prehistory*. Heinemann, 1981, pp. 1–6.

Klein, Jacob, and Tonia Sharlach. "A collection of model court cases from Old Babylonian Nippur (CBS 11324)." *Zeitschrift für Assyriologie und Vorderasiatische Archäologie*, 97, 2007, pp. 1–25.

Kneale, William, and Martha Kneale. *The Development of Logic*. Clarendon Press, 1962.

Kohler, Robert, and Kathryn Olesko. "Introduction: Clio meets science." *Osiris*, 27(1), 2012, pp. 1–16.

Kooijmans, Luuc. *Het orakel: De man die de geneeskunde opnieuw uitvond; Herman Boerhaave, 1668–1738*. Balans, 2011.

Kracke, E. A. *Civil Service in Early Song China*. Harvard University Press, 1953.

Kramer, Edna. "Maria Agnesi." In *Dictionary of Scientific Biography*. Edited by Charles Gillispie. Vol. 1. Charles Scribner's Sons, 1970–1980, p. 76.

Krisciunas, Kevin, and H. B. Paksoy. *The Legacy of Ulugh Beg*. Isis Press, 1992.

Kuhn, Thomas. *The Copernican Revolution*. Harvard University Press, 1957.

Kwa, Chunglin. *Kernthema's in de wetenschapsfilosofie*. Boom 2014.

La Vopa, Anthony. *The Labor of the Mind: Intellect and Gender in Enlightenment Cultures*. University of Pennsylvania Press, 2017.

Laird, Walter Roy. "Archimedes among the Humanists." *Isis*, 82(4), 1991, pp. 628–638.

Lakoff, Robin. Review of Herbert Brekle (ed.), *Grammaire générale et rassonnée ou La grammaire du Port Royal*, Fromann, 1966. *Language*, 45(2), 1969, pp. 343–364.

Lam, Joseph. "Chinese music: History and theory." In *The New Grove Dictionary of Music and Musicians*. Edited by Stanley Sadie. 2001, https://www.oxfordmusiconline.com/grovemusic.

Lam, Joseph. "Ming Music and Music History." *Ming Studies*, 1997(1), 1997, pp. 21–62.

Lamb, Evelyn. "Don't fall for Babylonian trigonometry hype: Separating fact from speculation in math history." *Scientific American*, August, 2017. https://blogs.scientificamerican.com/roots-of-unity/dont-fall-for-babylonian-trigonometry-hype/.

Lancaster-Brown, Peter. *Halley and His Comet*. Blandford Press, 1985.

Landsberger, Benno, et al. *Materialien zum Sumerischen Lexikon IV*. Pontificium Institutum Biblicum, 1956. http://ancientworldonline.blogspot.com/2012/09/early-materialien-zum-sumerischen.html.

Lanzarote-Guiral, José María. "The recognition of cave art in the Iberian Peninsula and the making of prehistoric archaeology (1878–1929)." In *The Making of the Humanities*. Edited by Rens Bod, Jaap Maat, and Thijs Weststeijn. Vol. 3: *The Modern Humanities*. Amsterdam University Press, 2014, pp. 359–377.

Law, Vivian. *The History of Linguistics in Europe*. Cambridge University Press, 2003.

Lefèvre, Wolfgang, Jürgen Renn, and Urs Schoepflin. *The Power of Images in Early Modern Science*. Springer, 2003.

Lehoux, Daryn. *What Did the Romans Know? An Inquiry into Science and Worldmaking.* University of Chicago Press, 2012.
Leppin, Hartmut. "Die Gesetzgebung Iustinian: der Kaiser und sein Recht." In *Erinnerungsorte der Antike: Die römische Welt.* Edited by Karl-Joachim Hölkeskamp and Elke Stein-Hölkeskamp. Beck, 2006, pp. 457–466.
Leung, Angela. "Organized medicine in Ming-Qing China: State and private medical institutions in the lower Yangzi region." *Late Imperial China*, 1987, 8(1), pp. 134–166.
Lewis, D. "Voyaging stars: Aspects of Polynesian and Micronesian astronomy." *Philosophical Transactions of the Royal Society of London, Series A, Mathematical and Physical Sciences*, 276(1257), 1974, pp. 133–148.
Li, X., G. Harbottle, J. Zhang, and C. Wang. "The earliest writing? Sign use in the seventh millennium BC at Jiahu, Henan Province, China." *Antiquity*, 77, 2003, pp. 31–44.
Li Zhizhen. *Compendium of Materia Medica.* Foreign Language Press, 2003.
Libbrecht, Ulrich. *Chinese Mathematics in the Thirteenth Century: The Shu-Shu-Chiu-Chang or Chin Chiu Shao.* Dover, 2006.
Liebenberg, Louis. *The Art of Tracking, the Origin of Science.* David Philip, 1990.
Liebs, Detlef. *Die Jurisprudenz im spätantiken Italien (AD 260–640).* Duncker & Humblot, 1987.
Lindberg, David. "The Greeks and the Cosmos." In *The Beginnings of Western Science: The European Scientific Tradition in Philosophical, Religious, and Institutional Context, Prehistory to A.D. 1450.* University of Chicago Press, 2007, pp. 21–44.
Lindberg, David. *Roger Bacon and the Origins of Perspectiva in the Middle Ages.* Clarendon Press, 1996.
Liu Zhiji. *Shitong tongshi.* Annotated by Pu Qilong. 2 volumes. Shanghai Guji Chubanshe, 1978.
Lloyd, Geoffrey. *Disciplines in the Making: Cross-Cultural Perspectives on Elites, Learning and Innovation.* Oxford University Press, 2009.
Lloyd, Geoffrey. *Early Greek Science: Thales to Aristotle.* W. W. Norton, 1970.
Lloyd, Geoffrey (ed.). *Hippocratic Writings.* Penguin Classics, 1984.
Lokin, J. H. A. "Codificatie in Constantinopel? Enige opmerkingen over de wetgevingstechniek van Theodosius II en Justinianus I." *Groninger Opmerkingen en Mededelingen*, 12, 1995, pp. 38–53.
Lowry, Joseph. *Early Islamic Legal Theory: The Risāla of Muḥammad ibn Idrīs al-Shāfiʿī.* Brill, 2007.
Lu, Yongxiang (ed.). *A History of Chinese Science and Technology.* Vol. 1. Springer, 2015.
Macrobius, Ambrosius Theodosius. *Commentary on the Dream of Scipio.* Translated by William Stahl. Columbia University Press, 1990.
Maddocks, Fiona. *Hildegard of Bingen: The Woman of Her Age.* Faber & Faber, 2013.
Maffei, Domenico. *La "Lectura super digesto veteri" di Cino da Pistoia.* Studio sui MSS Savigny. Giuffrè, 1963.
Major, John. *Heaven and Earth in Early Han Thought: Chapters Three, Four and Five of the Huainanzi.* State University of New York Press, 1993.
Makdisi, George. *The Rise of Humanism in Classical Islam and the Christian West.* Edinburgh University Press, 1990.
Mancosu, Paolo. *Philosophy of Mathematics and Mathematical Practice in the Seventeenth Century.* Oxford University Press, 1996.
Mancuso, Stefano, et al. "Experience teaches plants to learn faster and forget slower in environments where it matters." *Oecologia*, 175(1), 2014, pp. 63–72.

Manekin, Charles. *The Logic of Gersonides: An Analysis of Selected Doctrines, with a Partial Edition and Translation of "The Book of the Correct Syllogism."* Kluwer, 1992.

Mansfield, Daniel, and N. J. Wildberger. "Plimpton 322 is Babylonian exact sexagesimal trigonometry." *Historia Mathematica*, 44(4), 2017, pp. 395–419.

Manuel, Frank Edward. *Isaac Newton, Historian*. Harvard University Press, 1963.

Maris, Cees. *Legal Philosophy: The Most Important Controversies*. Springer, 2011.

Marshack, Alexander. *The Roots of Civilization: The Cognitive Beginnings of Man's First Art, Symbol and Notation*. McGraw-Hill, 1972.

Martianus Capella. *The nuptiis philologiae et Mercurii*. Translated by William Harris Stahl. In *Martianus Capella and the Seven Liberal Arts*. Vol. 2: *The Marriage of Philology and Mercury*. Columbia University Press, 1977.

Martens, Guus (trans.). *Huang Di nei jing su wen*. La Douze VOF, 2011.

Martzloff, Jean-Claude. *A History of Chinese Mathematics*. Springer, 1997.

Mason, Stephen. *A History of the Sciences*. MacMillan, 1962.

Mazliak, Paul. *Jean Fernel, premier physiologiste de la Renaissance*. Hermann, 2016.

Mazzotti, Massimo. *The World of Maria Gaetana Agnesi, Mathematician of God*. Johns Hopkins University Press, 2007.

McAllister, James. "The ontology of patterns in empirical data." *Philosophy of Science*, 77, 2011, pp. 804–814.

McClain, Ernest, and Ming Shui Hung. "Chinese cyclic cunings in late antiquity." *Ethnomusicology*, 23(2), 1979, pp. 205–224.

McClellan, James, III, and Harold Dorn. *Science and Technology in World History*. Johns Hopkins University Press, 2006.

McGee, Harold. *On Food and Cooking: The Science and Lore of the Kitchen*. Scribner, 2004.

McGinnis, Jon. *Avicenna*. Oxford University Press, 2010.

McGuire, J. E., and P. M. Rattansi. "Newton and the 'Pipes of Pan.'" In *Notes and Records of the Royal Society of London*, 21, 1966, pp. 108–143.

McNeely, Ian, and Lisa Wolverton. *Reinventing Knowledge: From Alexandria to the Internet*. W. W. Norton, 2008.

Meece, Stephanie. "A bird's eye view—of a leopard's spots: The Çatalhöyük 'map' and the development of cartographic representation in prehistory." *Anatolian Studies*, 56, 2006, pp. 1–16.

Mendell, Henry. "Reflections on Eudoxus, Callippus and their curves: Hippopedes and Callippopedes." *Centaurus*, 40, 1998, pp. 177–275.

Mensa i Valls, Jaume. *Arnau de Vilanova*. Rafael Dalmau, 1997.

Menski, Werner. *Comparative Law in Global Context*. 3rd edition. Cambridge University Press, 2009.

Mersenne, Marin. *Harmonie universelle, 1636–1637*. Translated into English by Roger Chapman. Martinus Nijhoff, 1957.

Merton, Robert. "The Matthew effect in science." *Science*, 159(3810), 1968, pp. 56–63.

Metzger, Bruce. *Der Text des Neuen Testaments: Einführung in die neutestamentliche Textkritik*. Kohlhammer, 1966.

Meyerhof, Max. "New light on universal Ibn Isḥâq and his period." *Isis*, 8(4), 1926, pp. 685–724.

Miao, Tian, and Zhang Baichun. "The development of knowledge on levers in ancient China." In *International Symposium on History of Machines and Mechanisms: Proceedings of HMM 2008*. Edited by Hong-Sen Yan and Marco Ceccarelli. Springer, 2009, pp. 251–268.

Michalowski, Piotr. "History as charter: Some observations on the Sumerian King List." *Journal of the American Oriental Society*, 103, 1983, pp. 237–248.
Miettinen, Olli. "The modern scientific physician: 1. Can practice be science?" *CMAJ*, 165(4), 2001, pp. 441–442.
Milton, John. *Paradise Lost*. Penguin Classics, 2003.
Minicka, Mary. "Towards a conceptualization of the study of Africa's indigenous manuscript heritage and tradition." *Tydskrif vir Letterkunde*, 2008, 45(1), pp. 143–163.
Mitford, Nancy. *Voltaire in Love*. Carroll and Graf, 1999.
Montheit, Michael. "Guillaume Budé, Andrea Alciato, Pierre de l'Estoile, Renaissance interpreters of Roman law." *Journal of the History of Ideas*, 58(1), 1997, pp. 21–40.
Moore, Sally Falk (ed.). *Law and Anthropology*. Blackwell, 2010.
Moretti, Franco. *Distant Reading*. Verso Books, 2013.
Morgan, Mary, and Margaret Morrison (eds.). *Models as Mediators*. Cambridge University Press, 1999.
Mosley, Adam. *Bearing the Heavens: Tycho Brahe and the Astronomical Community of the Late Sixteenth Century*. Cambridge University Press, 2007.
Most, Glenn. "Karl Lachmann (1793–1851): Reconstructing the Transmission of a Classical Latin Author." *History of Humanities*, 4(2), 2019, pp. 269–273.
Moyer, Ann. *Musica Scientia: Musical Scholarship in the Italian Renaissance*. Cornell University Press, 1992.
Muccillo, Maria. "Fabrici d'Acquapendente, Girolamo (detto l'Acquapendente o Fabrizio d'Acquapendente)." In *Dizionario biografico degli Italiani (DBI)*. Part 43. Edited by Fiorella Bartoccini. Istituto della Enciclopedia Italiana, 1993, pp. 768–774.
Muller, Jean-Claude. "Early stages of language comparison from Sassetti to Sir William Jones (1786)." *Kratylos*, 31, 1986, pp. 1–31.
Musgrave, Alan. *Essays on Realism and Rationalism*. Brill, 1999.
Nauta, Lodi. "Lorenzo Valla and Quattrocento skepticism." *Vivarium*, 44, 2006, pp. 375–395.
Needham, Joseph. *Science and Civilisation in China*. Vol. 3: *Mathematics and the Sciences of the Heavens and the Earth*. Cambridge University Press, 1959.
Needham, Joseph. *Science and Civilisation in China*. Vol. 4: *Physics and Physical Technology*. Cambridge University Press, 1962.
Needham, Joseph, Lu Gwei-Djen, and Nathan Sivin. *Science and Civilisation in China*. Vol. 6: *Biology and Biological Technology*. Cambridge University Press, 2000.
Nellen, Henk. *Hugo de Groot: Een leven in strijd om de vrede 1583–1645*. Balans, 2012.
Nellen, Henk. *Ismaël Boulliau (1605–1694): Astronome, épistolier, nouvelliste et intermédiaire scientifique; Ses rapports avec les milieux du libertinage érudit*. APA–Holland University Press, 1994.
Neugebauer, Otto. *The Astronomical Tables of al-Khwarizmi*. Munksgaard, 1962.
Neugebauer, Otto. *The Exact Sciences in Antiquity*. Munksgaard, 1951.
Neugebauer, Otto. *A History of Ancient Mathematical Astronomy*. Springer-Verlag, 1975.
Neugebauer, Ottom, and David Pingree. *The Pancasiddhantika of Varamihira*. 2 volumes. Munksgaard, 1970/1971.
Neugebauer, Otto, and A. J. Sachs. *Mathematical Cuneiform Texts*. American Oriental Series 29. American Oriental Society and the American Schools of Oriental Research, 1945.
Newton, Isaac. *The Correspondence of Sir Isaac Newton*. Vol. 2: *(1676–1687)*. Edited by H. W. Turnbull. Cambridge University Press, 1960.

Newton, Isaac. *The Principia*. A new translation by I. B. Cohen and A. Whitman. University of California Press, 1999.
Ng, On-cho, and Q. Edward Wang. *Mirroring the Past: The Writing and Use of History at Imperial China*. University of Hawai'i Press, 2005.
Nicomachus, *Enchiridion Harmonices*. Leipzig 1895.
Nissen, Hans, Peter Damerow, and Robert Englund. *Archaic Bookkeeping*. University of Chicago Press, 1993.
North, John. *Cosmos: An Illustrated History of Astronomy and Cosmology*. University of Chicago Press, 2008.
Norton, Scott. "Herbal medicines in Hawaii: From tradition to convention." *Hawaii Medical Journal*, 57, 1998, pp. 382–385.
Nothaft, Philipp. *Dating the Passion: The Life of Jesus and the Emergence of Scientific Chronology (200–1600)*. Brill, 2011.
Nowak, Stefan. *General Laws and Historical Generalizations in the Social Sciences*. Brill, 2009.
Nunn, John. *Ancient Egyptian Medicine*. British Museum Press, 1996.
O'Daly, Gerard. *Days Linked by Song: Prudentius' Cathemerinon*. Oxford University Press, 2012.
O'Grady, Patricia. *Thales of Miletus: The Beginnings of Western Science and Philosophy*. Western Philosophy Series 58. Ashgate, 2002.
O'Malley, Charles Donald. *Andreas Vesalius of Brussels, 1514–1564*. University of California Press, 1964.
Ogilvie, Marilyn Bailey. *Women in Science: Antiquity through the Nineteenth Century*. MIT Press, 1986.
Osita, Iroku. *A Day in the Life of God*. Enlil Institute. Dover, 2008.
Östling, Johan, Erling Sandmo, David Larsson Heidenblad, Anna Nilsson Hammar, and Kari Nordberg. *Circulation of Knowledge: Explorations in the History of Knowledge*. Nordic Academic Press, 2018.
Ouyang, Wen-chin. *Literary Criticism in Medieval Arabic-Islamic Culture: The Making of a Tradition*. Edinburgh University Press, 1997.
Owens, Jonathan. *Early Arabic Grammar Theory*. John Benjamins, 1990.
Palisca, Claude. *Humanism in Italian Renaissance Musical Thought*. Yale University Press, 1985.
Palisca, Claude. "Was Galileo's father an experimental scientist?" In *Music and Science in the Age of Galileo*. Edited by V. Coelho. Kluwer Academic, 1992, pp. 143–151.
Panini. *Ashtadhyayi*. Translated by Srisa Chandra Vasu. Nabu Press, 2011, reprint 1923.
Pankhurst, Richard. *An Introduction to the Medical History of Ethiopia*. Red Sea Press, 1999.
Passerin d'Entreves, Alexander. *Natural Law: An Introduction to Legal Philosophy*. Routledge, 2017.
Patota, Giuseppe (ed.). *L. B. Alberti, Grammatichetta e altri scritti sul volgare*. Salerno Editrice, 1996.
Perez, Éric. *Introduction au Shanghanlun: Historique, commentaire, diagnostic et traitement des maladies classées selon la nosographie des "Six méridiens."* Vitruve, 2014.
Peterson, Barbara Bennet (ed.). *Notable Women of China: Shang Dynasty to the Early Twentieth Century*. M. E. Sharpe, 2000.
Peterson, Willard. "The life of Ku Yen-Wu (1613–1682)." *Harvard Journal of Asiatic Studies*, 28, 1968, pp. 114–156.
Petschow, Herbert. "Zur Systematik und Gesetzestechnik im Codex Hammurabi." *Zeitschrift für Assyriologie*, 57, 1967, pp. 146–172.

Phillips, Denise. "Francis Bacon and the Germans: Stories from when 'science' meant 'Wissenschaft.'" *History of Science*, 53(4), 2015, pp. 378–394.
Piaget, Jean. *The Child's Conception of the World*. Routledge and Kegan Paul, 1928.
Picken, Laurence, and Noël Nickson (eds.). *Music from the Tang Court*. Cambridge University Press, 2000.
Pike, A., et al. "U-series dating of Paleolithic art in 11 caves in Spain." *Science*, 336(6087), 2012, pp. 1409–1413.
Pingree, D., et al. "The Paitamahasiddhanta of the Visnudharmottapurana." *Brahmavidya*, 21–22, 1967–1968, pp. 472–510.
Pingyi, Chu. "Scientific dispute in the imperial court: The 1664 calendar case." *Chinese Science*, 14, 1997, pp. 7–34.
Plato. *Politeia*. Athenaeum-Polak & Van Gennep, 1998.
Pliny the Elder. *Natural History*. 10 volumes. Translated by H. Rackham, W. Jones, and D. Eichholz. Loeb Classical Library, 1938–1962.
Plofker, Kim. *Mathematics in India*. Princeton University Press, 2009.
Plutarch. *Quaestiones Convivales*. In *Plutarch's Miscellanies and Essays: Comprising All His Works Collected under the Title of "Morals."* Edited by W. W. Goodwin. Little, Brown, 1898, pp. 1–216.
Poliziano, Angelo. *Miscellanea*. In *Opera omnia*. Edited by Ida Maïer. 3 volumes. Bottega d'Erasmo, 1970–1971.
Pollman, Karla. *Saint Augustine the Algerian*. Edition Ruprecht, 2003.
Pomeranz, Kenneth. *The Great Divergence: China, Europe, and the Making of the Modern World Economy*. Princeton University Press, 2000.
Popkin, Richard. *Isaac La Peyrère (1596–1676): His Life, Work and Influence*. Brill, 1987.
Powers, Harold S. *The Background of the South Indian Raga System*. PhD thesis, Princeton University, 1959.
Powers, Harold S. "The structure of musical meaning: A view from Banaras (a metamodal model for Milton)." *Perspectives of New Music*, 14(2), 1976, pp. 308–334.
Principe, Lawrence. *The Secrets of Alchemy*. University of Chicago Press, 2012.
Pringle, Heather. "The science of inequality: The ancient roots of the 1%." *Science*, 344(6186), 2014, pp. 822–825.
Proclus. *Commentary on the First Book on Euclid's Elements*. Translated by F. Morrow. Princeton University Press, 1970.
Psillos, Stathis. "Regularities, natural patterns and laws of nature." *Theoria*, 79, 2014, pp. 9–27.
Rademaker, C. S. M. *Leven en werk van Gerardus Joannes Vossius (1577–1649)*. Verloren, 1999.
Ragep, Jamil. *Nasir al-Din Tusi's Memoir on Astronomy (al-Tadhkira fi 'ilm al-hay'a)*. 2 volumes. Sources in the History of Mathematics and Physical Sciences. Springer-Verlag, 1993.
Raj, Kapil. "Beyond postcolonialism . . . and postpositivism: Circulation and the global history of science." *Isis*, 104(2), 2013, pp. 337–347.
Rajagopal, C. T., and M. S. Rangachari. "On medieval Keralese mathematics." *Archive for History of Exact Sciences*, 35, 1986, pp. 91–99.
Ramaswami Shastri, R. A. *A Short History of the Purva Mimamsa Shastra*. Annamalai University Sanskrit Series 3. Annamalai University, 1936.
Ramis de Pareia, Bartolomeo. *Musica Practica*. Commentary and translation by Clement Miller. Hänssler-Verlag, 1993.
Rappenglück, M. "The Pleiades in the 'Salle des Taureaux,' Lascaux Cave: Does a rock picture in the cave of Lascaux show the open star cluster of the Pleiades at the

Magdalénien era (ca 15,300 BC)?" *IVth SEAC Meeting Astronomy and Culture*, 1, 1997, pp. 217–225.
Rashed, Roshdi (ed.). *Al-Khwarizmi: The Beginnings of Algebra*. Saqi Books, 2010.
Rashed, Roshdi (ed.). *Encyclopedia of the History of Arabic Science*. Part 2. Routledge, 1996.
Rashed, Roshdi, and Angela Armstrong. *The Development of Arabic Mathematics: Between Arithmetic and Algebra*. Springer, 1994.
Rashed, Roshdi, and Christian Houzel. *Les "Arithmétiques" de Diophante: Lecture historique et mathématique*. De Gruyter, 2013.
Rau, Wilhelm. *Bhartṛharis Vākyapadīya I & II*. Steiner, 1977–1991.
Renn, Jürgen. "From the history of science to the history of knowledge—and back." *Centaurus*, 57(1), 2015, pp. 37–53.
Renn, Jürgen, and Matthias Schemmel. "Mechanics in the Mohist Canon and its European counterpart." In *Zhongguo ke ji dian ji yan jiu: Di san jie Zhongguo ke ji dian ji guo ji hui yi lun wen ji*. Edited by Hans Ulrich Vogel, Christine Moll-Murata, and Gao Xuan. Elephant Press, 2003, pp. 24–31.
Reynolds, L. D., and N. G. Wilson. *Scribes and Scholars: A Guide through the Transmission of Greek and Latin Literature*. Clarendon Press, 1991.
Richards, Paul. "Mary Tew Douglas (1921–2007)." *American Anthropologist*, 110(3), 2008, pp. 407–410.
Riedweg, Christoph. *Pythagoras: His Life, Teaching and Influence*. Cornell University Press, 2005.
Roberts, Victor, and Edward Kennedy. "The planetary theory of Ibn al-Shatir." *Isis*, 50, 1959, pp. 227–235.
Robins, Gay, and Charles Shute. *The Rhind Mathematical Papyrus: An Ancient Egyptian Text*. Dover, 1990.
Robinson, Chase. *Islamic Historiography*. Cambridge University Press, 2003.
Robson, Eleanor. *Mathematics in Ancient Iraq: A Social History*. Princeton University Press, 2008.
Robson, Eleanor. "Words and pictures: New light on Plimpton 322." *American Mathematical Monthly* (Mathematical Association of America), 109(2), 2002, pp. 105–120.
Rochberg, Francesca. "Empiricism in Babylonian omen texts and the classification of Mesopotamian divination as science." *Journal of the American Oriental Society*, 119(4), 1999, pp. 559–569.
Rochberg, Francesca. *The Heavenly Writing: Divination, Horoscopy, and Astronomy in Mesopotamian Culture*. Cambridge University Press, 2008.
Roebroeks, Wil, and Paola Villa. "On the earliest evidence for habitual use of fire in Europe." *Proceedings of the National Academy of Sciences (PNAS)*, 108(13), 2011, pp. 5209–5214.
Roero, Clara Silvia. "M. G. Agnesi, R. Rampinelli and the Riccati family: A cultural fellowship formed for an important scientific purpose, the Instituzioni analitiche." *Historia Mathematica*, 42(3), 2015, pp. 296–314.
Roller, Duane. *The "De Magnete" of William Gilbert*. Menno Hertzberger, 1959.
Romein, Jan. *Historische lijnen en patronen: Een keuze uit de essays*. Querido, 1971.
Rosen, Edward. "Alfonsine tables and Copernicus." *Manuscripta*, 20(3), 1976, pp. 163–174.
Rosenfeld, Boris. *A History of Non-Euclidean Geometry: Evolution of the Concept of a Geometric Space*. Springer, 1988.
Rosenthal, Franz. *The Classical Heritage in Islam*. Routledge, 1975.
Rossiter, Margaret. "The Matthew/Matilda effect in science." *Social Studies of Science*, 23, pp. 325–341.
Rota, Antonio. *Lo stato and diritto nella concezione di Irnerio*. Giuffre, 1954.

Roy, Ranjan. "Discovery of the series formula for π by Leibniz, Gregory, and Nilakantha." *Mathematics Magazine*, 63(5), 1990, pp. 291–306.
Rudman, Peter S. *How Mathematics Happened*. Prometheus Books, 2007.
Russo, Lucio. *The Forgotten Revolution: How Science Was Born in 300 BC and Why It Had to Be Reborn*. Springer, 2003.
Sabra, Abdelhamid. "Configuring the universe: Aporetic, problem solving, and kinematic modeling as themes of Arabic astronomy." *Perspectives on Science*, 6(3), 1998, pp. 288–330.
Salen, Katie, and Eric Zimmerman. *Rules of Play*, MIT Press, 2004.
Salutati, Coluccio. *Epistolario*. Edited by Francesco Novati. Fonti per la storia d'Italia, Istituto Storico Italiano, 1891–1911.
Sarma, Krishna Venkateswara. *A History of the Kerala School of Hindu Astronomy*. Hoshiarpur, 1972.
Sarton, George. *Introduction to the History of Science*. Vols. 1–3. Williams and Wilkins, 1931–1947.
Sasson, Jack (ed.). *Civilizations of the Ancient Near East*. Vol. 4. Charles Scribner's Sons, 1995.
Savage-Smith, Emilie. "Medicine." In *Encyclopedia of the History of Arabic Science*. Edited by Roshdi Rashed. Vol. 3. Routledge, 1996, pp. 903–962.
Schäfer, Dagmar. *The Crafting of the 10,000 Things: Knowledge and Technology in Seventeenth-Century China*. University of Chicago Press, 2011.
Schemmel, Matthias. "The transmission of scientific knowledge from Europe to China in the early modern period." In *The Globalization of Knowledge in History*. Edited by Jürgen Renn. Edition Open Access, 2010, pp. 269–293.
Scherzinger, Martin. "African Music and the History of Time." *History of Humanities*, 6(2), 2021, pp. 397–426.
Schiebinger, Londa. *The Mind Has No Sex? Women in the Origins of Modern Science*. Harvard University Press, 1991.
Schilt, Cornelis J. "Created in our image: How Isaac Newton was fashioned as a scientist and forgotten as a scholar." *History of Humanities*, 5(1), 2020, pp. 75–96.
Schipper, Mineke. *Imagining Insiders*. Cassell, 1999.
Schironi, Francesca. *Aristarco di Samotracia negli etimologici bizantini: Etymologicum genuinum magnum, Symeonis, Megalē grammatikē, Zonarae lexicon*. Vandenhoeck und Ruprecht, 2004.
Secord, James (ed.). "The big picture." Special issue, *British Journal for the History of Science*, 26, 1993, pp. 387–483.
Sengupta, Pradip Kumar. *Foundations of Indian Musicology*. Abhinav, 1991.
Settle, Thomas. "An experiment in the history of science." *Science* 133 (3445), 1961, pp. 19–23.
Seuren, Pieter. *Western Linguistics: An Historical Introduction*. Oxford University Press, 1998.
Shapin, Steven. *The Scientific Revolution*. University of Chicago Press, 1998.
Sharratt, Michael. *Galileo: Decisive Innovator*. Cambridge University Press, 1996.
Shen, Kangshen, John Crossley, and Anthony Lun. *The Nine Chapters on the Mathematical Art: Edition, Translation and Commentary*. Oxford University Press, 1999.
Shermer, Michael. "Scientia humanitatis: Reason, empiricism and skepticism are not virtues of science alone." *Scientific American* 312(6), 2015. https://www.scientificamerican.com/article/the-humanities-and-science-share-the-virtues-of-empiricism-and-skepticism/.
Shields, Christopher (ed.). *The Oxford Handbook of Aristotle*. Oxford University Press, 2012.
Shillington, Kevin (ed.). *Encyclopedia of African History*. Vol. 1. Routledge, 2004.
Shuchun, Guo, Chen Zaixin, and Guo Jinhai (eds.). *Zhu Shijie: Jade Mirror of the Four Unknowns*. Liaoning Education Press, 2006

Sibawayh. *Kitab*. Translated into German as *Sībawaihis Buch über die Grammatik übersetzt und erklärt* by Gustav Jahn. 2 volumes. Reuther & Reichard, 1895–1900, reprint Georg Olms, 1969.
Sigler, Laurence (ed.). *Fibonacci's Liber Abaci*. Springer, 2002.
Singh, Pamanand. "The so-called Fibonacci numbers in ancient and medieval India." *Historia Mathematica*, 12, 1985, pp. 229–244.
Sirén, Osvald. *The Chinese on the Art of Painting: Texts by the Painter-Critics, from the Han through the Ch'ing Dynasties*. Dover, 1936, reprint 2005.
Sivin, Nathan. *Cosmos and Computation in Early Chinese Mathematical Astronomy*. Brill, 1969.
Sivin, Nathan. "Science and medicine in imperial China: The state of the field." *Journal of Asian Studies*, 47(1), 1988, pp. 41–90.
Sivin, Nathan. *Science in Ancient China: Researches and Reflections*. Ashgate, 1995.
Slotsky, Alice. *The Bourse of Babylon: Market Quotations in the Astronomical Diaries of Babylonia*. CDL Press, 1997.
Smith, David, and Marcia Latham. *The Geometry of René Descartes*. Dover, 1954.
Smith, John D. "The remarkable Ibn al-Haytham." *Mathematical Gazette* (Mathematical Association), 76(475), 1992, pp. 189–198.
Snow, Charles Percy. *The Two Cultures and the Scientific Revolution*. Cambridge University Press, 1959.
Snyder, Laura. *Eye of the Beholder: Johannes Vermeer, Antoni van Leeuwenhoek, and the Reinvention of Seeing*. W. W. Norton, 2015.
Soll, Jacob. *The Reckoning: Financial Accountability and the Rise and Fall of Nations*. Basic Books, 2014.
Spedding, James, Robert Ellis, and Douglas Heath. *The Works of Francis Bacon*. 15 volumes. Houghton Mifflin, 1857–1874. http://onlinebooks.library.upenn.edu/webbin/metabook?id=worksfbacon.
Staal, Frits. "Greek and Vedic geometry." *Journal of Indian Philosophy*, 27, 1999, pp. 105–127.
Staal, Frits (ed.). *A Reader on Sanskrit Grammarians*. Motilal Banarasidass, 1985.
Steenbakkers, Piet. "Spinoza in the history of biblical scholarship." In *The Making of the Humanities*. Edited by Rens Bod, Jaap Maat, and Thijs Weststeijn. Vol. 1: *Early Modern Europe*. Amsterdam University Press, 2010, pp. 313–326.
Stein, Aurel. *Kalhana's Rajatarangini: A Chronicle of the Kings of Kashmir*. Saujanya Books, 2007.
Stevens, J. M. "Gynaecology from ancient Egypt: The papyrus Kahun; A translation of the oldest treatise on gynaecology that has survived from the ancient world." *Medical Journal of Australia*, 25–26, 1975, pp. 949–952.
Stigler, Stephen. *The History of Statistics*. Harvard University Press, 1990.
Stuurman, Siep. *François Poulain de la Barre and the Invention of Modern Equality*. Harvard University Press, 2004.
Tamayo, Daniel, Hanno Rein, Cristobal Petrovich, and Norman Murray. "Convergent migration renders TRAPPIST-1 long-lived." *Astrophysical Journal Letters*, 840(2), 2017, p. L19.
Taschow, Ulrich. *Nicole Oresme and the Frühling der Moderne*. 2 volumes. Halle, 2003.
Tavoni, Mirko. "The 15th-century controversy on the language spoken by the ancient Romans: An inquiry into Italian humanist concepts of 'Latin,' 'grammar,' and 'vernacular.'" *Historiographia Linguistica*, 9(3), 1982, pp. 237–264.
Taylor, Ken. *Celestial Geometry: Understanding the Astronomical Meanings of Ancient Sites*. Watkins, 2012.
Tedlock, Dennis. *Popol Vuh: The Definitive Edition of the Mayan Book of the Dawn of Life and the Glories of Gods and Kings*. Simon and Schuster, 1985.

Tedlock, Dennis. *2000 Years of Mayan Literature*. University of California Press, 2010.
Terrall, Mary. *The Man who Flattened the Earth: Maupertuis and the Sciences in the Enlightenment*. University of Chicago Press, 2002.
Thucydides. *The Peloponnesian War*. Translated by Martin Hammond. Oxford University Press, 2009.
Tolmacheva, Marina. *The Pate Chronicle: Edited and Translated from MSS 177, 321, 344, and 358 of the Library of the University of Dar es Salaam*. Michigan State University Press, 1993.
Tolsa, Cristian. "Ptolemy's law court analogy and Alexandrian philosophy." *Ancient Philosophy*, 36, 2016, pp. 465–477.
Tomlinson, Gary. *Culture and the Course of Human Evolution*. University of Chicago Press, 2018, pp. 4–18.
Toomer, Gerald. "Hipparchus and Babylonian astronomy." In *A Scientific Humanist: Studies in Memory of Abraham Sachs*, 9. Edited by Erle Leichty, Maria de J. Ellis, and Pamela Gerardi. Occasional Publications of the Samuel Noah Kramer Fund, 1988.
Toomer, Gerald. *Ptolemy's Almagest*. Princeton University Press, 1998.
Torkewitz, Dieter. *Das älteste Dokument zur Entstehung der abendländischen Mehrstimmigkeit*. Vol. 44 of *Beihefte zum Archiv für Musikwissenschaft*. Steiner, 1999.
Tosh, John (ed.). *Historians on History*. Routledge, 2008.
Tosi, Renzo. *I carmi greci di Clotilde Tambroni*. Pàtron, 2011.
Tripathi, V. N. "Astrology in India." In *Encyclopedia of the History of Science, Technology, and Medicine in Non-Western Cultures*. Edited by Helaine Selin. Springer, 2008, pp. 264–267.
Twitchett, Denis. *The Writing of Official History under the T'ang*. Cambridge University Press, 1992.
Unschuld, Paul (ed.). *Huang Di nei jing su wen*. University Press Group, 2011.
Unschuld, Paul. *Medicine in China: A History of Ideas*. University of California Press, 1985.
Unschuld, Paul. *What Is Medicine? Western and Eastern Approaches to Healing*. University of California Press, 2009.
Unschuld, Paul, and Zheng Jinsheng. *Chinese Traditional Healing: The Berlin Collections of Manuscript Volumes from the 16th through the Early 20th Century*. Vol. 1. Brill Academic, 2012.
Uriel, Heyd. *Studies in Old Ottoman Criminal Law*. Clarendon Press, 1973.
van Cranenburgh, Andreas. *Rich Statistical Parsing and Literary Language*. PhD thesis, University of Amsterdam, 2016. http://dare.uva.nl/search?metis.record.id=543163.
van den Bergh, G. C. J. J. *Geleerd recht*. 6th printing. Kluwer, 2011.
van den Bergh, Govaert. *Die holländische elegante Schule*. Klostermann, 2002.
van Dongen, Jeroen, and Herman Paul (eds.). *Epistemic Virtues in the Science and the Humanities*. Boston Studies in the Philosophy and History of Science. Springer, 2017.
van Gelder, G. J. H. *Beyond the Line: Classical Arabic Literary Critics on the Coherence and Unity of the Poem*. Brill, 1982.
van Helden, Albert. *Galileo Galilei, Sidereus Nuncius, or The Sidereal Messenger*. University of Chicago Press, 1989.
van Helden, Albert. *The Invention of the Telescope*. American Philosophical Society, 1977.
van Helden, Albert. "Willem Jacob 's Gravesande, 1688–1742." In *A History of Science in the Netherlands*. Edited by K. van Berkel, A. van Helden, and L. Palm. Brill, 1999, pp. 450–463.
van Leeuwen, Joyce. *The Aristotelian Mechanics: Text and Diagrams*. Springer, 2016.
van Miert, Dirk. *The Emancipation of Biblical Philology in the Dutch Republic, 1590–1670*. Oxford University Press, 2018.

van Nooten, B. "Binary numbers in Indian antiquity." *Journal of Indian Philosophy*, 21(1), 1993, pp. 31–50.

van Notten, Michael. *The Law of the Somalis: A Stable Foundation for Economic Development in the Horn of Africa*. Red Sea Press, 2005.

van Ruler, Han. *The Crisis of Causality: Voetius and Descartes on God, Nature and Change*. Brill, 1995.

Vansina, Jan. *Oral Tradition as History*. University of Wisconsin Press, 1985.

Vermij, Rienk. *Christiaan Huygens: De mathematisering van de werkelijkheid*. Veen, 2007.

VerSteeg, Russ. *Law in Ancient Egypt*. Carolina Academic Press, 2002.

Versteegh, Theo. *Greek Elements in Arabic Linguistic Thinking*. Brill, 1977.

Vesalius, Andreas. *On the Fabric of the Human Body*. Translated by W. F. Richardson and J. B. Carman. 5 volumes. Norman, 1998–2009.

Vitruvius. *Ten Books on Architecture*. Translated by Ingrid Rowland. Cambridge University Press, 1999.

Vogelsang, Ralf, et al. "New excavations of Middle Stone Age deposits at Apollo 11 Rockshelter, Namibia: Stratigraphy, archaeology, chronology and past environments." *Journal of African Archaeology*, 8(2), 2010, pp. 185–218.

Voltaire. *Lettres inédites de Voltaire*. Nabu Press, 2013.

von Erhardt, Rudolf, and Erika von Erhardt-Siebold. "Archimedes' sand-reckoner: Hippopedes and Callippopedes." *Centaurus*, 33, 1942, pp. 578–602.

von Staden, Heinrich. *Herophilos: The Art of Medicine at Early Alexandria*. Cambridge University Press, 2007.

Waithe, Mary Ellen. *Ancient Women Philosophers: 600 BC–500 AD*. Part 1. Martinus Nijhoff, 1987.

Walker, C. B. F. "Notes on the Venus tablet of Ammisaduqa." *Journal of Cuneiform Studies*, 36(1), 1984, pp. 64–66.

Walker, D. P. "Some aspects of the musical theory of Vincenzo Galilei and Galileo Galilei." *Proceedings of the Royal Musical Association*. 100, 1974, pp. 33–47.

Wang, Q. Edward. "Beyond East and West: Antiquarianism, evidential learning, and global trends in historical study." *Journal of World History*, 19(4), 2008, pp. 489–519.

Watts, Edward. *Hypatia: The Life and Legend of an Ancient Philosopher*. Oxford University Press, 2017.

Weatherall, Thomas. *Jus Cogens: International Law and Social Contract*. Cambridge University Press, 2015.

Weeks, Andrew. *Paracelsus: Essential Theoretical Writings*. Brill, 2008.

Weld, Daniel, and Johan De Kleer (eds.). *Readings in Qualitative Reasoning about Physical Systems*. Morgan Kaufmann, 1990.

Wenke, Robert. *Patterns in Prehistory*, 3rd edition. Oxford University Press, 1990.

West, John. "Ibn al-Nafis, the pulmonary circulation, and the Islamic golden age." *Journal of Applied Physiology*, 105(6), 2008, pp. 1877–1880.

West, M. L. "The Babylonian musical notation and the Hurrian melodic texts." *Music & Letters*, 75(2), 1994, pp. 161–179.

Westfall, Richard. *Never at Rest: A Biography of Isaac Newton*. Cambridge University Press, 1980.

White, Michael. *Isaac Newton: The Last Sorcerer*. Basic Books, 1999.

Whiteside, D. T. (ed.). *The Mathematical Papers of Isaac Newton*. Vol. 3: *1670–1673*. Cambridge University Press, 1969.

Wieacker, Franz. *Privatrechtsgeschichte der Neuzeit*. Vandenhoeck & Ruprecht, 1996.

Wiessner, Polly W. "Embers of society: Firelight talk among the Ju/'hoansi Bushmen." *Proceedings of the National Academy of Sciences (PNAS)*, 111(39), 2014, pp. 14027–14035.
Wilcke, Claus. "Der Kodex Urnamma (CU): Versuch einer Rekonstruktion." In *Riches Hidden in Secret Places: Ancient Near Eastern Studies in Memory of Thorkild Jacobsen*. Edited by Zvi Abusch. Eisenbrauns, 2002, pp. 291–333.
Wilkinson, Tony. "Hydraulic landscapes and irrigation systems of Sumer." In *The Sumerian World*. Edited by Harriet Crawford. Routledge, 2013, pp. 34–54.
Windelband, Wilhelm. *Geschichte und Naturwissenschaft*. 3rd edition. Heitz, 1904.
Witzel, Michael. "Autochthonous Aryans? The evidence from old Indian and Iranian texts." *Electronic Journal of Vedic Studies*, 7–3, 2001, pp. 1–115.
Wölfflin, Heinrich. *Principles of Art History: The Problem of the Development of Style in Later Art*. Translated by M. D. Hottinger. Dover, 1932.
Woolf, Cecil Nathan Sidney. *Bartolus of Sassoferrato: His Position in the History of Medieval Political Thought*. Cambridge University Press, 1913.
Wriggins, Sally. *Xuanzang: A Buddhist Pilgrim on the Silk Road*. Westview Press, 2003.
Wright, Owen. *Demetrius Cantemir: The Collection of Notations*. Vol. 2: *Commentary*. Ashgate, 2001.
Wright, Thomas. *Circulation: William Harvey's Revolutionary Idea*. Chatto, 2012.
Wujastyk, Dominik. *The Roots of Ayurveda: Selections from Sanskrit Medical Writings*. Penguin Books, 2003.
Wyles, Rosie, and Edith Hall. *Women Classical Scholars: Unsealing the Fountain from the Renaissance to Jacqueline de Romilly*. Oxford University Press, 2016.
Yang Guobin. *Dragon-Carving and the Literary Mind: An Annotated English Translation and Critical Study of "Wenxin Diaolong" by Liu Xie*. Foreign Language Teaching and Research Press, 2003.
Yar-Shater, Ehsan (ed.). *The History of al-Ṭabarī*. State University of New York Press, 1985–2007.
Yoder, Joella. *Unrolling Time: Christiaan Huygens and the Mathematization of Nature*. Cambridge University Press, 1989.
Yonglin, Jiang (trans.). *The Mandate of Heaven and the Great Ming Code*. University of Washington Press, 2010.
Young, Irwin. *The Practica Musicae of Franchinus Gafurius*. University of Wisconsin Press, 1969.
Zarlino, Gioseffo. *Istituzioni armoniche*. Translated by Oliver Strunk. In *Source Readings in Music History: From Classical Antiquity through the Romantic Era*. W. W. Norton, 1950, pp. 333–344.
Zha, Y.-L. "Research on Tsu Ch'ung-Chih's approximate method for π." In *Science and Technology in Chinese Civilization*. Edited by Cheng-Yih Chen, Roger Cliff and Kuei-Mei Chen. World Scientific, 1987, pp. 77–85.
Zhang, Jialong, and Fenrong Liu. "Some thoughts on Mohist logic." In *A Meeting of the Minds: Proceedings of the Workshop on Logic, Rationality and Interaction*. Edited by Johan van Benthem, Shier Ju, and Frank Veltman. College, 2007, pp. 85–102.
Zhihui Chen. "Scholars' recreation of two traditions of mathematical commentaries in late eighteenth-century China." *Historia Mathematica*, 44(2), 2017, pp. 105–133.
Zinkeisen, F. "The Donation of Constantine as applied by the Roman Church." *English Historical Review* 9(36), 1894, pp. 625–632.
Zinner, Ernst. *Regiomontanus: His Life and Work*. Translated by E. Brown. Elsevier, 1990.
Zohary, Daniel, Maria Hopf, and Ehud Weiss. *Domestication of Plants in the Old World*. Oxford University Press, 2013.

Index

Page numbers in **bold** refer to tables; those in *italics* refer to illustrations.

www.ingramcontent.com/pod-product-compliance
Ingram Content Group UK Ltd.
Pitfield, Milton Keynes, MK11 3LW, UK
UKHW040707260926
475585UK00010B/38/J

9 781421 443447